Lecture Notes in Computer Science 1803
Edited by G. Goos, J. Hartmanis and J. van Leeuwen

Springer

Berlin
Heidelberg
New York
Barcelona
Hong Kong
London
Milan
Paris
Singapore
Tokyo

Stefano Cagnoni et al. (Eds.)

Real-World Applications of Evolutionary Computing

EvoWorkshops 2000: EvoIASP, EvoSCONDI,
EvoTel, EvoSTIM, EvoRob, and EvoFlight
Edinburgh, Scotland, UK, April 17, 2000
Proceedings

 Springer

Series Editors

Gerhard Goos, Karlsruhe University, Germany
Juris Hartmanis, Cornell University, NY, USA
Jan van Leeuwen, Utrecht University, The Netherlands

Main Volume Editor

Stefano Cagnoni
University of Parma
Department of Computer Engineering
Parco delle Scienze 181/a, 43100 Parma, Italy
E-mail: cagnoni@ce.unipr.it

Cataloging-in-Publication data applied for

Die Deutsche Bibliothek - CIP-Einheitsaufnahme

Real world applications of evolutionary computing : proceedings /
EvoWorkshops 2000: EvoIASP ... , Edinburgh, Scotland, UK, April 17,
2000. Stefano Cagnoni ... (ed.). - Berlin ; Heidelberg ; New York ;
Barcelona ; Hong Kong ; London ; Milan ; Paris ; Singapore ; Tokyo :
Springer, 2000
  (Lecture notes in computer science ; Vol. 1803)
  ISBN 3-540-67353-9

CR Subject Classification (1998): C.2, I.4, F.3, I.2, G.2, F.2, J.2, J.1, D.1

ISSN 0302-9743
ISBN 3-540-67353-9 Springer-Verlag Berlin Heidelberg New York

Springer-Verlag is a company in the BertelsmannSpringer publishing group.
© Springer-Verlag Berlin Heidelberg 2000
Printed in Germany

Typesetting: Camera-ready by author
Printed on acid-free paper   SPIN: 10720173   06/3142   5 4 3 2 1 0

Volume Editors

Stefano Cagnoni
University of Parma
Department of Computer Engineering
Parco delle Scienze 181/a
43100 Parma, Italy
Email: cagnoni@ce.unipr.it

Riccardo Poli
The University of Birmingham
School of Computer Science
Edgbaston, Birmingham B15 2TT, UK
Email: R.Poli@cs.bham.ac.uk

George D. Smith
University of East Anglia
Department of Information Systems
Norwich, NR4 7TJ, UK
Email: gds@sys.uea.ac.uk

David Corne
University of Reading
P.O. Box 225, Whiteknights
Reading RG6 6AY, UK
Email: D.W.Corne@reading.ac.uk

Martin Oates
BT Labs at Adastral Park
Dept. Complex Systems Laboratory
Martlesham Heath
Suffolk IP5 3RE, UK
Email: moates@srd.br.co.uk

Emma Hart
University of Edinburgh
Division of Informatics
80 South Bridge
Edinburgh EH1 1HN, UK
Email: emmah@dai.ed.ac.uk

Pier Luca Lanzi
Milan Polytechnic
Department of Electronics
and Informatics
Piazza Leonardo da Vinci 32
Milan, Italy
Email: lanzi@elet.polimi.it

Egbert Jan Willem
National Aerospace Laboratory NLR
Data and Knowledge Systems Dept.
I.C.T. Division
Anthony Fokkerweg 2
1059 CM Amsterdam
The Netherlands
Email: boers@nlr.nl

Yun Li
University of Glasgow
Electronics and Electrical Engineering
Rankine Building
Glasgow G12 8LT, UK
Email: Y.Li@elec.glasgow.ac.uk

Ben Paechter
Napier University
School of Computing
219 Colinton Road
Edinburgh EH14 1DJ, UK
Email: benp@dcs.napier.ac.uk

Terence C. Fogarty
South Bank University
School of Computing,
Information Systems and Mathematics
103 Borough Road
London SE1 0AA, UK
Email: fogarttc@sbu.ac.uk

Preface

The increasingly active field of Evolutionary Computation (EC) provides valuable tools, inspired by the theory of natural selection and genetic inheritance, to problem solving, machine learning, and optimization in many real-world applications.

Despite some early intuitions about EC, that can be dated back to the invention of computers, and a better formal definition of EC, made in the 1960s, the quest for real-world applications of EC only began in the late 1980s. The dramatic increase in computer performances in the last decade of the 20th century gave rise to a positive feedback process: EC techniques became more and more applicable, stimulating the growth of interest in their study, and allowing, in turn, new powerful EC paradigms to be devised.

In parallel with new theoretical results, the number of fields to which EC is being applied is increasing day by day, along with the complexity of applications and application domains. In particular, industrially relevant fields, such as signal and image processing, computer vision, pattern recognition, industrial control, telecommunication, scheduling and timetabling, and aerospace engineering are employing EC techniques to solve complex real-world problems.

This volume contains the proceedings of EvoWorkshops 2000: six workshops on real-world applications of EC held concurrently on April 17th 2000 in Edinburgh. The workshops are: EvoIASP 2000, the second European Workshop on Evolutionary Computation in Image Analysis and Signal Processing, Evo-SCONDI 2000, the first European Workshop on Evolutionary Computation in Systems, Control and Drives in Industry, EvoTel 2000, the second European Workshop on Evolutionary Telecommunications, EvoStim 2000, the first European Workshop on Evolutionary Scheduling and Timetabling, EvoRob 2000, the third European Workshop on Evolutionary Robotics, and EvoFlight 2000, the first European Workshop on Evolutionary Aeronautics.

EvoWorkshops 2000 was held in conjunction with two other major European events: EuroGP 2000, the European Conference on Genetic Programming, held on April 15th and 16th, and ICES 2000, the Third International Conference on Evolvable Systems: from Biology to Hardware, held from April 17th to 19th.

We would like to thank the members of the international program committees of the six workshops which included both EC researchers and researchers in the specific fields of interest and ensured the high quality of the papers.

April 2000

Stefano Cagnoni, Riccardo Poli, Yun Li,
George Smith, David Corne, Martin Oates,
Emma Hart, Pier Luca Lanzi, Egbert J.W. Boers,
Ben Paechter, and Terence C. Fogarty

Organization

EvoWorkshops 2000 Organizing Committee

EvoIASP co-chair: Stefano Cagnoni (University of Parma, Italy)
EvoIASP co-chair: Riccardo Poli (University of Birmingham, UK)
EvoSCONDI chair: Yun Li (University of Glasgow, UK)
EvoTel co-chair: George Smith (University of East Anglia, UK)
EvoTel co-chair: David Corne (University of Reading, UK)
EvoTel co-chair: Martin Oates (British Telecom plc, UK)
EvoSTIM chair: Emma Hart (University of Edinburgh, UK)
EvoRob chair: Pier Luca Lanzi (Milan Polytechnic, Italy)
EvoFlight chair: Egbert J.W. Boers (NLR, The Netherlands)
Local chair: Ben Paechter (Napier University, UK)
Publications chair: Terence C. Fogarty (South Bank University, UK)

EvoWorkshops 2000 Program Committee

Panagiotis Adamidis, Technological Educational Institution of Thessaloniki, Greece
Giovanni Adorni, University of Parma, Italy
Marjan van den Akker, NLR, The Netherlands
Alistair Armitage, Napier University, UK
Wolfgang Banzhaf, University of Dortmund, Germany
Randall D. Beer, Case Western Reserve University, USA
Andrea Bonarini, Milan Polytechnic, Italy
Joachim E. Born, DaimlerChrysler AG, Germany
Alberto Broggi, University of Pavia, Italy
Stefano Cagnoni, University of Parma, Italy
Mridula Chakraborty, National Engineering Laboratory, UK
Ela Claridge, University of Birmingham, UK
Marco Colombetti, Milan Polytechnic, Italy
Luis Correia, New University of Lisbon, Portugal
Jason Daida, University of Michigan, USA
Marco Dorigo, Free University of Brussels, Belgium
Shang Y. Duan, Rover Ltd, UK
Agoston E Eiben, Leiden University, The Netherlands
Peter J Fleming, University of Sheffield, UK
Dario Floreano, EPFL, Switzerland
Oscar C. García, University of Granada, Spain
Georg Grübel, DLR, Germany
Darko Grundler, University of Zagreb, Croatia
Inman Harvey, University of Sussex, UK

Phil Husbands, University of Sussex, UK
Francisco Herrera, University of Granada, Spain
Masayuki Hirafuji, NARC Computational Modeling Laboratory, Japan
Daniel Howard, DERA, UK
Ken Hunt, University of Glasgow, UK
Nick Jakobi, University of Sussex, UK
Mario Koeppen, Fraunhofer IPK, Germany
John Koza, Stanford University, USA
Kay Chen Tan, National University of Singapore, Singapore
Yun Li, University of Glasgow, UK
Evelyne Lutton, INRIA, France
Jean Arcady Meyer, AnimatLab, France
Orazio Miglino, University of Naples, Italy
Julian Miller, University of Birmingham, UK
Stefano Nolfi, National Research Council, Italy
Peter Nordin, Chalmers University, Sweden
Ben Paechter, Napier University, UK
Riccardo Poli, University of Birmingham, UK
Jim Smith, University of the West of England, UK
Sathiaseelan Sundaralingam, Nokia, UK
Peter Swann, Rolls Royce plc, UK
Robert Smith, University of the West of England, UK
Kay Chen Tan, National University of Singapore, Singapore
Andrea G. B. Tettamanzi, Genetica, Italy
Andy Tyrrell, University of York, UK
Hans-Michael Voigt, GFAI, Germany
Q. Henry Wu, Liverpool University, UK
Xin Yao, University of Birmingham, UK
Ali Zalzala, Heriot-Watt University, UK

Sponsoring Institutions

Napier University, UK
EvoNet: the Network of Excellence in Evolutionary Computing

Table of Contents

EvoTel Talks

EvoTel Posters

EvoSTIM Papers

EvoRob Papers

EvoFlight Papers

Special Purpose Image Convolution with Evolvable Hardware

Joe Dumoulin[a], James A. Foster[b,c,d], James F. Frenzel[c,e], Steve McGrew[a]

[a]New Light Industries, Ltd., Spokane, WA
[b]Center for Secure and Dependable Software, U. Idaho, Moscow, ID
[c]Microelectronics Research and Communications Inst., U. Idaho, Moscow ID
[d]Dept. of Computer Science, U. Idaho, Moscow, ID
[e]Dept. of Electrical and Computer Engineering, U. Idaho, Moscow, ID

email: joe@bresgal.com, foster@cs.uidaho.edu, jff@mrc.uidaho.edu, stevem@iea.com

Abstract.

In this paper, we investigate a unique method of inventing linear edge enhancement operators using evolution and reconfigurable hardware. We show that the technique is motivated by the desire for a totally automated object recognition system. We show that an important step in automating object recognition is to provide flexible means to smooth images, making features more obvious and reducing interference. Next we demonstrate a technique for building an edge enhancement operator using evolutionary methods, implementing and testing each generation using the Xilinx 6200 family FPGA. Finally, we present the results and conclude by mentioning some areas of further investigation.

Introduction

Image edge enhancement is an important part of modern computerized object recognition methods. Edge enhancement typically requires convolution operators that produce weighted average transformations on individual pixels of some source image. Edge enhancement operators tend to be developed for particular types of problems and larger operators tend to be for very specific purposes. Large irregular pixel patterns require rather large and irregular edge enhancement operators.

The smoothing process assists an edge detection algorithm to distinguish between background noise and actual objects in an image.

Unfortunately, software implementations of large convolutions are extremely slow. Consequently, convolutions of any respectable size are typically implemented in specialized image processing hardware. Another problem with large convolutions is that it is often very difficult to "discover" good ones. The process of creating these convolution operators requires trial and error, and experience.

Recently, at New Light Industries, we needed special purpose convolution operators for use in low-cost, commercial off-the-shelf hardware devices. These convolutions needed to be very efficient and implementable in reconfigurable hardware. Using genetic algorithms and Xilinx XC6000 technology, we implemented an evolutionary system that designed convolution operators for our application. The genetic algorithm evaluated operators by comparing the

S. Cagnoni et al. (Eds.): EvoWorkshops 2000, LNCS 1803, pp. 1–11, 2000.

convolution of one original image with several prepared images that the convolution should be able to produce.

These experiments demonstrated both the viability, and some limitations, of creating edge enhancing convolution kernels using evolutionary techniques and prepared images. This paper describes our results and shows an evolved convolution that was used in an edge detecting algorithm. Finally, this paper presents some potential future research directions and applications.

Motivation

Computerized image processing provides many opportunities for implementing evolutionary problem-solving techniques. Our problem is to identify important features in a diffraction pattern that changes over time. This is analogous to identifying and tracking multiple objects moving through a series of frames. Consider a camera aimed at a moving object or an object illuminated by a moving light source. The problem is to identify the object as it moves through a series of still frames. Since we needed compact, inexpensive, efficient hardware to perform the identification, we chose to implement our solution in reconfigurable systems using XC6000 FPGAs from Xilinx.

Our target images were a series of diffraction patterns recorded over time with a moving light source. The diffraction patterns are optically derived Fourier Transforms of features on the surface of an object. The frames are captured in 8-bit in gray-scale and the objects to be recognized are geometric "blobs" that both move and change shape from frame to frame. We must identify each object as it moves and changes shape over a series of frames. We chose a simple edge-detection algorithm [8] as the method for extracting the border.

A key step in identifying object borders in a particular frame is applying a large convolution operator to the image. We used genetic methods to develop appropriate convolution operators for our training set. The following sections explain the approach and the results of our initial experiments.

Background

Image Processing Considerations
The literature on image processing contains many examples and techniques for performing image enhancement and image segmentation.   Generally, as one might expect with any computational task, there are tradeoffs between speed and accuracy in the segmentation process.   We used optically generated Fourier Transforms of surface features for this experiment. Identification of features in the frequency space of these images amounted to identifying the object being scanned. We needed to identify not just the features on many images, but the path of each feature through multiple consecutive frames. We investigated some well-known methods for image detection that we could use to facilitate segmentation and object tracking.

For the edge detection method, we looked at three algorithms: Marr-Hildreth [8], the Infinite Symmetric Exponential Filter (ISEF) [9], and morphological boundary extraction [10].

Marr-Hildreth edge detectors use a smoothing algorithm, usually a convolution with a Gaussian function, followed by a zero-crossing binarization of the image. We chose this method because it is simple, fast, and easy to implement in varied hardware and software.

The ISEF is a very high quality edge detector on images with high frequency uniform noise. The ISEF also computes zero crossings to find edges, but it performs a much more complex smoothing using a band-limited Laplacian operator.   The ISEF algorithm we implemented [11] used an edge following technique that consumed memory and probably could have been time-optimized as well. We found this algorithm to be very effective, but much slower than Marr-Hildreth. Since the latter was adequate for our task, we did not use the ISEF.

Morphological Boundary Extraction uses some simple set operations to outline edges on the boundaries of objects in an image. In the simplest case, the edge enhancement results from eroding a copy of the image using a simple kernel and then subtracting the resulting image from the original image. An image containing high frequency noise requires an additional opening step to clean the frame before the boundary extraction will work correctly. This algorithm was not as effective for extracting images from our images. The boundaries of different features in the image tend to bleed together if the erosion kernels are not of the proper size for the boundary. If the boundary is different in different parts of the image it may be necessary to erode different parts of the image with different kernels to get an accurate boundary.

We also investigated some thresholding methods, but found them to be very limited. One method we did not explore which might prove effective is Edge Level Thresholding, though this method proves to be slow relative to the methods we chose to investigate.

We decided to use the Marr-Hildreth method initially, because it was by far the fastest method given our environment. The ISEF technique proved more accurate in object segmentation, but the improvement was not warranted given that the process was significantly slower than Marr-Hilreth. The Morphological technique we investigated worked adequately only with a great deal of "hand-tweaking" the parameters. This actually made the morphological boundary extraction method a candidate for later experiments with genetic methods. For the current experiment, though, we found it difficult to control the results when we tried to automate the algorithm.

Convolution

Convolution operators are commonly used in edge enhancement techniques to increase contrast or emphasize features with particular shapes on digitized images. "Edges" are usually rapid changes in pixel values. One way to think about a convolution is as a differential operator that measures the rate of change in some direction along the image. Convolution operators for image processing are usually expressed as odd-valued square matrices. The operation is applied across the pixel field by multiplying individual pixel values by the matrix elements and then adding the results to get a new pixel value. The new value replaces the value for the center pixel in the field. The operation is applied repeatedly across the picture until the whole pixel field has been transformed.

Simple convolutions are often 3×3 or 5×5 matrices. One simple and straightforward convolution is the "directionless" or Laplacian operator. The Lapacian operator is directly analogous to the Laplacian of differential analysis: $\nabla^2 = \dfrac{\partial^2}{\partial x^2} + \dfrac{\partial^2}{\partial y^2}$

Using the definition of a difference, $f' \approx (f(x) - f(x-h))/h$, and letting h=1, the Laplacian can be expressed discretely as a 3×3 matrix as follows:

$$\begin{bmatrix} 0 & -1 & 0 \\ -1 & 4 & -1 \\ 0 & -1 & 0 \end{bmatrix}$$

This is only one common example of a convolution matrix. We will show examples created by genetic methods later in this paper.

When the convolution is applied to a 3×3 area on a pixel field, the operator gives a new value for the center pixel in a transformed image. The convolution is then moved across and down the pixel field until all the pixels have been converted. At the edges, there are a number of different techniques for operating on pixels that cannot be mapped to the center of a 3×3 matrix. In our experiments, we ignored the values at the borders of the pixel fields.

The convolution operation, though quite simple, can produce complex and often inscrutable results. We present a little example of convolution applied to a small array below. Suppose one has a section of a gray scale image such as that in Figure 1. This corresponds to some portion of an image.

0	2	10	6	4	0
1	5	12	6	3	2
0	2	8	5	4	1
1	3	12	4	2	2

0	0	0	0	0	0
0	6	19	0	0	0
0	0	17	0	9	0
0	0	0	0	0	0

Figure 1a. A section of a gray scale image. The convolution operates on incremental sections of the image, beginning with the highlighted section. Figure 1b. The result of the convolution with Laplacian of image 1a. Note that the edges of the image are 0 padded. Also any convolution that sums to a negative number is set to zero.

Convolution is defined as the sum of the element-by-element products of two matrices. So in the case of the Laplacian operator, and an N×N matrix, the convolution describes a new matrix whose elements are defined as: $A * \Delta_{i,j} = \sum_{p=-1}^{1} \sum_{q=-1}^{1} a_{i+p,j+q} \Delta_{p,q}$

After moving this convolution operator across the shaded area of the image above, one continues with the next 3×3 section of the matrix. Note how the convolution "enhances" one section of the matrix associated with local maxima that are not on edges.

The Laplacian is only one of many widely used convolution operators (or kernels). Usually, given a particular shape to recognize and a particular scale, there is a particular class of convolution kernels that will be most effective in enhancing the image to recognize features of the desired shape or scale. When shapes are large or irregular, it can be difficult to develop an effective (or efficient) convolution kernel to recognize the shape.

Evolutionary Computing with XC6000 FPGAs

We performed the image extraction and original image processing testing using a Matrox Genesis image grabber, which includes a TMS320C80 with parallel fixed-point MACs and a 32-bit data bus, allowing us to very handily evaluate the different edge detection methods discussed above. At the time of its purchase, the device approached $8,000 US, far from the COTS criteria that we were looking for. For this reason, and to facilitate the evolutionary aspect of our project, we chose the VCC Fat HOT PCI card for our development system. It was less than half the cost while providing the necessary computing power.

Developing programs for FPGAs using evolutionary methods requires some attention to the special constraints of the reconfigurable hardware. FPGA-based designs, evolutionary or not, are physically constrained by geometry and function unit capabilities. Function units within any FPGA provide basic binary logical operations (e.g., OR, AND, XOR), but beyond this, there is a wide variation in the particular functions that can be implemented within a single function unit on the FPGA.

Similarly, the geometry of the FPGA as a whole has an effect on design. The way in which individual function units are connected to other function units and to the edges of the device also constrains the routing of circuitry in the FPGA. We have chosen the Xilinx XC6000 series of FPGAs for our research for the following reasons:

- There is a significant body of research on evolutionary hardware design building around this class of FPGA devices.

- The chip design is open and configuration formats are accessible at the hardware level.
- Function unit routing is extremely flexible.
- The device has a well-defined PCI bus interface standard from the manufacturer; and this makes the device useful for experimenting with desktop computer interfaces in a consistent manner with other researchers and manufacturers.
- The manufacturer has developed API sets that make device programming easily accessible to the software developer using a desktop PC. Other API sets are available for different development environments.

The XC6000 gives us the flexibility to experiment with different evolutionary methods and to rapidly build and utilize new tools for our research. Many researchers describe the XC6200 architecture as it applies to our experiment. In particular, see [3], [4] and [5]. Many possible approaches exist for evolutionary design using FPGAs. We will discuss relevant approaches to one feature recognition subsystem that have emerged from our present research. What distinguishes different evolutionary design approaches is the representation and evaluation of trial solutions.

Evolutionary Hardware Design

Representation

The key to problem solving with genetic algorithms is to find a suitable representation of the problem. A good representation must encompass the full range of possible solutions, and must enable the genetic algorithm to generate new trial solutions with a relatively high probability of retaining beneficial features of old trial solutions. Mutation operators should change trial solutions in ways that lead to relatively small fitness changes, and recombination operators should tend to accumulate useful features of two or more trial solutions into a new trial solution. We investigated two different representations of FPGA configurations. One method uses a coded representation of possible functions and a coded representation of possible routings.

Functions	
0	Constant 0
1	Constant 1
2	X AND Y
3	X OR Y
4	X XOR Y
5	INV
6	BUF
Routing:	
1	North
2	East
3	West
4	South

Figure 2: Coded Representation Implemented as a Look-Up Table.

Coded representations are common (see [4], [6]) in evolutionary programming, partly because they are relatively easy to describe. In our representation, our chromosome is a 2D array of integers that defined a convolution kernel numerically. The fitness evaluation step converts the kernel into a set of bit values and then converts the bit values into a configuration. The configuration performs the convolution but allows us to hard-code (or, more accurately, hard-wire) the convolution kernel into the circuit description. This reduces the circuit size and speeds up the evaluation of the convolution.

The primary advantage of this representation is that it is abstracted from the particular FPGA architecture, so that it can be rewritten to configure different devices (as long as only common function unit/routing configurations are allowed). The primary disadvantages of this representation are:

- Only common FPGA function configurations are allowed, so special features of certain FPGAs (i.e., separable clocking, register protection in the XC6000) will not be available to the evolutionary process. Note that this can be an advantage in some circumstances, such as when it is desirable to constrain the FPGA to clocked circuits.
- The added level of abstraction (beyond a bit-level representation) adds a level of complexity to the testing process.

We also evaluated intrinsic evolution, in which we directly use the underlying structure of the XC6000 in the function unit encoding. (For examples of intrinsic evolution, see [3].) In this representation we pay particular attention to the address/data structure of the FPGA configuration. Programming the FPGA at this level is analogous to programming a standard CPU using native machine code. The complexity of creating working programs at this level is avoided in the evolutionary process. FPGA configurations evolve to approach the problem represented in the fitness stage.

Byte 0	Output Routing
Byte 1	Function Config/ Input
Byte 2	Function Config/ Input

Figure 3: Function Unit Configuration Bytes

A circuit configuration is represented as a string of bytes that together define the configuration of a function unit on the XC6000. The configuration of a single function unit comprises three bytes that define the local routing and the logic implemented in the function unit.

Our chromosome is a two-dimensional array of these three-byte units. Each element of the array represents a function unit. The index of each element represents the relative position of the function unit within the FPGA, thereby allowing us to construct an address for each element of the configuration. For details about what particular values indicate, see [1]. The primary advantages of this representation are:

- The representation may be directly interpreted as a "program" of the FPGA. Very little preliminary processing is needed to create a loadable configuration.
- All possible functions are allowed in principle, but they can be restricted at will by masking the appropriate bits of each function unit configuration.

The primary disadvantages of an intrinsic representation are:

- The resulting evolved configuration can be very difficult for a human FPGA programmer to understand.
- The configuration is always specific to a particular FPGA. A program generated via this method for one manufacturer's FPGA cannot be expected to run on a different manufacturer's FPGA.

The first representation discussed above is superior to the second for a number of practical reasons. Using the first method, we will not be tied to any particular FPGA for the purposes of development. It is, however, much more time-consuming to implement the first, so our experiments to date have concentrated on the second representation. There are, of course, other methods for creating FPGA configurations (See [6] in particular), but our methods are simple and adequate for the task.

Fitness Evaluation and Crossover

The evaluation of each solution created by the genetic algorithm is another key aspect of evolutionary design of programs for our experiment. We test each trial solution created by our genetic algorithm, and use the test results to assign a fitness value to the trial solution. Generally, the objective of the design process is to develop a design that meets certain performance criteria. In evolutionary design, however, it is important to use fitness values to *guide* evolution efficiently towards the final objective.

The basic flow of our FPGA evolution algorithm is:

1. Load a file that represents all the static configuration portions of the program (e.g., I/O Registers, Counters, and other control sections)
2. Create a random population of trial solutions.
3. For each trial solution, evaluate as follows:
   a.   Convert the trial solution to a loadable configuration.
   b.   Load the configuration.
   c.   Load the input registers.
   d.   Wait for some fixed amount of time.
   e.   Read the output register.
   f.   Compare the register with a calculated result based on the input to get the error.
   g.   Assign the fitness of the trial solution based on step f.
4. Select trial solutions according to fitness to generate a new population.
5. Generate new trial solutions by applying genetic operators (mutation and recombination) to the selected trial solutions.
6. Apply steps 3 through 6 repeatedly until a satisfactory solution emerges.

In step g above, changes can be made to emphasize or de-emphasize certain aspects of the calculated error at different stages in the evolution so that, for example, basic functions are evolved and optimized before they are combined into more complex structures. Fitness calculation can include weighting the error to give more influence to solutions that meet certain constraints or sub–criteria. This is extremely important in the case of our application since we will be comparing vector values representing patterns of features identified by the FPGA to reference vectors calculated by other means.

In order to maintain the integrity of working portions of each genome, we developed a crossover operator that preserves rectangular portions of the genome during crossover. By default, the 2D integer (3D Binary) genome in Gail will perform byte-level crossover. This has the effect of destroying local optimal areas in the genome from generation to generation. We tried to preserve local optimal behavior in the genome by treating each genome as a 2D array and each crossover component as a 2D sub-array of the genome.

For example, given two 9×9 genomes to cross over, the cross-over algorithm works as follows:

1. Randomly select a starting row and starting column. This will be the upper left corner of the crossover array.
2. Randomly select the number of rows and columns to cross over. This defines the size of the crossover array.

3. Select two genomes to cross over and remove a sub-array from the appropriate position and of the calculated size, from each of the genomes.
4. Exchange these sub-arrays between the genomes.
5. The evolved convolutions were created by genetically developing a convolution to attempt to "reproduce" an image that was enhanced with a general-purpose convolution (a pseudo-Laplacian) in special purpose image processing hardware.

   We evaluated fitness as follows:

1. Get a genome.
2. Convolve the original image with the genome.
3. Compare the resulting image with the image convolved with the 9×9 pseudo-Laplacian (See Figure 4 below) to get the fitness value. The evaluation compares each pixel in the resulting image with each pixel in the original image.
4. Compare fitnesses preserving the lowest value (highest fitness) individuals.

Results

The experiments we performed were carried out using a genetic algorithm defined using GALIB 2.4.2 and the XILINX API library for the 6200 parts. The GA was set up to evolve a 9×9 2-dimensional integer genome. We ran the GA with population size of 50, crossover probability of 20%, and mutation probability of 5%. We noticed that, under these conditions, the GA produced very stable results after approximately 100 generations.

As we see in the pictures below, low fitness members of the population actually produced "terrible" resulting images. We determined that this was due to numeric overflow during fitness evaluation. Because of this the fitness evaluation was re-examined and we decided to reward fitness values close to half the maximum value of a long integer. One "good" result is shown in Figure 6.

We present some examples of evolved convolutions and the enhanced images they produced below. The GA created and evaluated these images using the training set image in Figure 4.

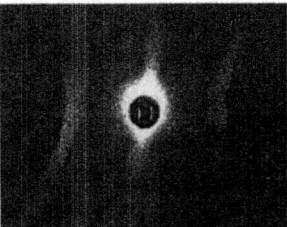 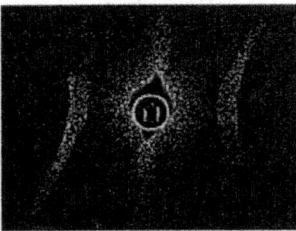 

$$\begin{bmatrix} -1 & 0 & 0 & 0 & -1 & 0 & 0 & 0 & -1 \\ 0 & 0 & 0 & 0 & 0 & 0 & 0 & 0 & 0 \\ 0 & 0 & 0 & 0 & 0 & 0 & 0 & 0 & 0 \\ 0 & 0 & 0 & 0 & 0 & 0 & 0 & 0 & 0 \\ -1 & 0 & 0 & 0 & 8 & 0 & 0 & 0 & -1 \\ 0 & 0 & 0 & 0 & 0 & 0 & 0 & 0 & 0 \\ 0 & 0 & 0 & 0 & 0 & 0 & 0 & 0 & 0 \\ 0 & 0 & 0 & 0 & 0 & 0 & 0 & 0 & 0 \\ -1 & 0 & 0 & 0 & -1 & 0 & 0 & 0 & -1 \end{bmatrix}$$

Figure 4: An original diffraction pattern from the training set (left), a pseudo Laplacian operator(right) and the convolved image used for comparison (center).

Evolved convolution operators produce some widely varying results. Below are a few examples of evolved operators and the images they produced from the original image in Figure 4.

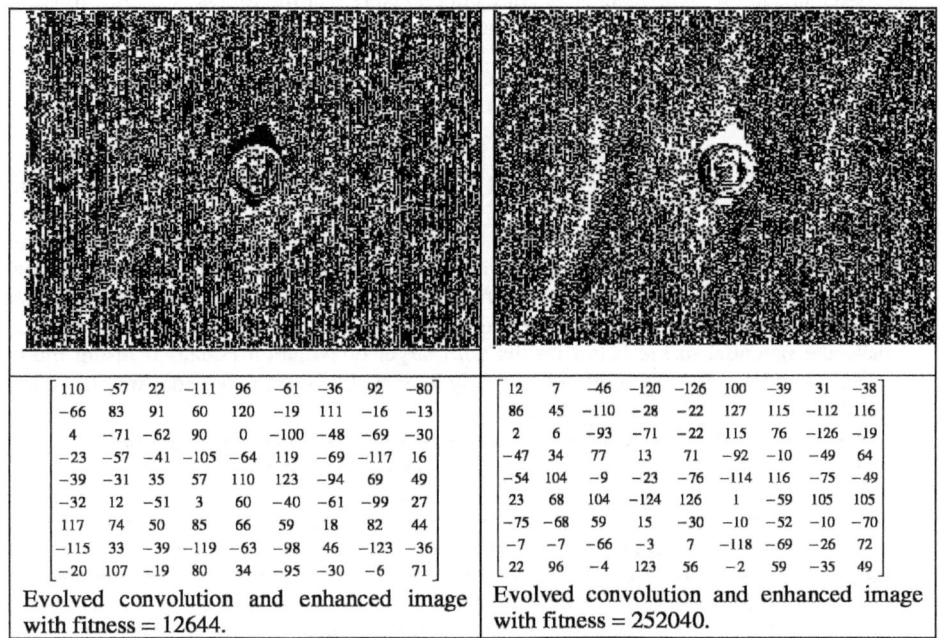

$\begin{bmatrix} 110 & -57 & 22 & -111 & 96 & -61 & -36 & 92 & -80 \\ -66 & 83 & 91 & 60 & 120 & -19 & 111 & -16 & -13 \\ 4 & -71 & -62 & 90 & 0 & -100 & -48 & -69 & -30 \\ -23 & -57 & -41 & -105 & -64 & 119 & -69 & -117 & 16 \\ -39 & -31 & 35 & 57 & 110 & 123 & -94 & 69 & 49 \\ -32 & 12 & -51 & 3 & 60 & -40 & -61 & -99 & 27 \\ 117 & 74 & 50 & 85 & 66 & 59 & 18 & 82 & 44 \\ -115 & 33 & -39 & -119 & -63 & -98 & 46 & -123 & -36 \\ -20 & 107 & -19 & 80 & 34 & -95 & -30 & -6 & 71 \end{bmatrix}$	$\begin{bmatrix} 12 & 7 & -46 & -120 & -126 & 100 & -39 & 31 & -38 \\ 86 & 45 & -110 & -28 & -22 & 127 & 115 & -112 & 116 \\ 2 & 6 & -93 & -71 & -22 & 115 & 76 & -126 & -19 \\ -47 & 34 & 77 & 13 & 71 & -92 & -10 & -49 & 64 \\ -54 & 104 & -9 & -23 & -76 & -114 & 116 & -75 & -49 \\ 23 & 68 & 104 & -124 & 126 & 1 & -59 & 105 & 105 \\ -75 & -68 & 59 & 15 & -30 & -10 & -52 & -10 & -70 \\ -7 & -7 & -66 & -3 & 7 & -118 & -69 & -26 & 72 \\ 22 & 96 & -4 & 123 & 56 & -2 & 59 & -35 & 49 \end{bmatrix}$
Evolved convolution and enhanced image with fitness = 12644.	Evolved convolution and enhanced image with fitness = 252040.

Figure 5: Two High Fitness Results

In the "high-fitness" convolution operators, we can see what appears to be a random distribution of pixel values through the dark areas of the original frame. These operators introduced a high-order randomness at the level of the convolution array size (9×9 pixels). This suggests that larger convolution arrays may provide more effective matching.

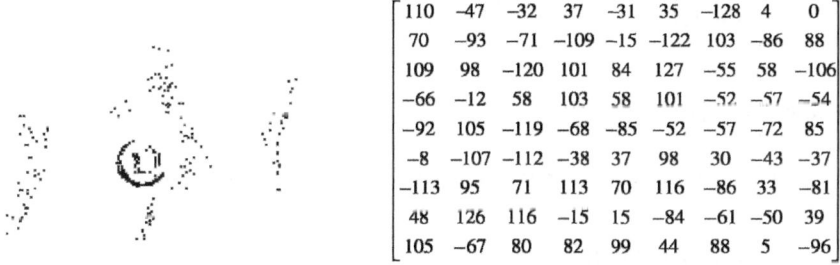

$$\begin{bmatrix} 110 & -47 & -32 & 37 & -31 & 35 & -128 & 4 & 0 \\ 70 & -93 & -71 & -109 & -15 & -122 & 103 & -86 & 88 \\ 109 & 98 & -120 & 101 & 84 & 127 & -55 & 58 & -106 \\ -66 & -12 & 58 & 103 & 58 & 101 & -52 & -57 & -54 \\ -92 & 105 & -119 & -68 & -85 & -52 & -57 & -72 & 85 \\ -8 & -107 & -112 & -38 & 37 & 98 & 30 & -43 & -37 \\ -113 & 95 & 71 & 113 & 70 & 116 & -86 & 33 & -81 \\ 48 & 126 & 116 & -15 & 15 & -84 & -61 & -50 & 39 \\ 105 & -67 & 80 & 82 & 99 & 44 & 88 & 5 & -96 \end{bmatrix}$$

Figure 6: Evolved convolution and enhanced image with fitness = 2018851897.

A low fitness result that provided better matching appears in Figure 6 above. This operator was generated in the 100th generation. It matches almost negatively with the template in Figure 4, yet it provides very good detail extraction and was in fact used in a Marr-Hildreth edge detector to find features in the sample frame (shown in Figure 4) and others produced in time-sequence in the same manner as our example. Note that the fitness is "close" to half the maximum value of an unsigned 64-bit integer.

Conclusions

We have a number of conclusions and possible paths of continued research from these initial results.

- **Improved Fitness Evaluation.** We need a fitness evaluation methodology more appropriate for the very large number of calculations required to compare these results pixel-by-pixel. One approach would average local results before doing the sum of squares.
- **Larger Convolution Kernels.** Since we are working with images of 640×480 pixels, we have the potential to see better results with larger convolution kernels or using multiple kernels and linearly combining the resulting images. Plans are now under way to 100×100 pixel kernels using a cluster computer.
- **The Utility of the FPGA vs. Clustered Systems.** The FPGA is very useful for fitness evaluation and execution of small kernels because the convolution mathematics can be designed into the circuit. The XC6200 in particular is very useful when we convolve a frame using multiple kernels. For large convolution kernels, however, multiple FPGAs are needed. This is because the data bus of the FPGA becomes a limitation when evaluating large sets of numbers. We will be exploring the use of matrix math operations built into modern COTS microprocessors (e.g., MMX, 3Dnow) to optimize calculation time for matrix operations. We hope to produce a performance comparison evaluating COTS techniques for integer matrix calculations in the future.
- **Applicability to Motion Detection.** For some very rudimentary types of objects, our techniques could prove useful for identifying "custom" kernels that correspond to particular objects. For example, a robotic vehicle could use this method to isolate unrecognized objects and save a library of discovered templates for future reference. We have no conclusive results along this line, only the suggestion of value.
- **Overall Viability.** The methodology described in this paper proved useful for our particular application. We hope to extend the methodology to provide useful generation of automatic convolution kernels for other general applications.

Acknowledgements

This work was funded by BMDO. James A. Foster was also partially funded by DOD/OST.

Bibliography and References

[1] XILINX XC6200 Field Programmable Gate Arrays, April 24, 1997, XILINX, Ltd

[2] R. Murgai, R. Brayton, A Sangiovanni-Vincentelli; Logic Synthesis for Field-Programmable Gate Arrays" 1995, Kluwer Academic Publishers

[3] A. Thompson, I. Harvey and P. Husbands; Unconstrained Evolution and Hard Consequences CSRP 397, (in Towards Evolvable Hardware, Springer-Verlag Lecture Notes in Computer Science, 1996)

[4] J. Koza, S. Bade, F. Bennett III, M. Keane, J. Hutchings, D. Andre; Rapidly Reconfigurable Field-Programmable Gate Arrays for Accelerating Fitness Evaluation in Genetic Programming, Published in Koza, John R. (editor). Late Breaking Papers at the

Genetic Programming 1997 Conference, Stanford University, July 13-16, 1997. Stanford, CA: Stanford University Bookstore. Pages 121 - 131.

[5] D. Montana, R. Popp, Suraj Iyer, and G. Vidaver; Evolvaware: Genetic Programming for Optimal Design of Hardware-Based Algorithms, 1998, BBN Technologies, Proc. Int. Conf. on Genetic Programming.

[6] J. Miller, P. Thomson; Evolving Digital Electronic Circuits for Real-Valued Function Generation using a Genetic Algorithm, 1998, Napier University, Proc. Int. Conf. on Genetic Programming.

[7] L. Pagie, P. Hogeweg; Evolutionary Consequences of Coevolving Targets, Evolutionary Computation 5(4):401-418, 1998.

[8] D. Marr, E. Hildreth; Theory of Edge Detection, Proceedings of the Royal Society of London, Series B, Vol. 207, pp.187-217, 1980.

[9] J. Shen, S.Castan; An Optimal Linear Operator for Step Edge Detection, Computer Vision, Graphics, and Image Processing: Graphical Models and Understanding, Vol.54, 2:pp. 112-133, 1992.

[10] J. Serra; Image Analysis and Mathematical Morphology, Academic Press, 1988.

[11] J.R. Parker; Algorithms for Image Processing and Computer Vision, John Wiley and Sons, 1997.

Stereoscopic Vision for a Humanoid Robot Using Genetic Programming

Christopher T.M. Graae, Peter Nordin and Mats Nordahl

Complex Systems Group, Institute of Physical Resource Theory, Chalmers University of Technology, S-412 96 Göteborg, Sweden
{nordin,tfemn}@fy.chalmers.se

Abstract. In this paper we introduce a new approach to adaptive stereoscopic Vision. We use genetic programming, where the input to the individuals is raw pixel data from stereo image-pairs acquired by two CCD cameras. The output from the individuals is the disparity map, which is transformed to a 3D map of the captured scene using triangulation. The used genetic engine evolves machine-coded individuals, and can thereby reach high Performance on weak computer architectures. The evolved individuals have an average disparity-error of 1.5 pixels, which is equivalent to an uncertainty of about 10% of the true distance. This work is motivated by applications to the control of autonomous humanoid robots – The Humanoid Project at Chalmers.

1 Introduction and Motivation

Man is the Standard for almost all interactions in our world where most environments, tools and machines are adapted to the abilities, motion capabilities and geometry of humans. Walking robots have a very large potential in environments created for humans as well as in more natural terrain. The largest potential is associated with robots of human-like dimensions walking on two legs – humanoid robots. It could be more efficient to control various machines by these robots than to rebuild all machines for direct Computer control [4].

This work is part of the Humanoid Project at Chalmers University of Technology. The project plans a series of humanoid experiments, all of which will be primarily controlled by evolutionary adaptive methods.

The final goal of the research is to build a human-sized robot based on a plastic human Skeleton to ensure geometric authenticity. The current status of the project is that a second-generation prototype of a small humanoid is being developed – ELVIS – with a height of about 60 cm.

Vision is the most important of our five senses. As an example a quarter of the human cerebral cortex is devoted to sight[1], but some animals devote over half of their

---

[1] According to a new estimate by neuroscientist David Van Essen of Washington University in St. Louis.

S. Cagnoni et al. (Eds.): EvoWorkshops 2000, LNCS 1803, pp. 12–21, 2000.

brain capacity to the task of interpreting signals from their eyes (this even though their eyes has build-in object detection and as in some frogs initiates reflexes such as catch-the-fly – independently of their brains)

There are two main reasons for using stereovision in robotics. It is a passive sensor without interference with the environment, nor with other sensor devices. It is also a reliable and effective way to extract range information, and it is easily integrated with other vision routines such as object recognition and tracking. For humanoid it can be motivated by its use in nature.

2   Method and Experiments

The genetic programming (GP) approach used in all experiments below is called Automatic Induction of Machine Code Genetic Programming (AIMGP) [5]. AIMGP is a linear GP-system that evolves binary machine code. The main benefit in this, compared to conventional systems, is a high increase in speed (about 60 times). The algorithm uses a steady state tournament selection, and the individuals are coded using a linear representation. The table below shows the most important settings that were used for all experiments.

Table 1. The most important parameter settings

Description	Value
Size of population	15000
No of demes	5
Size per deme	3000
Migration rate	1%
Crossover rate	90%
Mutation rate	90%
No of trainings	150 – 200
No of tests	150 – 200
Iterations	40 – 200
Constants	0.5, 1.0, 60.0

Each selected individual was trained on 150 – 200 training examples (corresponding or interest point pairs, see below). Each training example was iteratively run between 40 and 200 times, see Fig. 3. Finally the best individual was tested on 150 – 200 test examples.

The input for the GP individuals consisted of stereoscopic image data and the interest point in question. An interest point is the position in one of the images (here the left) in a stereoscopic image-pair, for which a corresponding point in the other image (the right) is sought.

The output is expected to be the corresponding interest point in the right image, thus giving information on the disparity of that point-pair. Some experiments expected the output to be the disparity value directly. Given the disparity and knowing the geometry of the set-up is enough to calculate the distance between the robots eyes (CCD cameras) and the real world position marked by the interest point-pair on the stereoscopic image-pair. This is done using triangulation.

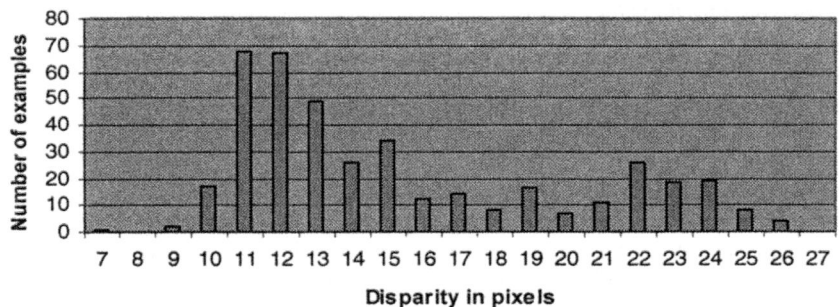

Fig. 1. One of the training sets

The scene used to capture stereoscopic image-pairs was made up of the white inside faces of a box, with the front side removed. One or more objects, placed in different positions in the scene, were used for each stereoscopic image-pair.

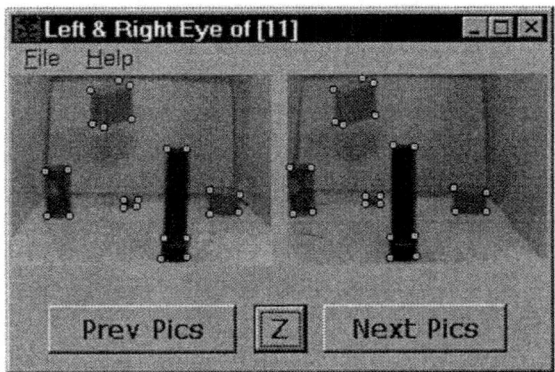

Fig. 2. A stereoscopic image-pair with marked interest points

Objects such as a black torch and a red floppy-disc box were used for the images in the first training, testing and validation sets.

Two different camera models were tried out, the QuickCam™ VC and the Kodak DVC 323. A resolution of 160x120 pixels with 256 colours[2] was used.

---

[2] 24-bit RGB values.

The training/evolution of the individuals in this first set of experiments was done off-line; i.e. on a stand alone PC instead of on the robot's on-board subnotebook. The captured images were saved in a bitmapped file format[3]. Interest points in these images were marked manually using a for-the-purpose-developed programme (see Fig. 2), which saved the coordinates of the interest point-pairs in a data-file. Each image-pair had its own data-file, usually with between 10 and 40 interest points-pairs.

2.1 Experiment series I – The Moving Points approach

The task of the first experiment was to find for each point in the left image, the corresponding point in the right image – i.e. to solve the correspondence problem. Each individual had control over two "measuring points", one in each image in the stereo image-pair, to be able to freely scan though the images in whatever way it found necessary. The area of an image available for an individual was restricted to a rectangle enclosing the interest point and its neighbourhood (a neighbourhood rectangle) - much the same as the sweep area window in Fig. 4. The individuals get the position of one interest point in the left image as input. By moving around the two measuring points in the two images, they search for the corresponding interest point in the right image.

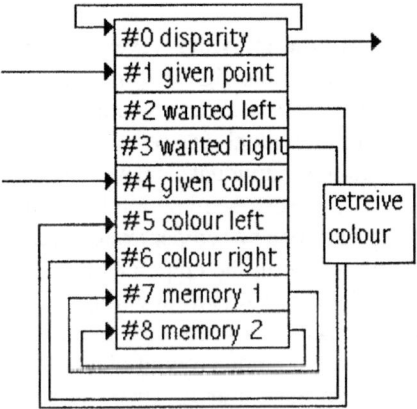

Fig. 3. An individual

Positions in the image were given as linear[4] coordinates inside the neighbourhood-rectangles. Origo were placed at the position of the left point (in the left image). The expected output was the disparity value for the two corresponding points, i.e. the horizontal distance between the two points. The early stage experiments expected a linear coordinate for the corresponding left point - the output - instead of the disparity.

To be able to solve this task, each individual was repeatedly executed – between 60 and 200 times. The individual got, at each iteration, the colour information of the two

---

[3] Microsoft Windows BMP, a device-independet bitmap (DIB) format.
[4] L = <width of img>*Y + X.

points defined in the previous iteration, the previous disparity estimation and the feed-back memories, with the hope that it step-by-step would improve the output, see Fig. 3.

2.2 Experiment series II – The Sweeping Box Approach

The approach in this experiment was to find for each point in the left image, the corresponding point in the right image through scanning/sweeping. The disparity is then the distance (or the horizontal shift) between these two points. The neighbouring pixel data to an interest point (i.e. a sub-image) in the left image were fed as input to the individual, together with the same amount of pixel data around a sweeping point in the right image. Table 2 below shows that the sub-images consisted of 25 pixels, arranged in a 5x5 square.

Table 2. The inputs and output for an individual

INPUTS (v[])	OUTPUTS (f[])
0 → 24 left image data	0 matching score
25 → 49 right image data	

The expected output from the individuals was a number that would describe how well the left sub-image matched the right sub-image. The individual was iteratively run for each pixel position in the sweep area window. See Fig. 4 for details. The highest peak in the output of an individual was recorded, and the corresponding sweeping point location was taken as the individual's answer of what should be used as data for the disparity calculation. All sweeps were parallel with the epipolar line, and the most common set-up was to do only one sweep per sweep area window. Two corresponding points should in theory be found on the epipolar line. This is not always the case, because of several reasons: inexact camera mounting, minor sampling errors and non-successful timing synchronisation (when taking the stereo-image snapshots). The offset error is not usually more than one row of pixels.

The closer an object were the cameras when the stereo-image pair was shot, the further apart the corresponding points (i.e. a higher disparity value). An object at infinity would have zero disparity. Due to this fact, the size (width) of the sweep area window was set to 40 pixels. This corresponded to the minimum distance - and therefore maximum disparity - an object could have in the chosen scene set-up.

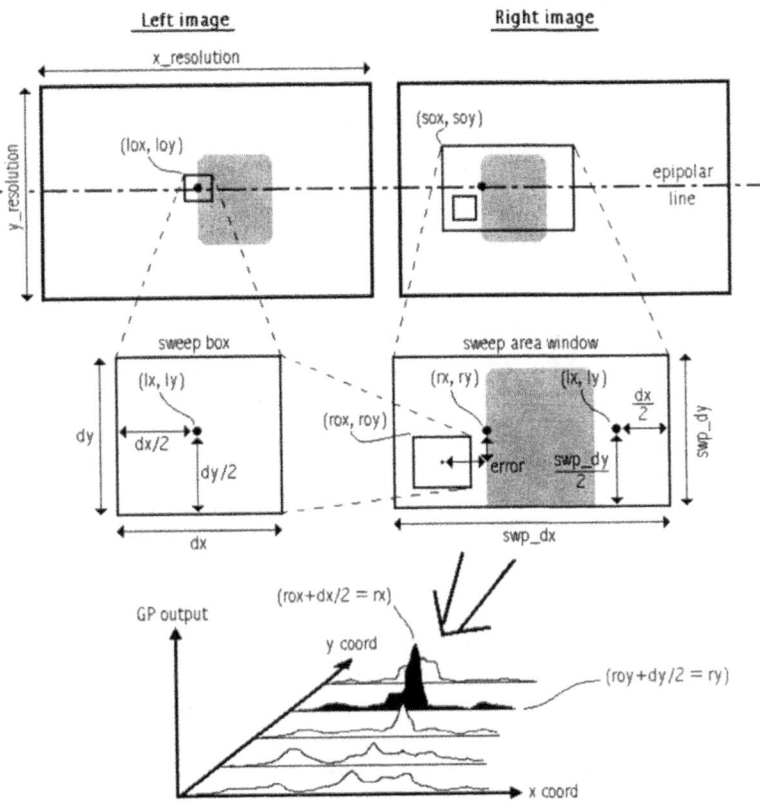

Fig. 4. The sweeping box approach.

Descriptions of the functions and values of the variables and constants in the figure above can be found in Table 3 below.

Table 3. Descriptions to Fig. 4.

Name	Description	Value
lx, ly	Coord. of the left interest point.	
rx, ry	Coord. of the right interest point.	
dx, dy	Width and height of sub-image used as input to the individuals.	5x5 pixels
Error	Distance from the correct position.	
swp_dx, swp_dy	Width and height of the sweeping area window.	40x5
lox, loy	Origo of the left box (sub-image).	
rox, roy	Origo of the right sweeping box.	
sox, soy	Origo of the sweeping area window.	
x coord	An x-axis along swp_dx.	0 – 40

y coord	An y-axis along swp_dy	$0-5$
GP output	The matching score output from an individual.	
x_resolution	Width of the stereoscopic images.	160 pixels
y_resolution	Height of the stereoscopic images.	120 pixels

3   Results

Below follows the results of the two applied approaches. First the less successful approach where the program is free to wander around in the pixels of the image and secondly the better working approach where a fixed scan path is used over the pixels in the image.

3.1 Experiment I

The evolution in this approach progressed slowly, and the majority of the individuals had an average disparity-error of around 14 pixels. That is about ten times more than is accepted to reach a useful accuracy in the distance estimation calculations.

The main problem was that the evolution got stuck in local minima. A disparity of 14 pixels is close to the average shift of the interest points in some of the training sets. Due to this fact, a majority of the individuals gave a constant output – a disparity of 14 pixels. The lazy nature of an evolutionary process kept itself in this local minimum. A couple of minor variations of the algorithm were tried out, in the hope of finding a way to force the individuals to search for structures in the images, but they did not improved the results. Filtering the points in the training set (Fig. 1) in a way that caused two groups of points - one representing "close" objects and the other objects "far away" - did not force the individuals to find a strategy to classify them as "far away" or "close".

A positive result from this experiment was that the genetic system evolved a sweeping behaviour for the control of its two freely movable points. But this did not help the individuals enough to find usable structures in the image data, retrieved from the position of the moving points.

3.2 Experiment II

This approach gave better results, and the evolution of useful individuals was really fast. The resulting average disparity-error of 1.5 pixels is acceptable and results in useful distances from the triangulation calculations. This disparity-error corresponds to an uncertainty of about 10% of the real world distance, which accurate enough for the first navigation experiments with our robot.

Variations in the experiment were tried out, and the evolutionary process progressed with the highest speed when giving the image-data in grayscales, instead of

RGB triplets. The conversion was done using the luminance/brightness part[5] in the YIQ colour model.

The figures below show the recorded behaviour of three individuals. These particular individuals are evolved over 400 generations, but good individuals arise already in generation 60. However, some of these early individuals tend to make infrequent large mistakes (up to eight pixels).

The blue graph (GP output - left axis) shows the individuals output value (their matching "probability") that they produced during one single sweep. The yellow graph (Error - right axis) shows the disparity-error in pixels that the individual would produce, if it used the current "Max output" (purple graph – left axis) as its final answer/output. The bottom scale shows the pixel-position of the sweeping box in the sweeping area window.

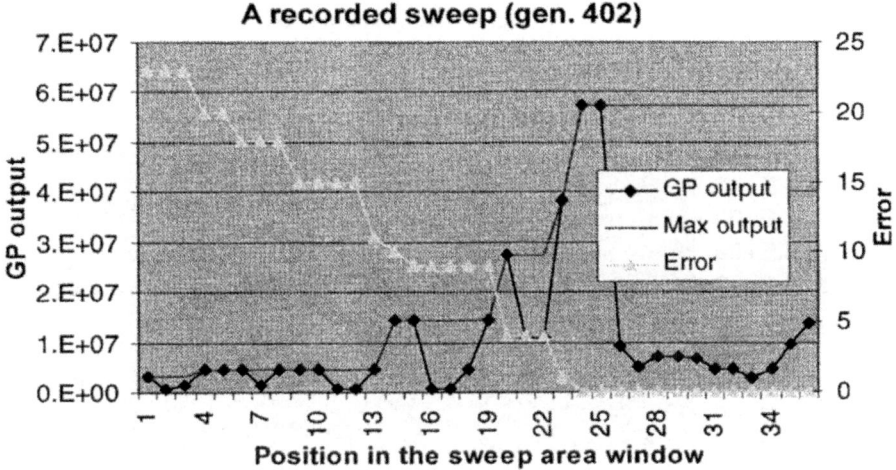

Fig. 5. Staircase behavior

---

[5] $Y = 0.299*R + 0.587G + 0.144B$. YIQ is used by the NTSC.

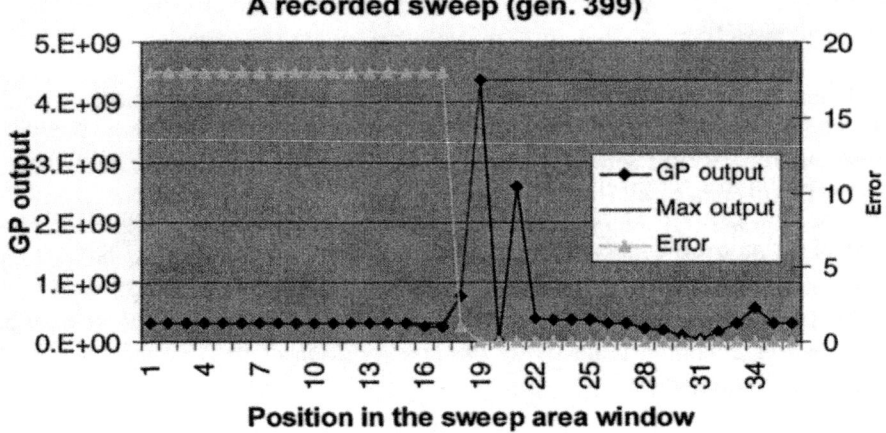

Fig. 6. Threshold behavior

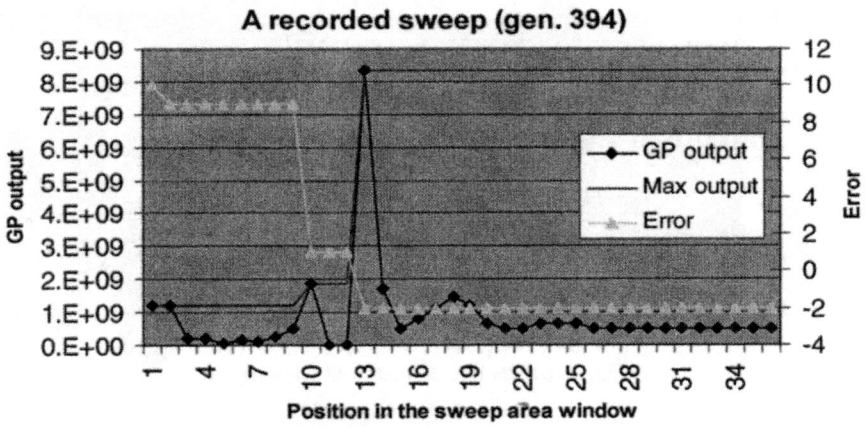

Fig. 7. A 2-pixel matching error

4   Future work

Object recognition is our next goal. One way would be to evolve 3D scenes using simple geometrical objects, e.g. spheres, boxes and planes. The fitness function would describe how well the evolved 3D scene matched the 3D scene calculated from the disparity map produced by the approach discussed in this article. This is true in the real world – even though you only see the front side (a "square") of a fridge, it is actually an object with a depth a "cube". This method would give a way of generalising from "flat" images into hypothesises of a full 3-D model of the world.

Future work on the sweeping box approach would be to include the shape of the sweeping box in the evolutionary process.

5   Summary and Conclusions

We have presented the first approach to stereovision using genetic programming and unprocessed pixel data. The resulting machine code individuals are efficient and enable real time 3-D map generation. However, much research remains from this *proof of concept experiment* but several directions of the approach seem promising. We also regard evaluations in the autonomous robot as an important line of research for the future.

Acknowledgements

Peter Nordin gratefully acknowledges support from TFR and NUTEK.

References

1.   Banzhaf Wolfgang, Nordin Peter, Keller Robert E., Francone Frank D. (1998) "Genetic Programming, An Introduction – On the Evolution of Computer Programs and Its Applications".
2.   Hearn Donald, Baker Pauline M. (1997) "Computer Graphics, C Version" 2:nd Edition, National Center of Supercomputing Applications, University of Illinois, USA.
3.   Iocchi Luca (1989) "Stereo Vision: "Multiresolution Stereo Vision System for Mobile Robots" Dipartimento di Informatica e Sistemistica, Università di Roma "La Sapienza", Italy.
4.   Nordin Peter, Nordahl Mats G. (1998) "An Evolutionary Architecture for a Humanoid Robot". Institute of Physical Resource Theory, Chalmers University of Technology, Göteborg, Sweden.
5.   Nordin, Peter. "Evolutionary Program Induction of Binary Machine Code and its Application. Krehl Verlag, Mnster, Germany.
6.   Stuart Rusell J., Norvig Peter (1995) "Artificial Intelligence, A Modern Approach". USA.
7.   The Humanoid Project WWW site http://humanoid.fy.chalmers.se

A Faster Genetic Clustering Algorithm

L Meng[1], Q H Wu[1], and Z Z Yong[2]

[1] Department of Electrical Engineering and Electronics
The University of Liverpool, Liverpool, L63 3GJ, U.K.
q.h.wu@liv.ac.uk
[2] Department of Electronic Engineering
Shenzhen University, Shenzhen 518060, P. R. China

Abstract. This paper presents a novel genetic clustering algorithm combining a genetic algorithm (GA) with the classical hard c-means clustering algorithm (HCMCA). It processes partition matrices rather than sets of center points and thus provides a new implementation scheme for the genetic operator - recombination. For comparison of performance with other existing clustering algorithms, a gray-level image quantization problem is considered. Experimental results show that the proposed algorithm converges more quickly to the global optimum and thus provides a better way out of the dilemma in which the traditional clustering algorithms are easily trapped in local optima and the genetic approach is time consuming.

1 Introduction

Clustering methods play a vital role in exploratory data analysis. In the field of pattern recognition [1], the conventional c-means clustering algorithms (CM-CAs) have been widely applied. Broadly speaking, CMCAs can be classified into *hard c-means* (HCM) and *fuzzy c-means* (FCM) clustering algorithms. Hard clustering deals with assigning each object point to exactly one of the clusters, whereas fuzzy clustering extends this concept to associate each object point to each of the clusters with a measure of belongingness. The belongingness of object points to the hard/fuzzy clusters is typically represented by a membership matrix called a hard/fuzzy partition, respectively. In general, CMCAs aim at finding the optimal partition and optimizing a clustering objective function using calculus-based methods. However, clustering objective functions are highly non-linear and multi-modal functions. As a consequence, by hill-climbing, CMCAs can be easily trapped into local extrema associated with a *degenerate* partition (i.e., a partition with one or more empty rows, meaning that fewer than c clusters were obtained in the final partition). Moreover, they are also significantly sensitive to the initial conditions.

A way to achieve both avoidance of local extrema and minimal sensitivity to initialization is to use stochastic optimization approaches, such as *evolutionary algorithms* (EAs). An EA is inspired by organic evolution and has been widely believed to be an effective global optimization algorithm. In [3] $\sim$ [6], genetically guided approaches were defined for the optimal clustering problems and

S. Cagnoni et al. (Eds.): EvoWorkshops 2000, LNCS 1803, pp. 22–33, 2000.

experiments were performed with different data sets. Results showed that an EA can ameliorate the difficulty of choosing an initialization for the CMCAs and provides a viable way to avoid local extrema. However, as stated in [3], an EA clustering approach takes up to two orders of magnitude more time than HCM/FCM clustering algorithms.

EAs are iterative schemes operating on a population of so-called individuals. Each individual in the population corresponds to a viable solution to the problem in hand. In previous work on genetically guided clustering ([4] ∼ [6]), each individual in the population has been designed to represent a matrix of cluster centers. Although under two sets of clustering criteria the mapping between a partition and the corresponding set of cluster centers is exclusive, the goal of a typical clustering approach is essentially the optimal partition rather than the positions of cluster centers. In this paper, a novel genetically guided clustering algorithm is defined, which uses partition matrices as the individuals in population. This hybrid algorithm combines a genetic algorithm (GA) with the classical hard c-means clustering algorithm (HCMCA) and is therefore termed as a genetic hard c-means clustering algorithm (GHCMCA). GHCMCA has been applied to a gray-level image quantization problem. Compared with HCMCA and GGA approaches, GHCMCA converges faster and always to an nondegenerate partition which is or is nearby the global optimum.

2 Problem Formulation

Consider a set of n vectors $X = \{\boldsymbol{x}_1, \boldsymbol{x}_2, \ldots, \boldsymbol{x}_n\}$ to be clustered into c groups of like data. Each $\boldsymbol{x}_i \in \Re^s$ is a feature vector consisting of s real-valued measurements describing the features of the object represented by $\boldsymbol{x}_i$. Hard or fuzzy clusters of the object points can be represented by a hard/fuzzy membership matrix called a hard/fuzzy partition. The set of all $c \times n$ nondegenerate hard partition matrices is denoted by $M_{c \times n}$ and defined as

$$M_{c \times n} = \{U \in \Re^{c \times n} \mid \sum_{i=1}^{c} U_{ik} = 1, 0 < \sum_{k=1}^{n} U_{ik} < n,$$
$$U_{ik} \in \{0, 1\}; 1 < i < c; 1 < k < n\} \tag{1}$$

where $U \in M_{c \times n}$ is a hard partition matrix. The number of possible U's, i.e. the number of ways of clustering n objects into c nonempty groups, is a Stirling number of the second kind given by (see [7])

$$(1/k!) \sum_{l=0}^{k} (-1)^{k-l} \binom{n}{k} l^n. \tag{2}$$

The clustering objective function for hard c-means (HCM) partition is the HCM function

$$J_1(U, V) = \sum_{i=1}^{c} \sum_{k=1}^{n} U_{ik} \cdot D_{ik}^2(\boldsymbol{v}_i, \boldsymbol{x}_k) \tag{3}$$

where $V = [v_1, v_2, \ldots, v_c]$ is a matrix of prototype parameters (cluster centers) $v_i \in \Re^s$; and $D_{ik}(v_i, x_k)$ is the Euclidean distance between the kth feature vector x_k and the ith cluster prototype v_i, which is of the form $D(x, y) = \sqrt{(x - y)'(x - y)}$. This objective function describes the accumulated squared error when replacing each feature vector by the center of cluster to which it belongs. Hence, it is actually a measure of distortion.

Minimizing the clustering objective function with respect to U leads to the following sets of clustering criteria. For each cluster i ($i \in [1, c]$), firstly,

$$v_i = \frac{\sum_{k=1}^{n} U_{ik} x_k}{\sum_{k=1}^{n} U_{ik}} \qquad (4)$$

i.e., the cluster centers are positioned at the center of mass of the feature vectors belonging to the cluster. The second set of criteria that minimizes the HCM function states that an objective point should be associated with the closest cluster center:

$$U_{ik} = \begin{cases} 1, & \text{if } i = \min_j[D_{jk}^2(v_j, x_k)]; \\ 0, & \text{otherwise.} \end{cases} \qquad (5)$$

In equations (1) $\sim$ (5), $i = 1, 2, \ldots, c$ denotes the index number of a cluster or a cluster center and $k = 1, 2, \ldots, n$ denotes the index number of an object point. HCMCA adopts both sets of these clustering criteria and minimizes the objective function $J_1(U, V)$ by alternatively updating V and U using equations (4) and (5).

3   A Novel Genetic Clustering Algorithm

Starting with an initial condition, a GA evolves a population towards successively better regions of the search space by means of randomized processes of *recombination, mutation,* and *selection.* The given optimization problem defines an environment that delivers a quality information (*fitness values*) for new search points, and the selection favors those individuals of higher quality to reproduce more often than worse individuals. The recombination mechanisms allows for mixing of parental information while passing it to their offspring, and mutation introduces innovation into the population.

In brief, a GA search for the optimal individual is typically implemented as follows.

1. Generate an initial population;
2. Evaluate the fitness value of each individual in the current population;
3. Select pairs of parents;
4. Generate offspring of the selected parents via recombination and mutation;
5. Replace the parents with their offspring and create a new generation;
6. Halt the process if a termination criterion is met. Otherwise, proceed to step 3.

To reduce the search space drastically, we introduce HCMCA into the typical GA implementation process. During every GA generation, with fitness evaluation taking place, a single HCMCA reallocation step is applied to each individual in the population. According to the partition matrix represented by an individual, the set of cluster centers are located using equation (4) and then the partition matrix is updated by equation (5), i.e. by assigning each object point to its closest cluster. After all the individual partitions are updated using HCMCA, the three basic genetic operations - selection, recombination and mutation - start.

In order to apply a genetic approach to a given problem, a number of fundamental issues must be addressed in advance. They are solution representation, fitness function, creation of the initial population and a succeeding new generation, implementation schemes of genetic operators, termination criterion, and the GA parameter settings. The rest of this section describes each of these issues in detail.

3.1   Solution Representation

As mentioned in section 1, each individual in the population is a hard partition matrix $U \in M_{c \times n}$. In hard clustering, any object point belongs to the closest cluster exclusively. There is only a 1 down any column of a hard partition matrix. Hereof, it is possible to simpify a $c \times n$ hard partition matrix U into an n-dimensional vector u with the ith element describes which row the 1 lies down the ith column of the original U. The possible values of the elements of u range from 1 to c. This simplification is adopted in the proposed HCMCA.

3.2   Fitness Function

To allow comparison of performance with the classical HCMCA as well as an existing genetic clustering approach - GGA [3]), HCM function $J_1(U, V)$ is used here as the objective function to be minimized. And the matrix V is calculated with respect to U using equation (4). To largely reduce the chance of GHCMCA becoming stuck at a degenerate partition, we have taken the number of empty clusters into consideration. Any value obtained using HCM function is scaled with a penalty factor. Different from the one in [3], our objective function is redefined as follows:

$$J'_1(U, V) = J_1(U, V) \times (1 + e/c) \tag{6}$$

where c is the total number of clusters and $0 \le e \le c$ denotes the number of empty clusters and is evaluated via counting the all-0 rows in U.

The goal of a clustering approach is to minimize the objective function while a GA favors fitter individuals. To compromise, we use the inverse of an individual's scaled HCM function value as its fitness value. In addition, a linear fitness scaling mechanism [8] has been introduced to maintain reasonable selection pressure.

3.3   Initialization

Consider that good choice for starting configurations should be free of overt biases. For the hard partition vectors u in the initial population, each element

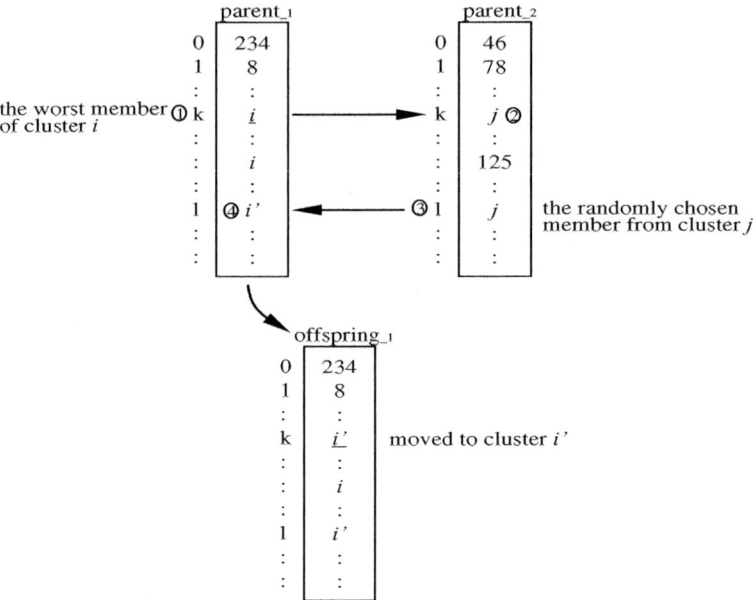

Fig. 1. Graphical description of the recombination strategy when it is applied to the ith cluster of parent_1's partition u_1.

is set to a randomly generated number in the range of $[1, c]$. By doing so, we actually randomly partition the object points to c initial clusters.

3.4   Genetic Operators

In every generation, a GA select parents from the current population. Theoretically, the probability of one individual being selected is proportional to its fitness value relative to the others' fitness values. After being selected, parents are mated to give birth to their offspring. Offspring are generated via the operations of recombination and mutation.

a. Selection. As to the selection operator, the *stochastic universal sampling* scheme is applied [9]. Based on the theoretical and empirical analysis, Baker concluded [9] that this scheme is an optimal sequential sampling scheme which, for the first time, assign offspring according to the theoretical specifications.

b. Recombination. A brand-new recombination strategy has been designed for the clustering problems. It reallocates the worst member in each cluster of a parent's partition according to its mating partner's partition. The worst member of a certain cluster is the farthest object point to the center of that cluster, among those belonging to it.

Particularly, for the ith cluster of parent$_1$'s partition u_1, recombination is carried out as follows:

1. find the worst member of the ith cluster of u_1 (suppose it's x_k);
2. according to the mating partner's partition vector u_2, find out to which cluster this worst member x_k belongs (suppose it's the jth cluster of u_2);
3. still according to u_2, randomly choose another object point from the jth cluster of u_2 (suppose it happens to be x_l);
4. back to u_1, check out to which cluster of u_1 the chosen object point x_l belongs and simply reallocate x_k to that cluster.

While still holding the randomness property, this recombination is well guided. In step 2, the mating partner's partition u_2 is referred to answer the following question: which object points should share the same cluster with the worst member of a certain cluster of partition u_1? Since there is at least one (itself) while may be more than one object points available, step 3 randomly choose one of them. Finally, in step 4, the worst member of partition u_1's ith cluster is reallocated to a new cluster such that it belongs to the same cluster as this randomly chosen object point. For better understanding, these implementation steps are described graphically in Figure 1.

The recombination process is applied to each parent of the mating pair cluster by cluster. That is to say, each cluster is adjusted independently of the others.

c. Mutation. After every recombination, mutation is imposed on each element of the newly constructed partition with a mutation probability p_m. Mutation sets the chosen elements to a randomly generated integer ranging from 1 to c.

3.5   Creation of A New Generation

Our genetic clustering algorithm is a steady-state GA, which replaces only a fraction of the population each generation. The motivation of introducing a steady-state GA is to keep a good balance between exploitation of the best regions found so far and continued exploration for potentially better payoff area.

However, a steady-state GA will increase the variance along the growth curves of individuals [10]. To reduce the variance, a First-In-First-Out (FIFO) deletion [10] is employed and thus permit the use of this steady-state GA with smaller populations. With FIFO deletion the population is simply a first-in-first-out queue with new individuals added to one end and deleted individuals removed from the other end.

Moreover, elitism is implemented to keep the best member of the population.

3.6   Parameter Settings

The genetic clustering approach is computational expensive. The dominant cost is that of calculating the fitness value for each individual in the population for every generation. This cost is a function of s, n, and c, i.e., the number of features, number of object points, and number of classes to cluster the object points. An increase in any of these parameters results in a linear increase in time per generation.

Since the total computational time is also proportional to the population size, small populations are preferred. Throughout the experiments with GHCMCA, we used a population size of 30.

In every generation, 60% individuals of the population will undergo the three basic genetic operations to generate offspring. The probabilities of succeeding recombination and mutation are 0.95 and 0.001, respectively [11]. These parameter settings have been found to offer best results.

The termination criterion used here is the number of generations. The approach stops when the required generation is created.

4   Experiments

As clustering is often applied to image processing and images are real-world domains of significant complexity in terms of number of object points to be clustered and number of classes, a gray level image quantization problem is considered. In this application case, the 256×256 black-and-white Lena image is firstly divided evenly into small blocks of 4×4 pixels. Then the gray levels of the pixels in each of these block compose a vector such that there are 4096 image vectors of 16 features (gray levels). The goal of this image quantization problem is to cluster these 4096 image vectors into 256 classes. Hence, $n = 4096$, $c = 256$, and $s = 16$.

To this image quantization problem, the classical HCMCA, a present genetically guided clustering algorithm - GGA [3], and our own GHCMCA have been applied independently. And the object points processed by all of these algorithms were the 4096 image vectors consisting of 16 pixel gray levels.

The classical HCMCA was implemented as follows:

1. Randomly partition the image vectors into initial clusters. Calculate the corresponding set of cluster centers using equation (4).
2. Equation (5) is applied such that according to its 16 pixel gray levels, each image vector is assigned to the closest cluster center. Meanwhile, the Euclidean distance between each image vector and its closest cluster center is summed to the accumulated squared error J_1.
3. Equation (4) is applied. Each cluster center is updated as the center of mass of all the image vectors belonging to it.

Steps 2 and 3 are applied alternatively until the relative difference in the accumulated squared error of two successive iterations is less than $\varepsilon = 0.005$ (i.e. abs$(J_{1\_last} - J_{1\_current})/J_{1\_current} < \varepsilon$). Step 2 is repeated using the last set of cluster centers. The resulting accumulated squared error J_1 and number of empty clusters are recorded.

Hall $et.$ al designed the GGA approach following the typical implementation steps of a GA and the HCMCA was also introduced in each generation. The three genetic operators were tournament selection, two-point recombination and random mutation. In GGA, individuals of the populations were sets of cluster centers instead of partition matrices.

For comparison, the GGA approach has been repeated. However, there are some differences between our experiments and theirs and these are highlighted as follows:

1. While they had used k-fold tournament selection with $k = 2$, we have used stochastic universal sampling scheme, as in our own GHCHCA.
2. While they had used a generational GA with an elitist strategy of passing the two fittest individuals to the next generation, we have used a steady-state GA with generation gap of 0.6 throughout. Elitism is implemented to keep the fittest individual.
3. While they had used a binary gray code representation for the individuals, we have used real value representation.

Like those used throughout the experiments for our genetic clustering algorithm, we have chosen the same parameter settings in all the repeated GGA approaches. They were a population size of 30, a recombination probability of 0.95, and a mutation probability of 0.001. Except that better solution can be obtained with larger population size, these value offer GGA the best performance [3]. Every approach stopped at the 40th generation.

5   Results

Tests on clustering image Lena by the classical HCMCA, the GGA approach, and our GHCMCA are undertaken respectively. The key measures for comparing algorithm performances are the mean squared error (MSE) and number of empty clusters, e, associated with a partition matrix. The mean squared error is the value of the HCM function J_1 averaged by the number of image pixels (MSE = $J_1/(256 \times 256)$), which indicates the distortion between the quantized image and the original image. During each run, the partition with the lowest accumulated squared error is traced and the MSE and e values associated with the optimal partition finally found are recorded.

The recorded values of MSE and e are averaged over the total number of runs, i.e. 100, 40 and 40 for HCMCA, GGA and GHCMCA respectively. The average values and the standard deviations of MSE and e are reported in Table 1. The MSE and e values associated with the very best partition ever obtained by each algorithm are also included.

Figure 2 shows the histogram of the recorded MSE and e values for HCMCA, GGA and GHCMCA runs, respectively. The histogram is a statistical representation of the distribution of local optimal obtained by the algorithm. Adding results of more runs did not change the distribution further.

From Table 1 and Figure 2, the following observations are drawn:

1. Using HCMCA, a broad distribution of local optimal is observed. The standard deviations of MSE and e are much higher than those of the genetic

Table 1. Results for trials with three clustering algorithms

	HCMCA	GGA	GHCMCA
average MSE	138.1317	81.5835	77.7275
st. dv. of MSE	5.3635	3.1002	1.2679
MSE of the best	125.1873	76.9357	74.8591
average e	184.0990	3.9500	0
st. dv. of e	6.5917	1.9994	0
e of the best	169	5	0

approaches. This indicates that classical HCMCA is sensitive to the initial distribution of cluster centers and easily trapped in local optima, while, on the other hand, shows the viability of a genetic approach to overcome these problems.

2. Both genetic guided clustering algorithms outperform the classical HCMCA in the sense that they end up in partitions associated with much lower distortion and significant fewer empty clusters. Partitions of similar MSE values were repeatedly found by GGA and GHCMCA. This indicates that these results are indeed nearly optimal.

3. In no case did GHCMCA result in a degenerate partition. On average, the best MSE value obtained by GHCMCA is slightly lower than that obtained by GGA. So is for the very best MSE ever found.

During each genetically guided clustering approach using either GGA or GHCMCA, the MSE value and empty cluster number of the best partition found up to and including each successive generation are recorded and then averaged over the total number of runs. Figure 3 shows the resultant average values of MSE and e with respect to the generation number for both GGA and GHCMCA. According to the two MSE curves, we see that, for both genetic approaches, the initial convergence rates are very high and as the generations progress convergence rate decreases rapidly. However, in the earlier generations, GHCMCA converges much faster than GGA and quickly reach the desired region where the nondegenerate partition matrices reside. GHCMCA can find an MSE value lower than the lowest ever found by HCMCA after the 5th generation, while GGA does it after generation 10. As stated, for clustering problems the fitness evaluation at every GA generation is time consuming. Although GGA soon catches up and from generation 18 the difference in MSE values between two genetic clustering algorithms is omittable, in the special cases where the speed as well as performance is required GHCMCA may provide a much faster way to find an acceptable solution. Furthermore, as HCMCA assures local optimality and due to its hill-climbing method converges much faster than any genetic approach, instead of waiting for the genetic approaches reach an exact optimal solution, we may stop genetic search after a necessary number of generations and use HCMCA to find the corresponding local optimum.

6   Conclusion

In this paper a novel genetic clustering algorithm is proposed, which combines a genetic algorithm (GA) with the classical hard c-means clustering algorithm (HCMCA). Unlike other clustering algorithms, GHCMCA processes partition matrices rather than sets of center points and thus allows a new implementation scheme for the genetic operator - recombination. For comparison of performance with other present clustering algorithms, experiments on a gray-level image quantization problem have been conducted. The results show that GHCMCA converges much quicker to the global optimum and provides a viable way to solve the dilemma where the classical HCMCA is found easily caught in local optima and a genetic approach requires large time consumption.

References

1. J. T. Tou and R. C. Gonzalez, *Pattern Recognition Priciples.* Addison-Wesley, Reading, Massachusetts, 1974.
2. P. Scheunders, "A genetic Lloyd-Max image quantization algorithm," *Pattern Recognition Lett.*, vol. 17, pp. 547-556, 1996.
3. L. O. Hall, I. B. zyurt and J. C. Bezdek, "Clustering with a genetically guided optimized approach," *IEEE Trans. Evolutionary Computation*, vol. 3, no. 2, pp. 103-112, 1999.
4. P. Scheunders, "A genetic c-means clustering algorithm applied to color image quantization," *Pattern Recognit.*, vol. 30, no. 6, pp. 859-866, 1997.
5. F. Klawonn, "Fuzzy clustering with evolutionary algorithms," in *Proc. Seventh IFSA World Congress*, vol. 2, pp. 312-323, 1997.
6. G. P. Babu and M. N. Murty, "Clustering with evolutionary strategies," *Pattern Recognit.*, vol. 27, no. 2, pp. 321-329, 1994.
7. M. Abramowitz and I. A. Stegun, eds., *Handbook of Mathemeitcal Functions*, U. S. Department of Commerce, National Bureau of Standards Applied Mathematical Series. 55, 1964.
8. D. E. Goldberg, *Genetic Algorithms in Search, Optimization, and Machine Learning.* Addison-Wesley Publishing Company, Inc, 1989.
9. J. E. Baker, "Reducing bias and inefficiency in the selection algorithm," *Proc. Second Int. Conf. Genetic Algorithms*, pp. 14-21, 1987.
10. K. A. De Jong and J. Sarma, "Generation gaps revisited," *Foundations of Genetic Algorithms 2*, D. Whitley (ed.), pp. 19-28. Vail, CO:Morgan Kaufmann, 1993.
11. N. N. Schraudolph and R. K. Belew, "Dynamic parameter encoding for genetic algorithms," *Machine Learning*, vol. 9, no. 1, pp. 9-21, 1992.

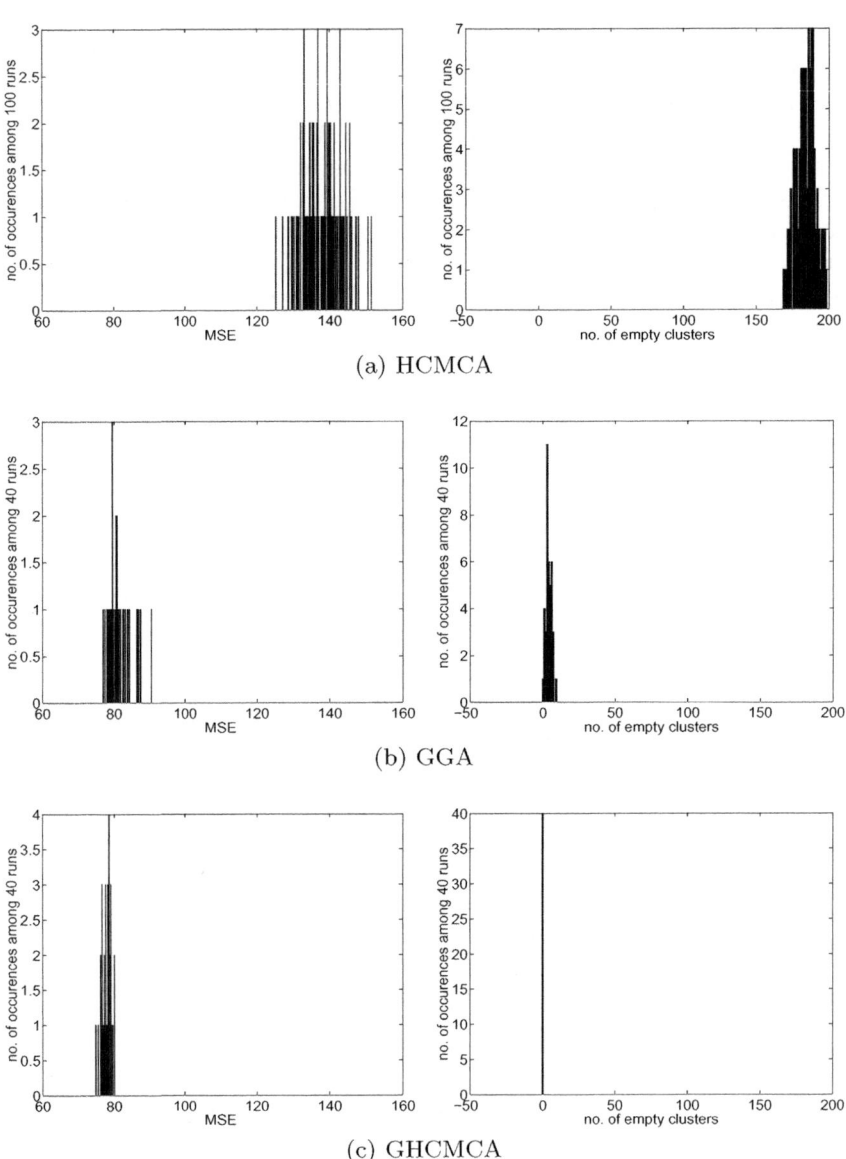

Fig. 2. Distribution of the mean square error and empty cluster number associated with the optimal partitions finally found when clustering the Lena image with HCMCA, GGA and GHCMCA, respectively.

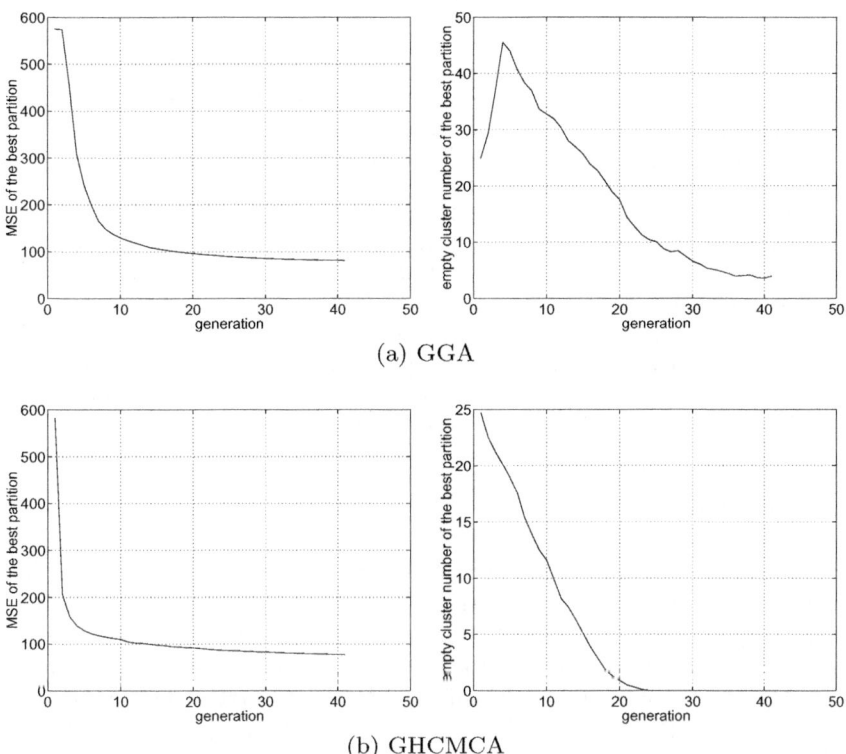

(a) GGA

(b) GHCMCA

Fig. 3. Convergence properties of GGA and GHCMCA with respect to the generation number

Scene Interpretation using Semantic Nets and Evolutionary Computation*

D. Prabhu[1], B. P. Buckles[2], and
F. E. Petry[2]

[1]*i2* Technologies, 1603 LBJ Freeway, Suite 780, Dallas TX 75234, USA and
[2]Department of Electrical Engineering & Computer Science, Tulane University, New
Orleans, LA 70118, USA

Abstract. The fitness function used in a GA must be measurable over
the representation of the solution by means of a computable function.
Often, the fitness is an estimation of the nearness to an ideal solution or
the distance from a default solution. In image scene interpretation, the
solution takes the form of a set of labels corresponding to the components
of an image and its fitness is difficult to conceptualize in terms of distance
from a default or nearness to an ideal. Here we describe a model in
which a semantic net is used to capture the salient properties of an ideal
labeling. Instantiating the nodes of the semantic net with the labels from
a candidate solution (a chromosome) provides a basis for estimating a
logical distance from a norm. This domain-independent model can be
applied to a broad range of scene-based image analysis tasks.

1 Introduction

We describe how a genetic algorithm (GA) can be employed to classify/label
objects in a scene for which no prior truth data exists. Relationships among the
objects in a typical scene from the domain of discourse are encapsulated within a
semantic net. The method was validated using a test suite of images captured by
satellites. Mainly, these included the infrared band of North Atlantic scenes and
two bands of AVHRR data depicting regions of the Western U.S. The objective
of the North Atlantic image analysis was to identify currents such as the Gulf
Stream and eddies. The objective of using AVHRR images was to detect and
identify clouds by type.

2 Background

Classification or labeling of segments is the focus of this paper. Labeling a segment of an image is a particularly difficult subtask because there must be an
automatic method of assigning a figure of merit to a candidate solution. A very

---

* This work was supported in part by a grant from NASA/Goddard Space Flight
Center, #NAG5-8570 and in part by DoD EPSCoR and the State of Louisiana
under grant F49620-98-1-0351.

S. Cagnoni et al. (Eds.): EvoWorkshops 2000, LNCS 1803, pp. 34–43, 2000.

general and computationally reasonable method for defining fitness function for image labeling is described that is based upon developing a semantic net for a typical scene from the domain. The classes of the semantic net are defined by the labels to be assigned to the segments. The predicates are relationships that exist among objects bearing the corresponding labels. Given a candidate solution (a labeling) for an image, a measure is described that estimates the conformance of the labeling to the semantic net.

3   Methodology

In this section, the formal approach and extensions to our prior ad hoc model using semantic nets. [1] is developed. A semantic network is a structure used to represent knowledge as a combination of nodes interconnected with arcs.

3.1   Description of Scene Properties with Semantic Nets

Let CL be the set of possible classification labels (or classes) specific to the application.

$$CL = \{c_1, c_2, \ldots, c_m\} \tag{1}$$

Let S be the set of segments/components in the given image:

$$S = \{s_1, s_2, \ldots, s_n\} \tag{2}$$

Each of these components is assumed to be completely characterized by T, the set of features specific to the application. Let,

$$T = \{t_1, t_2, \ldots, t_h\} \tag{3}$$

For any given image component, each of these features, t_i, takes values from some corresponding domain D_i.

Assume that every statement about such application-specific knowledge that is useful for the classification task can be described by using at most k_{max} number of components drawn from the image. First, consider only those statements which involve exactly k (for some $1 \leq k \leq k_{max}$) components of the image. Let C_k be the set of all k-tuples built from the indices of the elements of CL. i.e.,

$$C_k = \{ \langle j_1, \ldots, j_k \rangle \mid 1 \leq j_1, \ldots, j_k \leq m \} \tag{4}$$

Every image component $s \in S$ can be an instance of only one of the m classes drawn from the set CL. Denote this by ISA, an instance function which maps each image component to the index of a class.

$$ISA : S \longrightarrow \{1, \ldots, m\} \tag{5}$$

Thus, given a k-tuple of image components/segments, ISA can be used to generate a corresponding k-tuple $\langle j_1, \ldots, j_k \rangle \in C_k$. Let F_k be a set of feature-value comparator functions such that each function $f \in F_k$ maps a k-tuple of feature values drawn from the domain D_i corresponding to some feature $t_i \in T$, to an absolute comparison value.

$$F_k = \{ f \mid f : D_i^k \longrightarrow \mathcal{R} \} \tag{6}$$

Let P_k be the set of predicates such that each $p \in P_k$ is a fuzzy predicate mapping a given absolute comparison value to a fuzzy truth-value in the continuous range of $[0, 1]$.

$$P_k = \{ p \mid p : \mathcal{R} \longrightarrow [0, 1] \} \tag{7}$$

Predicates based on fuzzy logic are more appropriate than binary predicates for this task in view of the heuristic knowledge employed. In addition, $null \in P_k$, and $null(x) = 0$ for all x.

For *every* k-tuple of image components, *every* feature-value based relationship (inherited from the corresponding k-tuple of classes $\langle j_1, \ldots, j_k \rangle \in C_k$) among its elements can be quantified by using some specific feature-value comparator function and a corresponding predicate. For a given k-tuple of classes, $\langle j_1, \ldots, j_k \rangle \in C_k$, there may exist multiple relationships among the elements of the k-tuple of class-instances. Therefore, we define a mapping, $R1_k$ such that every given tuple of classes denotes a specific subset of the available predicates, *i.e.*,

$$R1_k : C_k \longrightarrow 2^{P_k} \tag{8}$$

We also define another functional mapping $R2_k$ such that every predicate is mapped to a feature comparator function. *i.e.*,

$$R2_k : P_k \longrightarrow F_k \tag{9}$$

Thus, every relationship involving k components can be uniquely denoted by the tuple $\langle \langle j_1, \ldots, j_k \rangle, p \rangle$ where, $\langle j_1, \ldots, j_k \rangle \in C_k$ and $p \in R1_k(\langle j_1, \ldots, j_k \rangle)$. Consider a k-tuple of image components/segments that is instantiated (by virtue of some ISA mapping) to be a tuple of class-instances corresponding to $\langle j_1, \ldots, j_k \rangle$. Every relationship $\langle \langle j_1, \ldots, j_k \rangle, p \rangle$ inherited by the image components can be quantified as follows: Suppose that $f = R2_k(p)$ and that f makes use of the feature $t_i \in T$ using k-tuple of values from the domain D_i. Now, this relationship can be quantified by the composition $p(f(x_1, \ldots, x_k))$ where $x_1, \ldots, x_k \in D_i$ are the feature values of the image segments.

The set of all such relationships constitutes a semantic net, SN_k. In other words,

$$SN_k = \{ \langle \langle j_1, \ldots, j_k \rangle, p \rangle \mid \langle j_1, \ldots, j_k \rangle \in C_k \wedge p \in R1_k(\langle j_1, \ldots, j_k \rangle) \} \tag{10}$$

This formal model for representing and utilizing the knowledge relating k elements can be summarized by FN_k, a fitness net.

$$FN_k = \langle CL, S, T, ISA, C_k, P_k, R1_k, F_k, R2_k, SN_k, W_k, E \rangle \tag{11}$$

where, W_k is a set of weights, one for each predicate $p \in P_k$, and E is a fitness function described in section 3.3.

This model can now be generalized to represent all possible statements from the knowledge base, *i.e.*, for all values of k, $1 \leq k \leq k_{max}$. Such a complete model is given by:

$$FN = \langle CL, S, T, ISA, \mathcal{C}, \mathcal{P}, R1, \mathcal{F}, R2, \mathcal{SN}, W, E \rangle \tag{12}$$

where,

$$\mathcal{C} = \bigcup_{1 \leq k \leq k_{max}} C_k \qquad \mathcal{P} = \bigcup_{1 \leq k \leq k_{max}} P_k$$
$$R1 = \bigcup_{1 \leq k \leq k_{max}} R1_k \qquad \mathcal{F} = \bigcup_{1 \leq k \leq k_{max}} F_k$$
$$R2 = \bigcup_{1 \leq k \leq k_{max}} R2_k \qquad \mathcal{SN} = \bigcup_{1 \leq k \leq k_{max}} SN_k$$
$$W = \bigcup_{1 \leq k \leq k_{max}} W_k$$

To illustrate the notions discussed above, consider a simple domain such as a chair shown in Figure 1. Here we have, $CL = \{s, b, a, l\}$, where the symbols de-

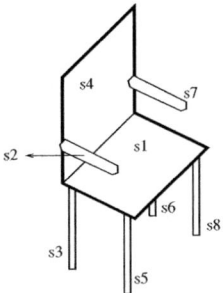

Fig. 1. A chair and its segments

note, and index, the classes "Seat", "Back", "Arm", and "Leg" respectively. Also, $S = \{s1, s2, s3, s4, s5, s6, s7, s8\}$ from Figure 1. There are two measures/features for each segment: angle of rotation from the horizontal, t_1, and surface area, t_2. While relationships of any degree (except zero) are permitted, for the sake of simplicity we assume only binary relationships for this example, *i.e.*, we consider only the case of $k = 2$ and $\mathcal{C} = C_2$. The set of feature comparator functions is given by $\mathcal{F} = F_2 = \{f_1, f_2\}$. f_1 takes the angle-measures of two segments as arguments and computes their relative orientation. f_2 computes the difference in the surface area of the given argument pair of segments. Further, let the set of predicates be $\mathcal{P} = P_2 = \{perpendicular\text{-}to, parallel\text{-}to, area\text{-}greater\text{-}than, null\}$. The predicates *perpendicular-to* and *parallel-to* return 1 if the segments are mutually perpendicular and parallel respectively. Otherwise, they return 0. Similarly, the predicate *area-greater-than* returns 1, if the difference computed by f_2 is positive and 0 otherwise. Obviously, fuzzy predicates can be used instead to assign values in the range $[0, 1]$. The mapping from class tuples to predicates, $\mathcal{R}1 = R1_2$, is simplified in this case since no class pair has more than one relationship. This is graphically shown in the top-half of Figure 2. The functional mapping, $\mathcal{R}2 = R2_2$, can be constructed easily since predicates *perpendicular-to* and *parallel-to* use the function f_1 and the predicate *area-greater-than* uses the function f_2. In this simple example, the semantic net $\mathcal{SN} = SN_2$ closely corre-

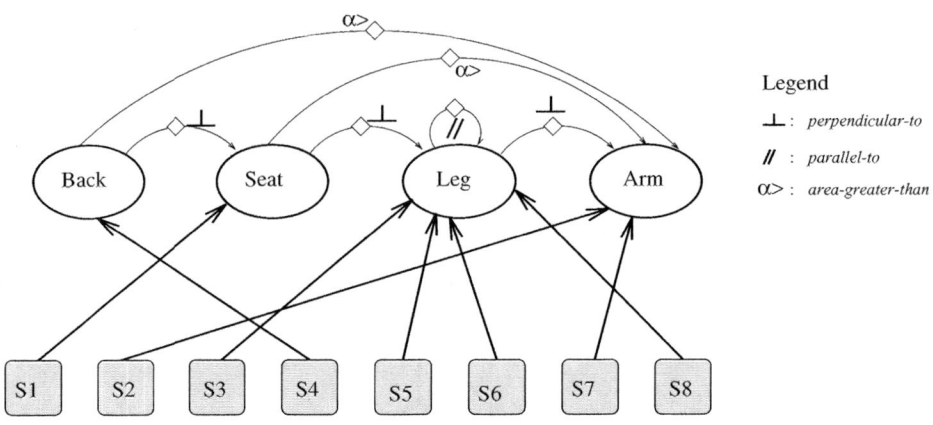

Fig. 2. A sample semantic net and a candidate ISA mapping for a scene consisting of a chair.

sponds to $R1_2$ and the Figure 2 shows the useful parts of the semantic net for a typical chair using the three predicates. Figure 2 also shows a particular instantiation, *i.e.*, an ISA mapping, of the segments from the scene shown in Figure 1. This mapping of $\{\langle s1,S\rangle, \langle s2,A\rangle, \langle s3,L\rangle, \langle s4,B\rangle, \langle s5,L\rangle, \langle s6,L\rangle, \langle s7,A\rangle, \langle s8,L\rangle\}$ results in the correct classification of the segments. Also, it is obvious that any other assignment of labels to the segments would result in a lower consistency evaluation. Formally, the semantic net shown in Figure 2 is given by

$$\mathcal{SN} = \{\ \langle\langle b,a\rangle, \alpha{>}\rangle,\ \langle\langle b,l\rangle, null\rangle,\ \langle\langle b,s\rangle, \bot\rangle,\ \langle\langle b,b\rangle, null\rangle,\ \langle\langle s,a\rangle, \alpha{>}\rangle,$$
$$\langle\langle S,L\rangle, \bot\rangle,\ \langle\langle s,s\rangle, null\rangle,\ \langle\langle l,a\rangle, \bot\rangle,\ \langle\langle l,l\rangle, //\rangle,\ \langle\langle a,a\rangle, null\rangle\ \}$$

where, the symbols "$\alpha{>}$", "$\bot$", and "$//$" denote the predicates *area-greater-than*, *perpendicular-to*, and *parallel-to* respectively.

3.2   Representing Candidate Solutions in GAs

A candidate solution for the classification task takes the form of a vector of indices $\langle j_1, \ldots, j_i, \ldots, j_n\rangle$, containing one element for each segment $s_i \in S$. This vector of indices represents a possible ISA mapping for the segments in the set S. In other words, $ISA(s_i) = j_i$ where, $1 \le i \le n$ and $1 \le j_i \le m$. For example, the vector $\langle S, A, L, B, L, L, A, L\rangle$ represents the labeling shown in Figure 2. Similarly, the vector $\langle L, A, S, B, A, L, L, L\rangle$ represents another candidate solution, albeit of inferior quality.

3.3   Computing Fitness Using Semantic Net Description

Fitness is a quantitative measure of the consistency of an ISA relationship that a candidate solution represents. The procedure described below may appear to be computationally expensive. However, we have found that, in practice, there are very few relationships involving more than two components/segments and that a large number of high-order relationships are reduced to null predicates. Further, all the predicates can be precomputed for the segments in a given image and the repeated fitness computations need only do the summation of the various predicate values using a table look-up.

$$E = \sum_{k=1}^{k_{max}} \sum_{\langle\langle c_k\rangle, p\rangle \in SN_k} \sum_{\langle s_k\rangle} w \times p(f(x_1, \ldots, x_k)) \qquad (13)$$

where $\langle c_k \rangle$ abbreviates $\langle j_1, \ldots, j_k \rangle \in C_k$ and $\langle s_k \rangle$ represents any k-tuple of image segments instantiated to $\langle j_1, \ldots, j_k \rangle$ via the ISA mapping. Also, $f = R2_k(p)$, $w \in W_k$ is the weight corresponding to predicate p, and $x_1, \ldots, x_k \in D_i$ are the feature values of the image segments in $\langle s_k \rangle$ corresponding to some feature $t_i \in T$, depending on the feature comparator function f. For domains in which knowledge is unevenly distributed in the semantic net, practice may dictate that normalization over the set of predicates for each class or class-tuple be performed.

To illustrate fitness, examine the best solution $\langle S, A, L, B, L, L, A, L \rangle$ taken from Figure 2. Since there are four and two instances of the classes l and a respectively, we need to sum twenty-three predicate values, ignoring the *null* predicate. Assuming unit weights for all predicates, on summation, the fitness value for the ideal solution can be seen to be twenty three, since all the predicates have a value of 1 in this case. In contrast, consider the obviously sub-optimal solution $\langle L, A, S, B, A, L, L, L \rangle$. Here, only ten out of the twenty three predicate values have a value of 1 resulting in a fitness value of ten.

4   Experiments and Analysis

The goal in the oceanographic problem is to label the mesoscale features of the North Atlantic from satellite images using a known set of classes. The classes in this case are "Gulf Stream North Wall"(N), "Gulf Stream South Wall"(S), "Warm Eddy"(W), "Cold Eddy"(C), and "Other"(O), $i.e.$, $CL = \{N, S, W, C, O\}$. We use edge-segmented images of the region in the infrared band (10.3-11.3 μm). An infrared satellite image and its companion segmented image are shown in Figures 3 and 4. For the image shown, we have, $S = \{s1, s2, \ldots, s35\}$. We use only two measures for each segment in the image – its position and length. They are computed by using coordinates of the centroid (based on the mass of the segment) and the two end-points.

The set of predicates $\mathcal{P}$ and the set of functions $\mathcal{F}$ are informally shown in Table 1. The semantic net $\mathcal{SN}$ for the domain is shown in Table 2. The mappings $\mathcal{R}1$ and $\mathcal{R}2$ are implicit in these tables. We use unit weights for all predicates.

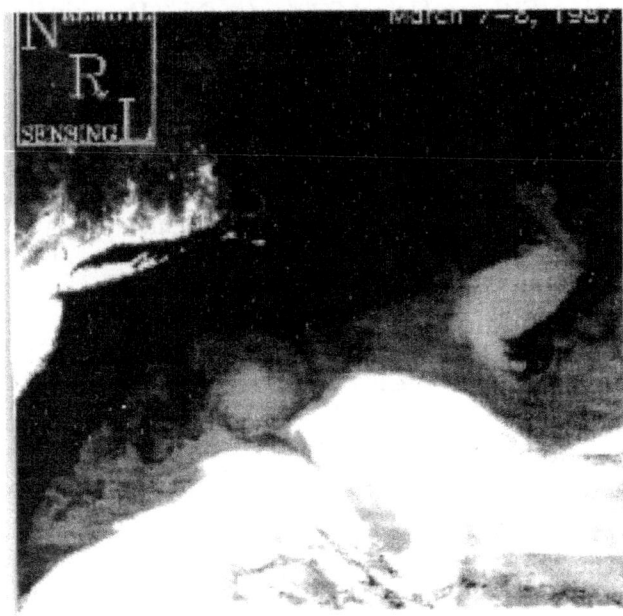

Fig. 3. Original Infrared Image of the Gulf Stream

Table 1. Description of predicates and functions for oceanic labeling

Predicate	Function	Comments
Is-North-Of(i,j)	If $AvgLat(i) \geq AvgLat(j)$, $= 1$; Otherwise, $= 0$.	Segment i is north of segment j
Is-Near(i,j)	$\exp(-\beta \times X_{i,j})$	$X_{i,j}$ is distance between segments i and j.
Is-Not-Near(i,j)	$1 - Is\text{-}Near(i, j)$	Fuzzy complement of *Is-Near*.
*Is-North-Of-*And *-Fifty-Km-From(i,j)*	$\min\{$ *Is-North-Of*(i, j), $\exp(-\beta \times \lvert X_{i,j} - 50\rvert)$ $\}$	Segment i is north of segment j and is 50 km from it.
*Arcs-Of-Circle-*And *-Less-Than-Hundred* *-Km-Distant(i,j)*	$\min\{$ *Arcs-Of-Circle*(i, j), *Less-Than-Hundred- Km-Distant*(i, j) $\}$	*Arcs-Of-Circle* is estimated based on intersection of cords from segments. Second predicate is computed as $= 1$, if $X_{i,j} \leq 100$, and $= \exp(-\beta \times \lvert X_{i,j} - 100\rvert)$, otherwise

Also, it is noted that the predicates are computed *a priori* for all the segment pairs and stored in a lookup table.

Candidate solutions are represented as vectors of labels. For the image shown in Figure 4, any label vector $\langle c_1, c_2, \ldots, c_{35} \rangle$ such that $c_i \in \{n, s, w, c, o\}$ for all

Table 2. A semantic net for oceanic segment labeling

Class tuple	Predicate Name
$\langle w, w \rangle$	$Is\text{-}Near(i,j)$
$\langle w, w \rangle$	$Arcs\text{-}Of\text{-}Circle\text{-}\text{And}$ $\text{-}Less\text{-}Than\text{-}Hundred\text{-}Km\text{-}Distant(i,j)$
$\langle w, n \rangle$	$Is\text{-}North\text{-}Of(i,j)$
$\langle w, n \rangle$	$Is\text{-}Near(i,j)$
$\langle n, n \rangle$	$Is\text{-}Near(i,j)$
$\langle n, s \rangle$	$Is\text{-}North\text{-}Of\text{-}\text{And-}Fifty\text{-}Km\text{-}From(i,j)$
$\langle s, s \rangle$	$Is\text{-}Near(i,j)$
$\langle s, c \rangle$	$Is\text{-}North\text{-}Of(i,j)$
$\langle s, c \rangle$	$Is\text{-}Near(i,j)$
$\langle c, c \rangle$	$Is\text{-}Near(i,j)$
$\langle c, c \rangle$	$Arcs\text{-}Of\text{-}Circle\text{-}\text{And}$ $\text{-}Less\text{-}Than\text{-}Hundred\text{-}Km\text{-}Distant(i,j)$
$\langle o, w \rangle$	$Is\text{-}Not\text{-}Near(i,j)$
$\langle o, n \rangle$	$Is\text{-}Not\text{-}Near(i,j)$
$\langle o, s \rangle$	$Is\text{-}Not\text{-}Near(i,j)$
$\langle o, c \rangle$	$Is\text{-}Not\text{-}Near(i,j)$
$\langle o, o \rangle$	$Is\text{-}Near(i,j)$
Other tuples	$null$
Legend: w = Warm Eddy; c = Cold Eddy; o = Other n = North Wall of Gulf Stream s = South Wall of Gulf Stream	

i, constitutes a feasible candidate solution. Such label vectors are encoded as bit strings suitable for GA search. Table 3 shows the parameters used for the GA runs. Each run with these settings was repeated 10 times, each starting with a different initial random population. The accuracy of the best solution generated by the GA in each run with respect to the fairly difficult image shown in Figures 3 and 4 is compared with that of manual labeling and is listed in Table 4. Figure 4 shows the best labeling obtained over all the runs.

5   Conclusions

Here we describe a domain-independent framework for labeling image segments for scene interpretation. This approach is based on the abstract representation of a typical scene from the domain of discourse. The abstraction form, *i.e.*, semantic network, permits encoding the descriptions of relationships of arbitrary degree among the instances of scene objects. A GA is used in searching the space of candidate solutions for the best labeling. Fitness of a candidate solution is

Table 3. Parameters of GA runs for oceanic labeling

Description	Value
Population size	200
Number of generations	200
Selection operator	Proportional selection using stochastic remainder sampling with replacement
Crossover operator	Uniform crossover (allele level)
Probability of crossover	0.600
Mutation operator	Bit mutation
Probability of mutation	0.005

Table 4. Accuracy of GA-generated oceanic labeling

Run #	Accuracy %	Run #	Accuracy %
1	80	6	63
2	57	7	71
3	66	8	77
4	83	9	71
5	83	10	69

estimated by evaluating the conformance of the solution to the relationships depicted in the semantic net.

References

1. C. A. Ankenbrandt, B. P. Buckles, and F. E. Petry, "Scene recognition using genetic algorithms with semantic nets", *Pattern Recognition Letters*, vol. 11, no. 4, pp. 285–293, 1990.
2. B. P. Buckles and F. E. Petry, Eds., *Genetic Algorithms*, IEEE Computer Society Press, 1992.
3. B. Bhanu, S. Lee, and J. Ming, "Self-optimizing image segmentation system using a genetic algorithm", in *Proceedings of the Fourth International Conference on Genetic Algorithms*, R.K. Belew and L.B. Booker, Eds., San Mateo, CA, 1991, pp. 362–369, Morgan Kaufmann.
4. S. M. Bhandarkar and H. Zhang, "Image segmentation using evolutionary computation", *IEEE Trans. on Evolutionary Computation*, vol. 3, no. 1, pp. 1–21, apr 1999.
5. R. Tönjes, S. Growe, J. Bückner, and C.-E. Liedtke, "Knowledge-based interpretation of remote sensing images using semantic nets", *Photogrammetric Engineering & Remote Sensing*, vol. 65, no. 7, pp. 811–821, jul 1999.
6. J. Bala, K. DeJong, and P. Pachowicz, "Using genetic algorithms to improve the performance of classification rules produced by symbolic inductive methods", in

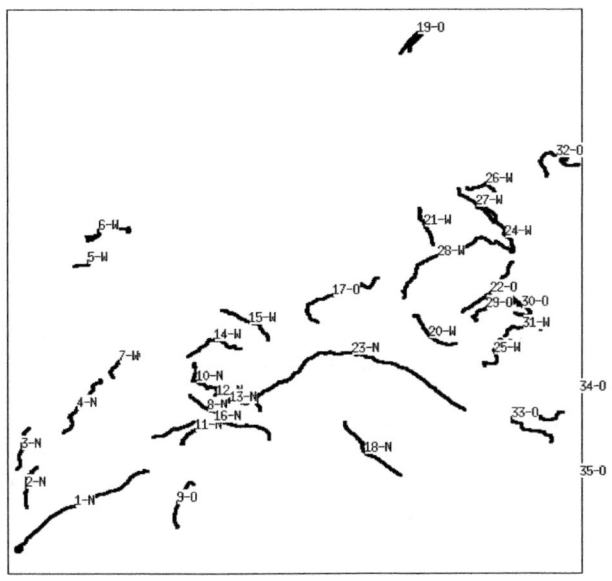

Fig. 4. Best Labeling of the Gulf Stream found by the GA

Proceedings of 6th International Symposium Methodologies for Intelligent Systems ISMIS'91, Z. W. Ras and M. Zemankova, Eds., Charlotte, NC, 16-19 Oct 1991, pp. 286–295, Springer-Verlag, Berlin, Germany.

7. S. Truve, "Using a genetic algorithm to solve constraint satisfaction problems generated by an image interpreter", in *Theory and Applications of Image Analysis. Selected Papers from the 7th Scandinavian Conference, Aalborg, Denmark*, P. Johansen and S. Olsen, Eds. Aug, 13-16 1991, pp. 133–147, World Scientific.

8. A. Hill and C. J. Taylor, "Model-based image interpretation using genetic algorithms", *Image and Vision Computing*, vol. 10, no. 5, pp. 295–300, Jun 1992.

9. D. B. Fogel, "Evolutionary programming for voice feature analysis", in *Proceedings of 23rd Asilomar Conference on Signals, Systems, and Computers*, oct 1989, pp. 381–383.

Evolutionary Wavelet Bases in Signal Spaces

Adelino R. Ferreira da Silva

Universidade Nova de Lisboa,
Dept. de Eng. Electrotécnica,
2825 Monte de Caparica, Portugal
afs@mail.fct.unl.pt

Abstract. We introduce a test environment based on the optimization of signals approximated in function spaces in order to compare the performance of different evolutionary algorithms. An evolutionary algorithm to optimize signal representations by adaptively choosing a basis depending on the signal is presented. We show how evolutionary algorithms can be exploited to search larger waveform dictionaries for best basis selection than those considered in current standard approaches.

1 Introduction

In order to facilitate an empirical comparison of the performance of different evolutionary algorithms a test environment must be provided. Traditionally, sets of test functions with specific topological properties, commonly known as *fitness landscapes*, have been proposed by several authors to be used in performance benchmarking. In particular, the De Jong's test function set has been a standard for genetic algorithm benchmarks since 1975. In most cases, the optimization objective is formulated as a global function minimization problem.

In this paper, we depart from this view by considering the optimization of functions approximated in function spaces. Series expansions of continuous-time signals go back at least to Fourier's original expansion of periodic functions. A basis is a set of linearly independent functions that can be used to produce all admissible functions $f(t)$. The idea of representing a signal as a sum of elementary basis functions, or equivalently to find orthogonal bases for certain function spaces, is very powerful. However, classic approaches have limitations, in particular there are no "good" local Fourier series that have both time and frequency localization. An alternative is the construction of wavelet bases, which use scaling instead of modulation in order to obtain an orthonormal basis for $\mathbb{L}_2(\mathbb{R})$. An entropy-based algorithm for best basis selection has been proposed in the literature [6]. Under the specific conditions of its application, the standard best basis (SBB) algorithm finds the optimum basis decomposition according to a specified cost functional. We show that this algorithm can be used to benchmark evolutionary algorithms.

A second objective of this paper, is to show how evolutionary algorithms can be exploited to search larger waveform dictionaries for best basis selection than those considered in current standard approaches. We extend the scope of

S. Cagnoni et al. (Eds.): EvoWorkshops 2000, LNCS 1803, pp. 44–53, 2000.

the SBB algorithm by searching larger waveform dictionaries in order to find better adaptive signal representations. In Sect. 3, we present an evolutionary algorithm for best basis selection. Adapted waveform analysis uses libraries of orthonormal basis and an efficient functional to match a basis to a given signal or family of signals. Two often used libraries are wavelet-packets and localized trigonometric functions, since they support the expansion of the waveforms in orthonormal basis whose elements have good time-frequency localization properties. These libraries constitute huge collections of basis from which we can pick and choose the best matching basis. Flexible decompositions are important for representing time-frequency atoms whose time-frequency localizations vary widely. In this article, we propose the use of evolutionary algorithms [10] as the main searching tool for best basis selection. The proposed approach generates a population of solutions based on basis expansions of multi-filter, time-shifted, wavelet-packet libraries. An evolutionary algorithm operates on the population to evolve the best solution according to a given objective function. Libraries of bases represent the population from which we want to select the best-fit individuals. This optimization approach permits more flexibility in searching for best basis representations than traditional approaches.

2   Wavelet Spaces

In this section, we briefly review the framework of wavelet basis selection. Let $x \in \mathbb{R}^n$ be an input signal and let $\Omega_{0,0} = \mathbb{R}^n$ represent the signal space and $B_{0,0} = (e_1, \ldots, e_n)$ be the standard basis of $\mathbb{R}^n$. Wavelet packets split this original space into two mutually orthogonal subspaces smoothly and recursively, i.e.,

$$\Omega_{j,k} = \Omega_{j+1,2k} \oplus \Omega_{j+1,2k+1} \tag{1}$$

for $j = 0, 1, \ldots, J$, $k = 0, \ldots, 2^j - 1$, and $J(\leq \log_2 n)$ is the maximum i level of recursions specified by the user. Here, we have $n_j \triangleq \dim \Omega_{j,.} = n/2^j$. The wavelet packet transforms, recursively split the frequency domain via the so-called conjugate quadrature filters. These splits naturally generate a set of subspaces with the binary tree structure. Let $\alpha = (j, k)$ be an index to specify a node (i.e., a subspace with its basis set) of this binary tree. The index j specifies the depth of the binary tree; this is an index of the *width* of frequency bands for wavelet packets. The index k specifies the *location* of the frequency bands for wavelet packets. Let $\Omega = \Omega_\alpha$ be such a collection of subspaces and let $B = \{B_\alpha\}$ be the corresponding set of basis vectors where $B_\alpha = (\omega_{\alpha 1}, \ldots, \omega_{\alpha n_j})$ is a set of basis vectors that spans Ω_α. Each basis vector in B is called a time-frequency atom, and the whole set B is referred to as a *time-frequency dictionary* or a *dictionary of orthonormal bases*. These dictionaries contain many orthonormal bases. If the depth of the tree is J, each dictionary contains more than $2^{2^{(J-1)}}$ different bases.

An important question is how to select from a large number of bases in the dictionary a basis which performs "best" for one's task. In order to measure the performance of each basis, we need a measure of efficiency or fitness of a basis for the task at hand. For this purpose, several so-called *information cost functionals* have been proposed. A commonly used information cost functional is *entropy*. The entropy of a vector $d = \{d(k)\}$ is defined by,

$$\mathcal{H}(d) = -\sum_k | \frac{d_k}{\| d \|_2} |^2 \, log_2 \, | \frac{d_k}{\| d \|_2} |^2 . \tag{2}$$

This cost functional was used as the objective function to drive the evolutionary optimization approach outlined in Sect. 3. The goal of the optimization approach is to find an optimal, or *quasi-optimal* (in some sense) basis representation for a given dataset.

The above considerations are the fundamentals behind the SBB algorithm [6]. However, it is possible to extend the library of bases in which the best representations are searched for, by introducing additional degrees of freedom that adjust the time-localization of the basis functions [2]. One such extension is the shift-invariant wavelet packet transform. Actually, one well-known disadvantage of the discrete wavelet and wavelet packet transforms is the lack of shift invariance. The added dimension in the case of shift-invariant decompositions is a *relative shift*, between a given parent-node and its relative children nodes. Shifted versions of these transforms for a given input signal, represent new bases to be added to the library of bases, which may further improve our ability to find the "best" adapted basis. These modifications of the wavelet transform and wavelet packet decompositions lead to orthonormal best-basis representations which are shift-invariant and characterized by lower information cost functionals [5]. Wavelet packet trees may be extended as joint wavelet packet trees to profit from enlarged libraries of bases, thus increasing our chances of getting truly adapted waveform representations. However, enlarged search spaces entail combinatorial explosion problems. We rely on evolutionary optimization approaches to guide us on the search process.

3   Evolutionary Formulation

The two major major steps in applying any heuristic search algorithm to a particular problem are the specification of the representation and the evaluation (fitness) function. When defining an evolutionary algorithm one needs to choose its components, such as mutation and recombination that suit the representation, selection mechanisms for selecting parents and survivors, and an initial population. Each of these components have parameter values which determine whether the algorithm will find an near-optimum solution and whether it will find a solution efficiently.

3.1   Representation

In the work reported here, a variable length integer sequence is used as the basic genotype. The objective is to evolve wavelet decomposition trees through the evolution of genetic sequences. The technique used to initialize the population is based on generating an initial random integer sequence, according to the values of the allele sets specified for the individual genes. The initial genotype sequence which codifies the wavelet tree matches the breadth-first (BF) sequence required to generate a complete binary tree, up to a pre-specified maximum depth. We refer to these sequences as tree-mapped sequences. A well-built decomposition tree for wavelet analysis purposes, is generated by imposing appropriate constraints to the genotype sequence as specified in Sect. 3.2. The imposition of the constraints yields variable length code sequences after resizing. An alphabet $\mathbb{A} = \{0, 1, 2\}$ is used to codify the wavelet tree nodes according to their types as specified in Sect. 3.3, thus enabling us to map any tree structure into a code sequence. The mapping of a code sequence to a complete BF tree traversal yields an initial sequence with length $l = 2^L - 1$, for a tree of depth L. The length l is also the number of nodes in a complete binary tree of depth L. When coding a complete binary tree using a complete BF sequence the last level of terminal nodes is redundant. Therefore, we have used a codification based on resized complete BF sequences to code genetic sequences. The chromossomes are constructed as follows. The first gene assumes integer values $g_0 \in \mathbb{F}$, where $\mathbb{F}$ is the set of possible filter types used in the implementation, as explained in Sect. 3.3. The remaining genes are used to codify the wavelet decomposition tree.

3.2   Constraints

There are several methods for generating trees which can be used to initialize the population. The *full, grow* and *ramped half-and-half* methods of tree generation were introduced in the field of genetic programming [9]. These methods are based on tree depth. The ramped half-and-half method is the most commonly used method of generating random parse trees because of its relative higher probability of generating subtrees of varying depth and size. However, these methods do not produce a uniform sampling of the search space. In this work, we use constrained genetic sequences for genome initialization. Two types of constraint operators are used to guarantee that valid tree-mapped genetic sequences are generated: (1) top-down operator, and (2) bottom-up operator. In addition, by applying these operators we look for a uniform sampling of the tree search space. In terms of binary tree data structures, the top-down constraint guarantees that if a node t_i has null code $c_i = 0$ then its two sons $t_{i,0}$ and $t_{i,1}$ must have null code $c_{i,0} = 0$ and $c_{i,1} = 0$. The bottom-up constraint guarantees that if at least one of the sons $t_{i,0}$ and $t_{i,1}$ of a node t_i has non-null code, then the parent $t_{i,0}$ must have non-null code $c_i \neq 0$. These constraint operators are biased in opposite ways. Starting from a uniform random code sequence, the bottom-up constraint operator constructs valid genetic sequences which are biased towards complete, full-depth trees. By he same token, the top-down constraint operator constructs valid genetic sequence which is biased towards null, minimum depth trees. To get a more uniform sampling of the sequence space, for sequences of maximum $s = 2^L - 1$, we use the following initialization procedure:

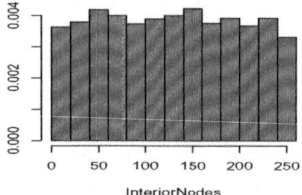 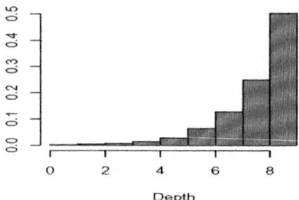

Fig. 1. Histogram for the distribution of interior nodes in generated sequences (left panel); histogram for the distribution of trees with specified depth (right panel)

1. initialize sequence s to terminal code value $c_i = 0 \in \mathbb{A}$,
2. get a random value $r \in [1, s]$,
3. initialize subsequence $s_1 = [1, r) \in s$ with random interior code values, $c_i = \{1, 2\} \in \mathbb{A}$,
4. randomly select one of the two constraint operators, bottom-up or top-down, to apply to s.

By resizing (pruning) constrained code sequences we allow for genetic sequences of variable length, hence tree representations of variable depth. The left panel in Fig. 1 shows a histogram for the number of interior nodes generated by the initialization procedure, for a maximum specified depth $L = 8$ of the equivalent tree, and 3000 stochastic genetic sequences. The right panel in Fig. 1 presents a histogram for the distribution of trees with specified depth, generated from the same stochastic samples with resizing.

3.3   Specification

The approach of organizing libraries of bases as a tree has been extended to construct a joint tree, to guide the process of generating shifted wavelet packet transforms. Libraries of bases represent the population from which we want to select the best-fit individuals. In the current formulation, the genotype sequence $\mathcal{G}$ allows for three optimization parameters: best filter, best wavelet packet basis and best shifted basis. The genetic representation is used to create an initial population and evolve potential solutions to the optimization problem. The genotype is made up of the genes which guide the discrete wavelet decomposition of each waveform, in accordance with the joint tree representation. A cost functional is then applied to the wavelet coefficients, and its value is used to derive the fitness of the individual. In terms of entropy, the optimization problem amounts to evolve a minimum-entropy genotype. Therefore, the best individual is the one with minimum evolved entropy in a given library space. Since we were able to formulate the three subtasks, wavelet packet decomposition, shifted wavelet transform, and wavelet filter to be applied in a common data structure, the original multiple optimization problem can be solved in terms of a single aggregate functional. The first gene g_0 in $\mathcal{G}$ is responsible for the optimization of the filter used in the decomposition. We have used in the implementation 16 possible types of different filters, thus $g_0 = \{0, \ldots, 15\}$. The filters considered in the implementation were the Haar filter, the Daubechies filters D4, D6 and D8, and several biorthogonal filters commonly used in image analysis as implemented in [7]. In particular, the filter set included the 7/9 spline filter referred to in [1], the 7/9, 11/13, 10/6, 3/5, 6/2, and 3/9 filters defined in [11], the 7/9 "FBI-fingerprint" filter, and the 10-tap filter listed in [3].

The analysis phase of the (discrete) shift wavelet packet transform is codified in the genetic sequence $\mathcal{G}$. The collection of wavelet packets comprises a library of functions with a binary tree structure. To obtain the wavelet packet analysis of a function, or data set in the discrete case, we first find its coefficient sequence in the root subspace, then follow the branches of the wavelet packet coefficient tree to find the expansion in the descendent subspaces. Assigning to each tree node a wavelet split value $s_i \in \{0, 1, 2\}$ we may enumerate all possible binary tree structures. The value $s_w = 1$ references unshifted interior nodes, i.e., nodes with left and right children subtrees associated with unshifted decompositions. The value $s_w = 2$ references time shifted interior nodes. The value $s_w = 0$ references the leaves.

4   Spaces of Test Signals

A well-known implementation of the SBB algorithm which can be used for comparison purposes is contained in the *WaveLab* package [4]. We reference by *evolutionary best basis* (EBB), the evolutionary formulation presented in Sect. 3 for best basis selection using multifilter, time shifted wavelet packet bases. By *canonical best basis* (CBB), we mean an algorithm which is able to reproduce (approximate) the results of the SBB algorithm, using optimization methodologies for best basis selection different from those conceived for the SBB algorithm. In this sense, we may map the EBB algorithm into a CBB algorithm. The SBB algorithm is based on building an information cost tree in order to minimize some cost measure on the transformed coefficients [6]. The evolutionary algorithm proposed in Sect. 3 was applied to a set of test signals and the results compared with the results produced by SBB, based on the entropy minimization criterion. For evaluation purposes, we will use the test signals depicted in figure 2. Two of these signals are artificial signals. The other two signals are built from collected data. The signal *HypChirps* includes two hyperbolic chirps. The signal *MishMash* includes a quadratic chirp, a linear chirp, and a sine, as used in the WaveLab package. The signal *Seismic* is distributed throughout the seismic industry as a test dataset. Finally, the signal *Sunspots* represent the monthly sunspot numbers. The following basic parameters have been used in the *steady-state* evolutionary algorithm [8, 12]: (1) population size: 50; (2) crossover probability: 0.95; (3) mutation probability: 0.02; (4) replacement percentage: 0.6.

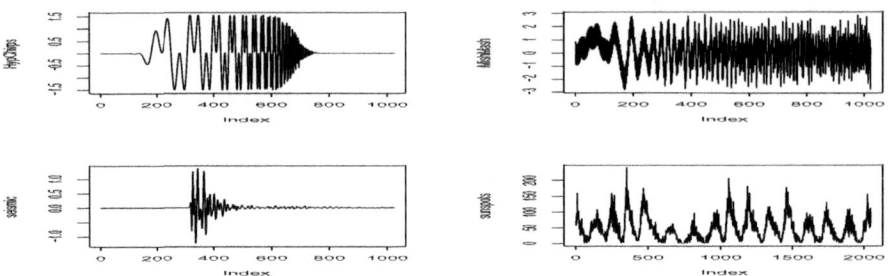

Fig. 2. Test signals

4.1   Standard Wavelet Spaces

To reproduce the application conditions of the SBB algorithm, the EBB approach was restricted to handle unshifted wavelet packet decompositions and use a specific filter. Hence, the CBB algorithm is a restricted version of the EBB algorithm, which is used to reproduce the SBB results using a different methodology. Both the EBB and the CBB algorithms are evolutionary. The minimum entropy values associated with the best basis selected by the SBB (WaveLab) algorithm for these signals, using the Daubechies $D8$ filter and $L = 9$ decomposition levels, are reported in Table 1. The SBB entropy values are also depicted by the dashed lines in Fig. 3. The CBB algorithm was applied to the same set of test signals to evolve the best basis using the same entropy cost functional. Fig. 3 shows the evolution of the median minimum entropy (median best value) with the number of generations for each of the test signals. The entropy values represent the median over 30 runs of the CBB algorithm. The median entropy values after $ngen = 80$ generations are shown in Table 1 as well. The optimal entropy values and the rate of convergence can be used as benchmarks to compare the performance of different evolutionary algorithms, or simply to tune the value of the control parameters.

Table 1. Comparative median minimum entropy values and reconstruction errors

Signal	Median Minimum Entropy (L=9)			Reconst. Error
	SBB (D8)	CBB (D8, ngen=80)	EBB (ngen=80)	EBB (ngen=80)
HypChirps	3.7908	3.7908	2.7707	1.0 e-6
MishMash	4.4805	4.5162	3.5910	2.6 e-6
Seismic	2.6022	2.6025	1.9735	2.7 e-7
Sunspots	3.0059	3.0059	2.7749	3.9 e-8

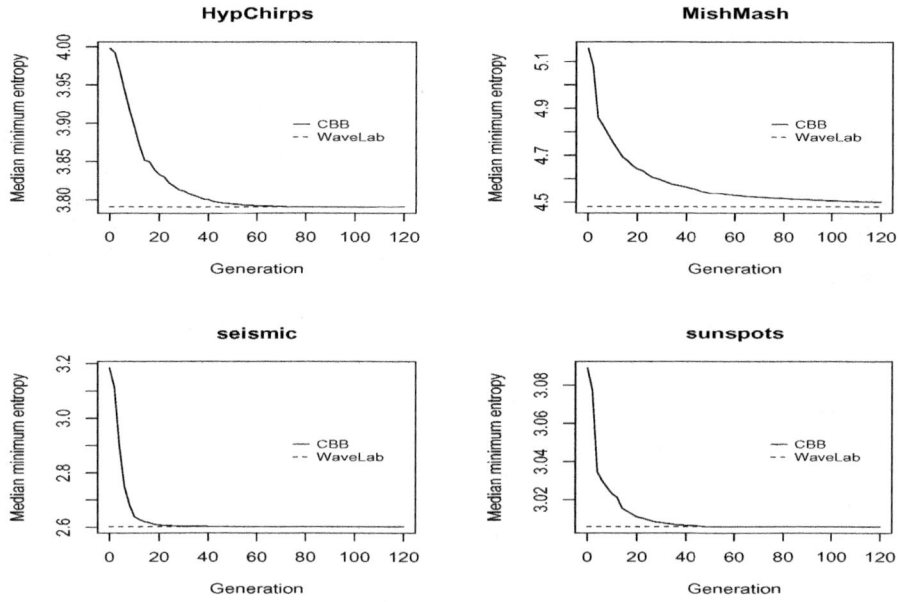

Fig. 3. Evolution of the median minimum entropy by the CBB algorithm

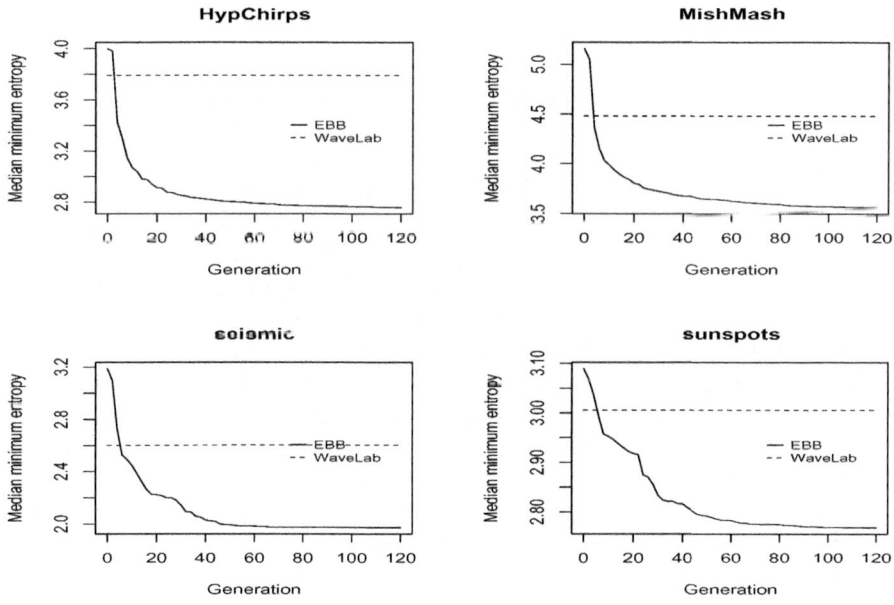

Fig. 4. Evolution of the median minimum entropy by the EBB algorithm

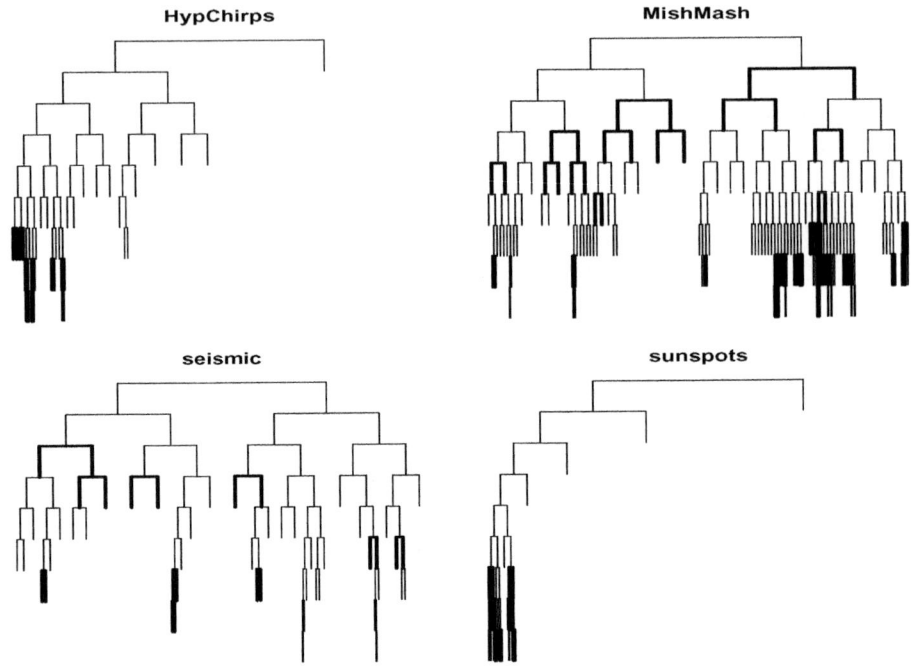

Fig. 5. Evolved trees by the EBB algorithm

4.2   Enlarged Wavelet Spaces

To show the application of the EBB technique we have applied it to evolve basis
for the optimization of three parameters: wavelet packet decomposition, shiftable
decomposition and filter to use. Fig. 4 presents the evolution of the minimum
entropy values generated by the EBB algorithm for the signals in the test set. The
values are the median values over 30 runs of the EBB algorithm. The dashed
horizontal line in Fig. 4 is the value of the best basis entropy generated by
SBB. We notice that the EBB algorithm is able to greatly reduce the minimum
entropy value used to assess best basis adaptability, compared to both the SBB
and the CBB algorithms. Table 1 references the numeric median values for the
EBB minimum entropy after $ngen = 80$ generations. The most selected filters
among the best evolved filters for each test signal, were the following: Brislawn
10-tap filter for signals HypChirps and MishMash, Villasenor 3/9 filter for signal
Seismic, and Villasenor 6/2 filter for signal Sunspots. Fig. 5 depicts typical best
evolved trees for a sample run of the EBB algorithm. Darker lines represent
shifted wavelet packet transforms. Thinner lines represent unshifted transforms.

Another important evaluation factor is the reconstruction error. Given a signal f, we reconstruct an approximate signal $\hat{f}$ from the transformed coefficients
by applying the inverse shifted wavelet packet transform, and calculating the l_2
error between these two signals $\| f - \hat{f} \|_2$. Table 1, presents the numeric median

values of the reconstruction errors for each of the test signals, using the EBB algorithm for 80 generations over 30 runs.

5 Conclusion

The approximation of signals in functions spaces was used to introduce a test environment aimed at the comparative performance of different evolutionary algorithms. We have considered *entropy* as the optimization fitness criterion to be used. However, other cost functions may prove useful to extend the range of the test environment. In particular, measures directed to the optimization of multi-objective criteria may be incorporated in the proposed framework. On the other hand, the proposed test environment may be easily extended to incorporate two dimensional signal spaces. In terms of signal processing, well-adapted signal expansions are important, for instance, in signal compression. For orthonormal basis and additive cost measures, the standard algorithm for best basis selection is efficient. However, with the introduction of overcomplete waveform dictionaries the algorithm has increasing difficulty in finding well-adapted signal representations. The proposed evolutionary approach offers more flexibility in searching for well-adapted signal representations than standard approaches.

References

1. M. Antonini, M. Barlaud, P. Mathieu, and I. Daubechies. Image coding using wavelet transform. *IEEE Trans. on Image Process.*, 1(2), April 1992.
2. G. Beylkin. On the representation of operators in bases of compactly supported wavelets. *Society for Industrial and Applied Mathematics*, 6(6):1716–1740, December 1992.
3. C. M. Brislawn. Two-dimensional symmetric wavelet transform tutorial program. Technical report, Los Alamos National Laboratory, December 1992.
4. J. Buckheit and D. L. Donoho. Wavelab and reproducible research. Technical report, Department of Statistics, Stanford University, 1995.
5. I. Cohen, S. Raz, and D. Malah. Orthonormal shift-invariant wavelet packet decomposition and representation. *Signal Processing*, 57(3):251–270, March 1997.
6. R. R. Coifman and M. V. Wickerhauser. Entropy based methods for best basis selection. *IEEE Trans. on Inf. Theory*, 38(2):719–746, 1992.
7. G. Davis. *Baseline Wavelet Transform Coder Construction Kit*. Mathematics Department, Dartmouth College, January 1997.
8. D. E. Goldberg. *Genetic Algorithms in Search, Optimization, and machine learning and Filter Banks*. Addison-Wesley, Reading, Massachusetts, 1989.
9. John R. Koza. *Genetic Programming - On the Programming of Computers by Means of Natural Selection*. MIT Press, Cambridge, MA, 1992.
10. Z. Michalewicz. *Genetic algorithms + data structures = evolution programs*. Artificial Intelligence. Springer-Verlag, New York, 1992.
11. J. Villasenor, B. Belzer, and J. Liao. Wavelet filter evaluation for image compression. *IEEE Trans. on Image Process.*, 4(8):1053–1060, August 1995.
12. M. Wall. *GAlib: A C++ Library of Genetic Algorithm Components*. Mechanical Engineering Department, Massachusetts Institute of Technology, August 1996.

Finding Golf Courses: The Ultra High Tech Approach

Neal R. Harvey, Simon Perkins, Steven P. Brumby, James Theiler,
Reid B. Porter, A. Cody Young, Anil K. Varghese, John J. Szymanski and
Jeffrey J. Bloch

Space and Remote Sensing Sciences Group,
Los Alamos National Laboratory, Los Alamos, NM 87545, USA

Abstract. The search for a suitable golf course is a very important issue
in the travel plans of any modern manager. Modern management is also
infamous for its penchant for high-tech gadgetry. Here we combine these
two facets of modern management life. We aim to provide the cutting-
edge manager with a method of finding golf courses from space!
In this paper, we present GENIE: a hybrid evolutionary algorithm-based
system that tackles the general problem of finding features of interest in
multi-spectral remotely-sensed images, including, but not limited to, golf
courses. Using this system we are able to successfully locate golf courses
in 10-channel satellite images of several desirable US locations.

1 Introduction

There exist huge volumes of remotely-sensed multi-spectral data from an ever-
increasing number of earth-observing satellites. Exploitation of this data requires
the extraction of features of interest. In performing this task, there is a need for
suitable analysis tools. Creating and developing individual algorithms for specific
feature-detection tasks is important, yet extremely expensive, often requiring a
significant investment of time by highly skilled analysts. To this end we have
been developing a system for the automatic generation of useful feature-detection
algorithms using an evolutionary approach.

The beauty of an evolutionary approach is its flexibility: if we can derive a
fitness measure for a particular problem, then it might be possible to solve that
problem. Many varied problems have been successfully solved using evolutionary computation, including: optimization of dynamic routing in telecommunications networks [1], optimizing image processing filter parameters for archive film
restoration [2], designing protein sequences with desired structures [3] and many
others.

When taking an evolutionary approach, a critical issue is how one should
represent candidate solutions in order that they may be effectively manipulated.
We use a genetic programming (GP) method of representation of solutions, due
to the fact that each individual will represent a possible image processing algorithm. GP has previously been applied to image-processing problems, including:

S. Cagnoni et al. (Eds.): EvoWorkshops 2000, LNCS 1803, pp. 54-64, 2000.

edge detection [4], face recognition [5], image segmentation [6], image compression [7] and feature extraction in remote sensing images [8–10]. The work of Daida et al. Brumby et al. and Theiler et al. is of particular relevance since it demonstrates that GP can be employed to successfully evolve algorithms for real tasks in remote-sensing applications.

2   System Overview

We call our feature detection system "GENIE" (GENetic Image Exploitation) [9, 10] GENIE employs a classic evolutionary paradigm: a population of individuals is maintained and each individual is assessed and assigned a fitness value. The fitness of an individual is based on an objective measure of its performance in its environment. After fitness determination, the evolutionary operators of selection, crossover and mutation are applied to the population and the entire process of fitness evaluation, selection, crossover and mutation is iterated until some stopping condition is satisfied.

2.1   Training Data

The environment for each individual in the population consists of a set of training data. This training data consists of a data "cube" of multi-spectral data together with some user-defined data defining "ground-truth". Ground-truth, in this context, is not what is traditionally referred to as ground-truth (this being in-situ data collected at, or as close as possible to, the time the image was taken). Here, ground-truth refers to what might normally be referred to as "analyst-supplied interpretation" or "training data". This training data for our system is provided by a human analyst, using a Java-based tool called ALADDIN. Through ALADDIN, the user can view a multi-spectral image in a variety of ways, and can "mark up" training data by "painting' directly on the image using the mouse. Training data is ternary-valued with the possible values being "true", "false", and "unknown". *True* defines areas where the analyst is confident that the feature of interest **does** exist. *False* defines areas where the analyst is confident that the feature of interest **does not** exist. Fig. 1 shows a screen capture of an example session. Here the analyst has marked out golf courses as of interest.

2.2   Encoding Individuals

Each individual *chromosome* in the population consists of a fixed-length string of *genes*. Each gene in GENIE corresponds to a primitive image processing operation, and so the whole chromosome describes an algorithm consisting of a sequence of primitive image processing steps.

Genes and Chromosomes   A single gene consists of an operator name, plus a variable number of input arguments, specifying where input is to come from;

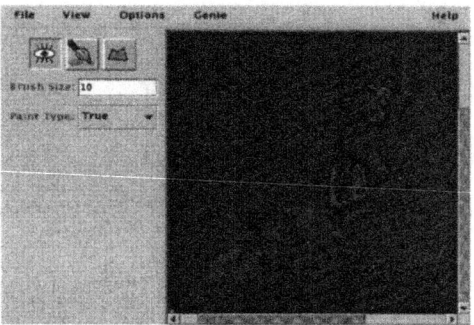

Fig. 1. GUI for Training Data Mark-Up. Note that ALADDIN relies heavily on color, which does not show up well in this image. The light colored patches in the center-right and upper-right parts of the image are two golf courses that have been marked up as "true". Most of the rest of the image has been marked up as "false", except for a small region around the golf courses which has been left as "unknown".

output arguments, specifying where output is to be written to; and operator parameters, modifying how the operator works. Different operators require different numbers of parameters. The operators used in GENIE take one or more distinct image planes as input, and generally produce a single image plane as output. Input can be taken from any data planes in the training data image cube. Output is written to one of a small number of *scratch planes* — temporary workspaces where an image plane can be stored. Genes can also take input from scratch planes, but only if that scratch plane has been written to by another gene positioned earlier in the chromosome sequence.

The image processing algorithm that a given chromosome represents can be thought of as a directed acyclic graph where the non-terminal nodes are primitive image processing operations, and the terminal nodes are individual image planes extracted from the multi-spectral image used as input. The scratch planes are the 'glue' that combines together primitive operations into image processing pipelines. Traditional GP ([11]) uses a variable sized (within limits) tree representation for algorithms. Our representation differs in that it allows for reuse of values computed by sub-trees since many nodes can access the same scratch plane, i.e. the resulting algorithm is a graph rather than a tree. It also differs in that the total number of nodes is fixed (although not all of these may be actually used in the final graph), and crossover is carried out directly on the linear representation.

We have restricted our "gene pool" to a set of *useful* primitive image processing operators. These include spectral, spatial, logical and thresholding operators. Table 1 outlines these operators. For details regarding Laws textural operators, the interested reader is referred to [12, 13].

The set of morphological operators is restricted to function-set processing morphological operators, i.e. gray-scale morphological operators having a flat structuring element. The sizes and shapes of the structuring elements used by

Table 1. Image Processing Operators in the Gene Pool

Code	Operator Description	Code	Operator Description
ADDP	Add Planes	MEAN	Local Mean
SUBP	Subtract Planes	VARIANCE	Local Variance
ADDS	Add Scalar	SKEWNESS	Local Skewness
SUBS	Subtract Scalar	KURTOSIS	Local Kurtosis
MULTP	Multiply Planes	MEDIAN	Local Median
DIVP	Divide Planes	SD	Local Standard Deviation
MULTS	Multiply by Scalar	EROD	Erosion
DIVS	Divide by Scalar	DIL	Dilation
SQR	Square	OPEN	Opening
SQRT	Square Root	CLOS	Closing
LINSCL	Linear Scale	OPCL	Open-Closing
LINCOMB	Linear Combination	CLOP	Close-Opening
SOBEL	Sobel Gradient	OPREC	Open with Reconstruction
PREWITT	Prewitt Gradient	CLREC	Close with Reconstruction
AND	And Planes	HDOME	H-Dome
OR	Or Planes	HBASIN	H-Basin
CL	Clip Low	CH	Clip High
LAWB	Laws Textural Operator $S3^T \times L3$	LAWC	Laws Textural Operator $L3^T \times E3$
LAWD	Laws Textural Operator $E3^T \times E3$	LAWE	Laws Textural Operator $S3^T \times E3$
LAWF	Laws Textural Operator $L3^T \times S3$	LAWG	Laws Textural Operator $E3^T \times S3$
LAWH	Laws Textural Operator $S3^T \times S3$		

these operators is also restricted to a pre-defined set of primitive shapes, which includes, square, circle, diamond, horizontal cross and diagonal cross, and horizontal, diagonal and vertical lines. The shape and size of the structuring element are defined by operator parameters. Other local neighborhood/windowing operators such as mean, median, etc. specify their kernels/windows in a similar way. The spectral operators have been chosen to permit weighted sums, differences and ratios of data and/or scratch planes.

We use a notation for genes that is most easily illustrated by an example: the gene [ADDP rD0 rS1 wS2] applies pixel-by-pixel addition to two input planes, read from data plane 0 and from scratch plane 1, and writes its output to scratch plane 2. Any additional required operator parameters are listed after the input and output arguments.

Note that although all chromosomes have the same fixed number of genes, the *effective size* of the resulting algorithm graph may be smaller than this. For instance, an operator may write to a scratch plane that is then overwritten by another gene before anything reads from it. GENIE performs an analysis of chromosome graphs when they are created and only carries out those processing steps that actually affect the final result. Therefore, in some respects, we could refer to the fixed length of the chromosome as a "maximum" length.

2.3 Backends

Complete classification requires that we end up with a single binary-valued output plane from the algorithm. It would be possible to treat, say, the contents of scratch plane 0 after running the chromosome algorithm, as the final output from the algorithm (thresholding would be required to obtain a binary result). However, we have found it to be of great advantage to perform the final classification using a non-evolutionary algorithm.

To do this, we first select a subset of the scratch planes and data planes to be *answer planes*. Typically in our experiments this subset consists of just the scratch planes. We then use the provided training data and the contents of the answer planes to derive the *Fisher Discriminant*, which is the linear combination of the answer planes that maximizes the mean separation in spectral terms between those pixels marked up as "true" and those pixels marked up as "false", normalized by the "total variance" in the projection defined by the linear combination. See [14] for details of how this discriminant works.

The output of the discriminant-finding phase is a gray-scale image. This is then reduced to a binary image by using Brent's method [15] to find the threshold value that minimizes the total number of misclassifications (false positives plus false negatives) on the training data.

2.4   Fitness Evaluation

The fitness of a candidate solution is given by the degree of agreement between the final binary output plane and the training data. This degree of agreement is determined by the Hamming distance between the final binary output of the algorithm and the training data, with only pixels marked as true or false contributing towards the metric. The Hamming distance is then normalized so that a perfect score is 1000. To put this in a more formal/mathematical context. Let H be the Hamming distance between the final binary output of the algorithm and the training data, with only pixels marked as true or false contributing towards the metric, let N be the number of classified pixels in the training image (i.e. pixels marked as either "true" or "false") and let F be the fitness of the candidate solution.

$$F = (1 - (H/N)) \times 1000 \qquad (1)$$

2.5   Software Implementation

The genetic algorithm code has been implemented in object-oriented Perl. This provides a convenient environment for the string manipulations required by the evolutionary operations and simple access to the underlying operating system (Linux). Chromosome fitness evaluation is the computationally intensive part of the evolutionary process and for that reason we currently use RSI's IDL language and image processing environment. Within IDL, individual genes correspond to single primitive image operators, which are coded as IDL procedures, with a chromosome representation being coded as an IDL batch executable. In the present implementation, an IDL session is opened at the start of a run and communicates with the Perl code via a two-way unix pipe. This pipe is a low-bandwidth connection. It is only the IDL session that needs to access the input and training data (possibly hundreds of Megabytes), which requires a high-bandwidth connection. The ALADDIN training data mark-up tool was written in Java. Fig. 2 shows the software architecture of the system.

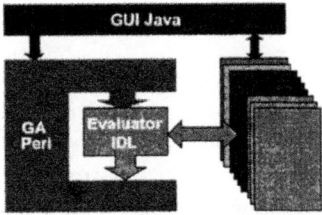

Fig. 2. Software Architecture of the System Described. Note that the feature depicted on the right of this diagram represents the input data, training data and scratch planes

3   Why Golf Courses?

The usefulness of devising algorithms for the detection of golf courses may not, at first, seem apparent (except to a manager, perhaps!). However, due to the nature of golf courses and their characteristics in remotely-sensed data, they are of great use in testing automatic feature-detection systems, such as described here. They possess distinctive spectral and spatial characteristics and it is the ability of feature-detection algorithms to utilize both these "domains" that we seek to test. It is also useful that there exists a great deal of "ground truth" data available: a great many golf courses, for the benefit of low-tech managers, are marked on maps. In addition, golf courses usually possess a well-known, particular type of vegetation and it is rare to find information regarding specific vegetation types on maps. Fig. 3 (a) shows a map of NASA's Moffet Field Air Base, clearly showing the position of a golf course. Fig. 3 (b) shows a false col-

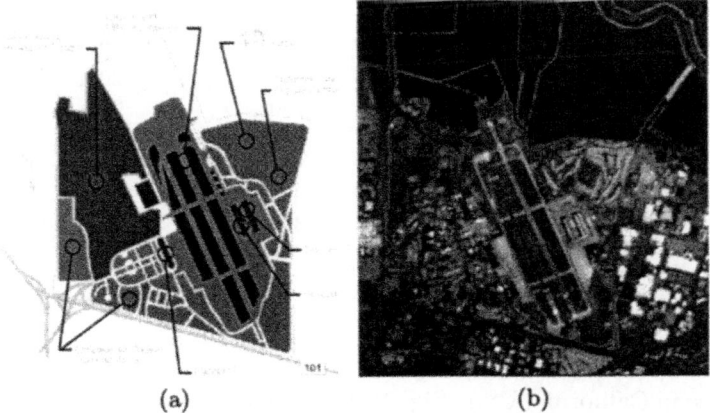

Fig. 3. (a) Map of NASA's Moffet Field Air Base, showing a golf course (available at http://george.arc.nasa.gov/jf/mfa/thesite2.html) (b) Image from remotely-sensed data of NASA's Moffet Field Air Base

our image of some remotely sensed data of the same region. The airfield and golf course are clearly visible.

4   Remotely-Sensed Data

The remotely-sensed images referred to in this paper are 10-channel simulated MTI data, produced from 224-channel AVIRIS data, each channel having 614 × 512 pixels. The images displayed are false-color images (which have then been converted to gray-scale in the printing process). The color mappings used are the same for all images shown (an exception being Fig. 1 where the false-color image has had a red and green overlay, corresponding to "false" and "true" pixels, as marked by the human analyst). The particular color mappings used here involve averaging bands A and B for the blue component, bands C and D for the green component and bands E and F for the red component. In addition, the images have been contrast enhanced. The choice of color mappings was arbitrary, in that it was a personal decision made by the analyst, made in order to best "highlight" the feature of interest, from his/her perspective and thus enable him/her to provide the best possible training data. This choice of color-mappings, together with a contrast-enhancement tool, are important and very useful features of ALADDIN. Table 2 provides details about MTI data.

Table 2. MTI Band Characteristics

Band	Wavelength (μm)	Color	SNR	Ground Sample Distance
A	0.45-0.52	blue/green	120	5m
B	0.52-0.60	green/yellow	120	5m
C	0.62-0.68	red	120	5m
D	0.76-0.86	NIR	120	5m
E	0.86-0.89	NIR	500	20m
F	0.91-0.97	NIR	300	20m
G	0.99-1.04	SWIR	600	20m
H	1.36-1.39	SWIR	4	20m
I	1.55-1.75	SWIR	700	20m
O	2.08-2.35	SWIR	600	20m
J	3.50-4.10	MWIR	250	20m
K	4.87-5.07	MWIR	500	20m
L	8.00-8.40	LWIR	800	20m
M	8.40-8.85	LWIR	1000	20m
N	10.2-10.7	LWIR	1200	20m

Figs. 3(a), 4(a) and 5(a) are data taken over an area of NASA's Moffet Field Air Base in California, USA. Fig. 3(a) is a sub-set of the data shown in Fig. 4(a). Figs. 3(a) and 5(a) are non-adjacent regions of the original data. These sub-sets of the data contain a lot of different features, but, of course, have a common feature of interest: golf courses.

5 Searching for Golf Courses

We reserve the data described above (Fig. 3(a)) for testing an evolved golf-
course finder algorithm and set the system the task of finding a golf course on
some other data. This data, showing the "truth" as marked out by an analyst,
is shown in Fig. 1. The golf course area has been marked as "true" and most
of the remaining data has been marked as "false". The system was run for 400
generations, with a population of 100 chromosomes, each having a fixed length of
20 genes. At the end of the run the best individual had a fitness of 966 (a perfect
score would be 1000). This fitness score actually translates into a detection rate
of 0.9326 and a false alarm rate of 0.00018. The results of applying the best
overall algorithm found during the run to the data used in the training run are
shown in Fig. 4.

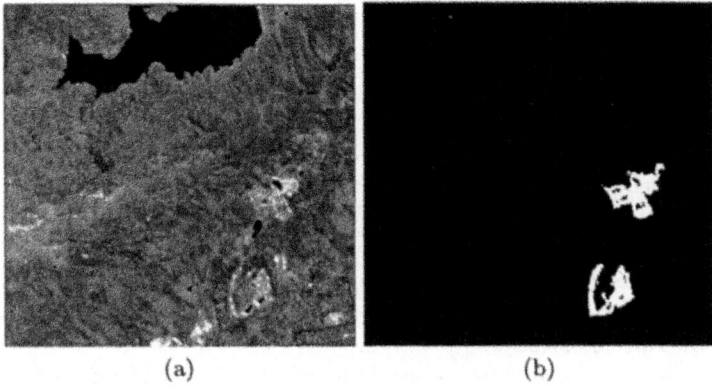

(a)   (b)

Fig. 4. (a) Image of training data (b) Result of applying algorithm found to training
data

It can be seen that the algorithm has been able to successfully detect the
golf course and has not detected any of the other features within the image.

In order to test the robustness of the algorithm found, it was applied to out-
of-training-sample data, as described previously, and shown in Fig. 3 (b). The
results are shown in Fig. 5.

It should be noted that the data shown in Fig. 5 covers a greater area than
shown by the map in Fig. 3 (a). It can be seen that the algorithm has successfully
found the golf course shown on the map. It can also be seen that the algorithm
has detected other golf courses. On closer examination of the data, it would
appear that further golf courses do, in fact, exist at those locations. It can also
be seen that the algorithm has not found any spurious features.

The "short" (redundant genes stripped out) version of the chromosome found
is detailed below.

[LAWG rD2 wS0] [OPREC rD3 wS3 5 1] [ADDP rS0 rS3 wS1] [ADDP rS1
rD6 wS1] [LAWE rD6 wS4] [LAWG rD6 wS0] [OPCL rS4 wS3 1 1] [DIL rS1 wS1
1 0] [OPREC rS1 wS1 5 0] [MEDIAN rS1 wS2 1] [LAWH rD2 wS4]

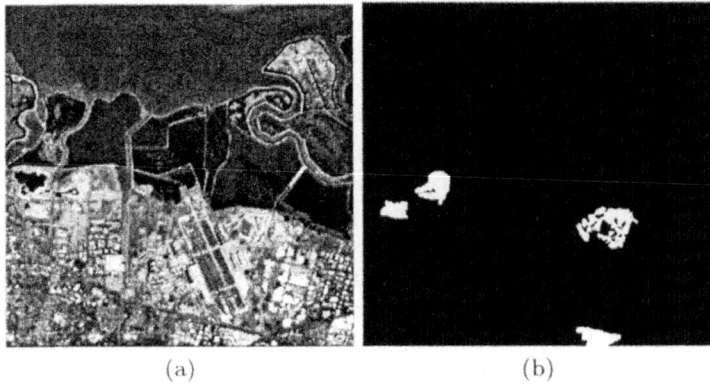

(a)   (b)

Fig. 5. (a) Image of out-of-training-sample data (b) Result of applying algorithm found to out-of-training-sample data

A graphical representation of the algorithm found is shown in Fig. 6. Note that the circles at the top of the graph indicate the data planes input to the algorithm (in this case only 3 data planes out of a possible 10 have been selected), the 5 circles in the center represent the scratch planes and the circle at the bottom represents the final, binary output of the overall algorithm. The operations above the line of scratch planes represent that part of the overall algorithm incorporated in the chromosome. The operations below the line of scratch planes represent the optimal linear combination of scratch planes and intelligent thresholding parts of the overall algorithm.

It is interesting to have some kind of objective measure of the algorithm's performance on the out-of-training-sample data. To this end an analyst marked up training data (i.e. true and false) for this data, with respect to the golf courses present. This enabled determination of a fitness for the algorithm on this data as well as detection and false alarm rates. The fitness of the algorithm was 926.6, the detection rate was 0.8532 and false-alarm rate was 3.000E-05.

6   Comparison with Other Techniques

In order to compare the feature-extraction technique described here to a more conventional technique, we used the Fisher discriminant, combined with the intelligent thresholding, as described previously, to try and extract the golf courses in the images shown/described. This approach is based purely on spectral information. On application to the data used in the training run (Fig. 4(a)), this "traditional" approach produced a result having a fitness of 757.228 (with respect to the training data/analyst-supplied interpretation), which translates into a detection rate of 0.5159 and a false-alarm rate of 0.00141. On application to the out-of-training-sample data, the result had a fitness of 872.323, which translates into a detection rate of 0.7477 and false-alarm rate of 0.00305. Both of these results are significantly below the performance of the results produced by the GENIE system described here.

7   Conclusions

A system for the automatic generation of remote-sensing feature detection algorithms has been described. This system differs from previously described systems in that it combines a hybrid system of evolutionary techniques and more traditional optimization methods. It's effectiveness in searching for useful algorithms has been shown, together with the robustness of the algorithms discovered. It has also been shown to significantly out-perform more traditional, purely-spectral approaches.

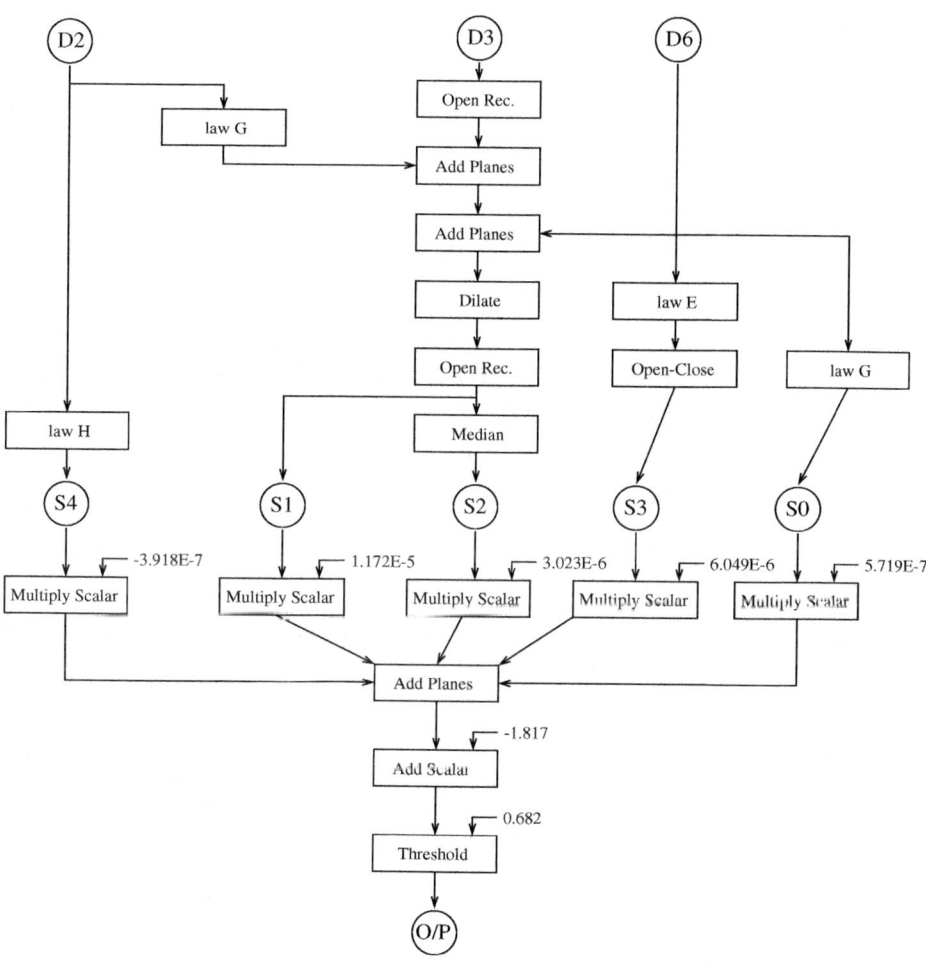

Fig. 6. Graphical representation of algorithm found

References

1. Cox, L.A., Jr., Davis, L., Qiu, Y.: Dynamic anticipatory routing in circuit-switched telecommunications networks, in Handbook of Genetic Algorithms, L. Davis, ed., pp. 124-143, Van Nostrand Reinhold, New York, 1991.
2. Harvey, N.R., Marshall, S.: GA Optimization of Spatio-Temporal Grey-Scale Soft Morphological Filters with Applications in Archive Film Restoration. In: Poli, R., Voigt, H.-M., Cagnoni, S., Corne, D., Smith, G.D., Fogarty, T.C. (eds.): Evolutionary Image Analysis, Signal Processing and Telecommunications (1999) pp. 31–45
3. Dandekar, T., Argos, P.: Potential of genetic algorithms in protein folding and protein engineering simulations, Protein Engineering 5(7), pp. 637-645, 1992.
4. Harris, C., Buxton, B.: Evolving edge detectors, Research Note RN/96/3, University College London, Dept. of Computer Science, London, 1996.
5. Teller, A., Veloso, M.: A controlled experiment: Evolution for learning difficult image classification, in 7th Portuguese Conference on Artificial Intelligence, Volume 990 of Lecture Notes in Computer Science, Springer-Verlag, Berlin, 1995.
6. Poli, R., Cagoni, S.: Genetic programming with user-driven selection: Experiments on the evolution of algorithms for image enhancement, in Genetic Programming 1997: Proceedings of the 2nd Annual Conference, J. R. Koza, et al., editors, Morgan Kaufmann, San Francisco 1997.
7. Nordin, P., Banzhaf, W.: Programmatic compression of images and sound, in Genetic Programming 1997: Proceedings of the 2nd Annual Conference, J. R. Koza, et al., editors,, Morgan Kaufmann, San Francisco, 1996.
8. Daida, J.M., Hommes, J.D., Bersano-Begey, T.F., Ross, S.J., Vesecky, J.F.: Algorithm discovery using the genetic programming paradigm: Extracting low-contrast curvilinear features from SAR images of arctic ice, in Advances in Genetic Programming 2, P. J. Angeline and K. E. Kinnear, Jr., editors, chap. 21, MIT, Cambridge, 1996.
9. Brumby, S.P., Theiler, J., Perkins, S.J., Harvey, N.R., Szymanski, J.J., Bloch J.J., Mitchell, M.: Investigation of Image Feature Extraction by a Genetic Algorithm in Proc. SPIE 3812, pp. 24–31, 1999.
10. Theiler, J., Harvey, N.R., Brumby, S.P, Szymanski, J.J., Alferink, S., Perkins, S., Porter, R., Bloch, J.J.: Evolving Retrieval Algorithms with a Genetic Programming Scheme in Proc. SPIE 3812, in Press.
11. Koza, J.R.: Genetic programming: On the Programming of Computers by Means of Natural Selection MIT Press, 1992
12. Laws, K.I.: Texture energy measures in Proc. Image Understanding Workshop, Nov. 1979, pp. 47–51.
13. Pietikainen, M., Rosenfeld, A., Davis, L.S.: Experiments with Texture Classification using Averages of Local Pattern Matches IEEE Trans. on Systems, Man and Cybernetics, Vol. SMC-13, No. 3, May/June 1983, pp. 421–426.
14. Bishop, C.M.: Neural Networks for Pattern Recognition, pp. 105–112, Oxford University Press, 1995.
15. Press, W.H., Teukolsky, S.A., Vetterling, W.T., Flannery, B.P.: Numerical Recipes in C, 2nd Edition, Cambridge University Press, 1992, pp. 402–405..

Sound Localization for a Humanoid Robot by Means of Genetic Programming

Rikard Karlsson, Peter Nordin and Mats Nordahl

Complex Systems Group, Chalmers University of Technology, S-412 96, Göteborg, Sweden, Email: nordin,tfemn@fy.chalmers.se

Abstract. A linear GP system has been used to solve the problem of sound localization for an autonomous humanoid robot, with two microphones as ears. To determine the angle to the sound source, an evolved program was used in a loop over a stereo sample stream, where the genetic program gets the latest sample pair plus feedback from the previous iteration as input. The precision of the evolved programs was dependent on the experimental setup. For a sawtooth wave from a fixed distance the smallest error was $8°$. When letting the distance to the same source vary the error was $23°$. For a human voice at varying distances the error was up to $41°$.

1 Introduction

The purpose of this paper is to investigate sound localization using GP. The system is intended to be used in a humanoid robot equipped with two microphones. Sound localization is performed by a small binary machine code program on an unprocessed stereo stream of sampled sound. Since the system is intended for a humanoid robot, some degree of similarity to human sound localization is desired (see [1] for another study of this problem in a robot context). Because of the limited CPU power on-board the robot the computational requirements need to be minimized

The GP system needs to generalize from a limited set of training data. The system also needs to be able to localize many different kinds of sounds, sounds from all directions in the horizontal plane, sounds with different intensity, sounds from all distances, sounds when echoes are present, and sounds with background noise present.

As a comparison, we first give a short description of the human auditory system and its sound localization capabilities.

1.1 The auditory system

The ability to localize sounds is an important part of the auditory system and has been essential to our survival. Sound propagates through a medium as a longitudinal wave. On its way to the eardrum it passes the outer ear, called the pinna, and the auditory canal. The longitudinal wave in the air is transferred

S. Cagnoni et al. (Eds.): EvoWorkshops 2000, LNCS 1803, pp. 65–76, 2000.

to the eardrum, or tympanic membrane. The signal is amplified on its way by the ossicles, which work as levers from the tympanic membrane to the smaller area of the oval window. The movement of the oval window is transferred into a longitudinal wave in the liquid-filled cochlea.

The longitudinal wave propagates down the long stretched basilar membrane in the cochlea. The membrane is narrower and thinner, and thus more sensitive to higher frequencies, near the oval window and responds to lower frequencies further down the membrane. Between 35 000 and 50 000 neurons are distributed along the basilar membrane. Each has a characteristic frequency related to the frequency causing maximal displacement of the basilar membrane at its location. This organization, where neighbouring neurons have similar response, is called tonotopy. It has a frequency range from 200 Hz to 20 000 Hz. Frequencies lower than 200 Hz are coded in terms of time of neural firing, by phase locking between the sound wave and the neuron firing. Between 200 Hz and 4 000 Hz both phase locking and tonotopy are used, while only tonotopy is used above 4 000 Hz.

1.2   Human sound localization

Let us now investigate the cues for sound localization present right at the eardrums. Consider a dummy head in an anechoic (echo-free) chamber with one sound source. The sound wave from the sound source propagates through the air with a velocity of 340 m/s and reaches the dummy head and its pinnae, auditory canals and eardrums. If the dummy head is facing the sound source at an angle it will take the sound longer to reach one ear than the other. This difference in time of arrival is known as the interaural time delay (ITD). If the sound is continuous the ITD can instead be determined by comparing phases of the sound signals at the eardrums.

If we know the relative phases and the wavelengths of the waves, we can only determine uniquely which wave lags the other if we know that one wave is always delayed by less than half a wavelength with respect to the other. This puts a limit on the frequency. When the sound comes from the left or right the delay is maximal. The distance between the ears is approximately 20 cm, which means that the wavelength needs to be longer than 40 cm. This corresponds to frequencies smaller than 850 Hz, for air with a sound velocity of 340 m/s.

Another cue that can be used to determine the direction to a sound source is the interaural intensity difference (IID), since the head casts a sound shadow and reduces the amplitude of a passing sound wave. Sound waves longer than the width of the head are strongly diffracted, and frequencies larger than 1700 Hz are strongly shaded.

The pinnae also filter the acoustic spectrum in a direction-dependent way. This filter is called head-related transfer function (HRTF). It can be seen as a function that maps intensity to intensity and takes the frequency of the sound and the direction to the sound source as arguments. The HRTF provides another potential cue for sound localization.

For localization in the horizontal plane ITD and IID are major cues for humans, while the HRTF is a cue for localization in the vertical planes. In humans the ITD is used for frequencies smaller than 2 kHz, while the IID is used for frequencies above 2 kHz. These results agree fairly well with the arguments above.

More details on human sound localization and its accuracy can be found in [2]. In the horizontal plane, humans make an error of 4° to 10° depending on the direction to the sound source. Sounds straight ahead of us are easiest to localize, while it is harder to localize sounds in the median plane. It is even more difficult to localize sounds if one is deaf on one ear. But it is possible to use only the HRTF as cue.

2   Experimental setup

Two microphones were used to record sounds in stereo from different directions in the horizontal plane. For sound localization to be as simple and efficient as possible we evolved programs that could localize sound using only a raw sampled sound signal as input. The program was iterated with feedback of the output from the previous iteration and the next pair of sound samples as input.

2.1   Recording of sound files

Sound recordings in stereo were made with two microphones placed in the dummy head of the robot. The signal from the microphones was preamplified, and sampled and recorded in stereo by a PC sound card. All recordings were made in an office with background noise mainly from computer fans.

The choice of sampling frequency and resolution limits which cues can be used for sound localization, since the sampling leads to a discretization error. This choice also determines the frequency range. The hardware restricts us to 8 or 16 bits representation. The discretization error is at most half the resolution in time and sample value. To calculate the ITD and IID a difference between signals is involved, so the maximal error is the resolution in time and sample value.

The use of ITD for sound localization requires high time resolution; the sample resolution is less important. If the sound arrives from infinity the ITD depends on the angle to the sound approximately as $ITD = \frac{D}{v}sin\varphi$ where D is the distance between the ears, v is the sound velocity and φ is the angle to the sound source. The error is then given by: $\Delta\varphi = \frac{180}{\pi}\frac{v}{Dcos\varphi}\Delta ITD$

To obtain an error of 10° for a sound from straight ahead the sampling time must be as short as $4.6 \cdot 10^{-4}$ s, corresponding to a sampling frequency of 22 kHz. Frequencies up to 850 Hz can be used to calculate ITD for humans, while the limit is 1900 Hz for a robot with 9 cm between its ears. The IID on the other hand is mainly sensitive to the sample resolution. In practice, frequencies above 1700 Hz can be used to calculate IID for humans, while the limit for the robot is 3800 Hz. To use the HRTF a large range of frequencies must be present and a high resolution in the sample value is needed.

Taking these facts into account, a 16 kHz sampling frequency was decided upon, based on a compromise between reducing the discretization error, and reducing the amount of input data for the genetic program.

Several sounds were recorded using different experimental setups. A human singing voice was recorded with an empty head of the robot (a cubic aluminum box, open in the top and back). Two sets of 1600 Hz sawtooth waves were recorded with different pinnae. Six commands spoken by a human voice were also recorded. These recordings are refered to as recordings One, Saw I and II and Commands in the rest of the paper. The samples were recorded using 16 bits.

In the setup for recording One the head was empty. The pinnae were made of plastic, and were similar to human pinnae. No auditory canals were present. A male human singing a tone was recorded from 16 evenly distributed directions in the horizontal plane. For the recording Saw I the head was filled with insulating material. Auditory canals with a diameter of 7 mm and a depth of 10 mm were made out of plastic. So were the small pinnae. A 1600 Hz sawtooth wave was transmitted from a loudspeaker 1.2 meters from the head. Sounds were recorded from 16 directions equally spaced in the horizontal plane.

In the recording Commands the head was filled with insulating material, and the robot was equipped with ears made of modeling clay and modeled after human ears. The auditory canals were approximately 12 mm long. By accident, the ear canals differed somewhat in width. Sounds were recorded with the dummy head placed in five different positions in the room. In each recordings were made from 16 evenly distributed directions in the horizontal plane. The sound source was placed either one, two, three or four meters from the dummy head. The recorded sounds were a male voice giving the commands walk, stand, forward, back, stop, right, left and grab.

In the recording Saw II the same head as in recording Commands was used (except that the ear canals were made equally wide). Sawtooth waves with a frequency of 1600 Hz were recorded with the head placed at three different locations in the room, 1 m , 2 m and 3 m from the sound source. At 1 m sounds from 16 directions evenly distributed in the horizontal plane were recorded. In the other two positions sounds from nine directions was recorded.

2.2   Genetic programming for sound localization

The programs evolved used a raw sampled sound signal as input. The program was iterated with feedback of the output from the previous iteration and the next pair of sound samples as input. An individual program can be viewed as a function whose inputs and outputs are arrays. The input consists of two memories, the previous real outputs, two samples and a constant. The output consists of the two memories and two real outputs.

The memories represent the output from the individual used as input in the next iteration. The 'real outputs' are the outputs used to evaluate the angle to the sound source. These are used as feedback in the same way as the memories. The samples are a pair of input samples from the left and right sound channel.

Initially the inputs in form of memories and previous real outputs are set to zero. During subsequent iterations, memories and previous real outputs, are used as feedback. The first time an individual is run it is given the first pair of samples in the fitness case as input, and a new pair of samples is then supplied in each iteration.

We can describe how the individuals are used in pseudo-code, see figure 1.

```
mem1 = mem2 = out1 = out2 = 0
L sample = L fitness case_i( 1 )
R sample = R fitness case_i( 1 )

for( k = 2 to 40){
    ( mem1, mem2, out1, out2 )
        = individual( mem1, mem2, out1, out2, L sample, R sample, const )
    L sample = L fitness case_i( k )
    R sample = R fitness case_i( k )
}

calculate angle to sound source from previous real outputs
```

Fig. 1. The use of an individual to determine the direction to a sound source is shown in pseudo-code.

The two real outputs 'out1' and 'out2' are interpreted as the x- and y-coordinate in a cartesian coordinate system, and the angle is calculated from this information. This representation is suggested by the geometry of the problem (an angular representation was also tried).

The number of iterations needed can be estimated from the physics of the problem. To use the ITD the individual needs to see at least a number of samples corresponding to the time delay. The maximal time delay is approximately the time it takes the sound to travel the distance between the ears, approximately 0.26 ms for a 9 cm robot head. This corresponds to 4.2 samples for a sampling frequency of 16 000 Hz. To use the IID the individual needs to see at least one wave length. For a sound of 400 Hz this corresponds to 40 samples. 400 Hz is well below the frequencies of highest intensity in the human speaking voice. Since higher frequencies are most interesting for the HRTF no more samples than to evaluate the IID are likely to be needed.

2.3   GP system

A linear GP system evolving binary machine code was used [3]. The system operates with a four individual tournament selection using steady state. Ten demes where used, together with an arithmetic function set.

A fitness case consists of a number of pairs of samples. These were chosen randomly from a longer array of samples from a recording corresponding to a specific direction to the sound source. All directions were given equal weight in the training set. New fitness cases were chosen each time an individual was evaluated, to improve generalization.

The test set was chosen randomly from the larger set of samples in the same way as the training set. The evaluation was done on new fitness cases chosen the same way as the ones used for training. A validation set was used to check how well the genetic program generalizes to a different set than the training set.

The error of an individual was calculated as difference, the shortest way around the circle, between the correct angle and the angle calculated from the output of the genetic program. The fitness was defined by the following expression:

$$fitness = \sqrt{\frac{1}{N-1} \sum_{i=1}^{N} error^2}$$

where N is the number of fitness cases in the training set and the index i corresponds to a specific fitness case.

3   Results

The success of the evolved program varied significantly with the experimental setup.

The results from experiments I to V are shown in figures 2 to 7. Figures 2 to 5 show the fitness of the best individual when evaluated on the training, test and validation sets. They also show the average and median fitness for the population as a whole, and the program length of the best genetic program and the average length of the genetic programs in the population.

The nine diagrams in figure 6 and figure 7 show the angle calculated after each iteration during the execution of the genetic program in experiments IV and V. In figure 6 one graph is shown in each diagram and in figure 7 three graphs corresponding to sounds coming from different distances are shown. In the nine diagrams the desired answer ranges from -90° in the first to +90° in the last with a spacing of 22.5°. The evaluations were done on a test set that was identical to the training set.

Table 1 shows the parameters that were varied between experiments.

The following parameters were identical in all experiments:

(1) The genetic programs were trained on sounds coming from the directions $\varphi = 0°$, $\pm22.5°$, $\pm45°$, $\pm67.5°$ and $\pm90°$.

(2) All samples were amplified in all five experiments in such a way that the maximal sample amplitude was identical in all fitness cases. After this amplification a second amplification was performed in all experiments except for experiment V. The last amplification was chosen randomly between 0.8 and 1 for each fitness case, to avoid that the genetic program learned to recognize each recorded sound file.

Exp. nr.	Data for: Training	Test	Val.	Err. Train.	Err. Test	Err. Val.	Settings: #loops	#gen.
I	Saw I (9)	Saw I (9)	-	12°	11° - 17°	-	20	400
II	human (27)	human (18)	Saw I (9)	41°	34° - 62°	30° - 50°	40	150
III	Saw I (9)	Saw I (18)	human (9)	8°	6° - 11°	40° - 56°	40	380
IV	Saw I (9)	Saw I (9)	human (18)	10°	13°	42° - 63°	[36, 45]	270
V	Saw II (27)	Saw II (27)	Saw I (9)	23°	24° - 28°	45° - 53°	[36, 45]	320
			human (45)			70°		

Table 1. The table is a summary of parameter settings and results for experiments I to V. Column two to four show the sound files used as training, test and validation sets, and the number of sound files used. In column five to seven the fitness of the best individual calculated on the training, test and validation sets are shown. The following column shows the number of iterations of the genetic programs. In experiment IV and V the number of iterations was chosen randomly in the given interval. The numbers in the last column show the number of generations each experiment was run.

(3) The population consisted of 40 000 individuals divided into 40 demes.

(4) A 90 % crossover probability was used. The probability for mutations to occur was the same. The probability for homologous crossover was 40 % and there was no migration between demes. The rate of crossover between demes was 1 %.

(5) The only constant available to the genetic programs was -413 608 020. No individual was allowed to be longer than 128 bytes (32 blocks).

(6) In experiment I and II the fitness was evaluated on ten fitness cases from each recorded sound file. In the other experiments that number was five. When the fitness was evaluated on the test set and training set twenty fitness cases were taken from each recorded sound file in the test and validation sets.

No real success were obtained when the training set were from Recording One. The error for sound localization on the training set was never smaller than s 21°. This result was obtained when all individuals were evaluated on exactly the same fitness cases. An error of 42° was obtained when the fitness cases were changed between each evaluation of an individual. This indicates that the genetically evolved program overtrained and memorized the fitness cases instead of solving the real problem of sound localization.

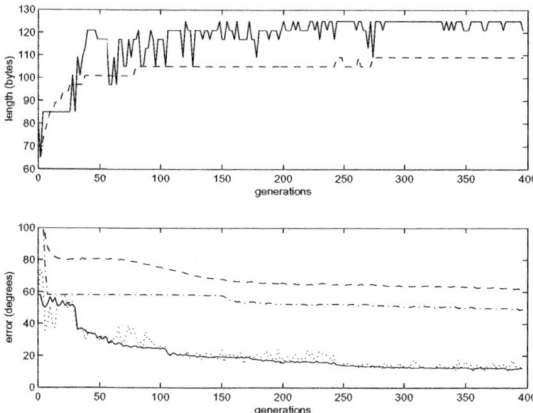

Fig. 2. Results from experiment I. The upper diagram shows the length in bytes of the best individual (solid) and the average length (dashed). The lower diagram shows the fitness of the best individual on the training set (solid) and the test set (dotted), the average fitness (dashed) and the median fitness (dashdot).

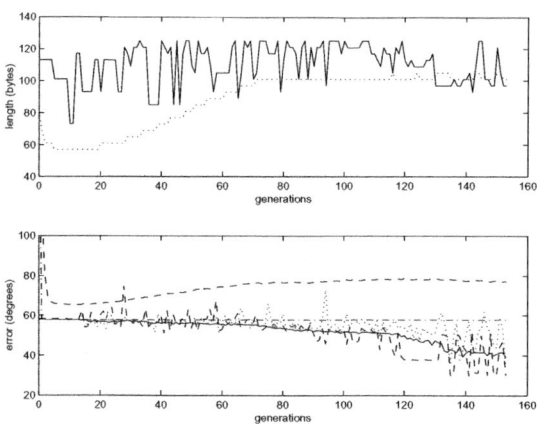

Fig. 3. Results from experiment II. The upper diagram shows the length in bytes of the best individual (solid) and the average length (dotted). The lower diagram shows the fitness of the best individual on the training set (solid), the test set (dotted) and the validation set (dashed lower). It also shows the average fitness (dashed upper) and the median fitness (dashdot).

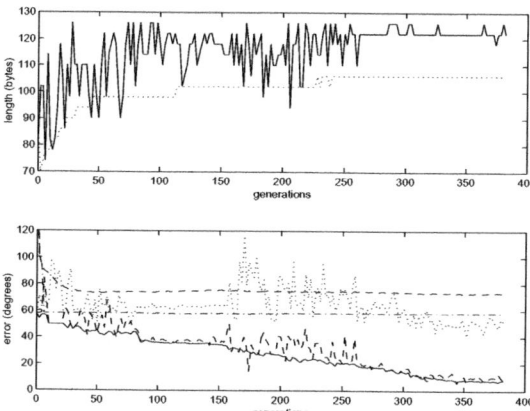

Fig. 4. Results from experiment III. The upper diagram shows the length in bytes of the best individual (solid) and the average length (dotted). The lower diagram shows the fitness of the best individual on the training set (solid), the test set (dotted) and the validation set (dashed lower). It also shows the average fitness (dashed upper) and the median fitness (dashdot).

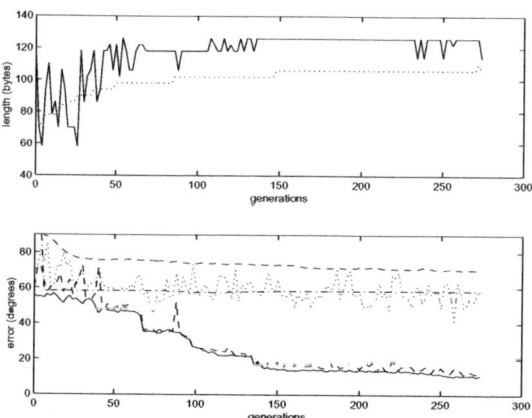

Fig. 5. Results from experiment IV. The upper diagram shows the length in bytes of the best individual (solid) and the average length (dotted). The lower diagram shows the fitness of the best individual on the training set (solid), the test set (dotted) and the validation set (dashed lower). It also shows the average fitness (dashed upper) and the median fitness (dashdot).

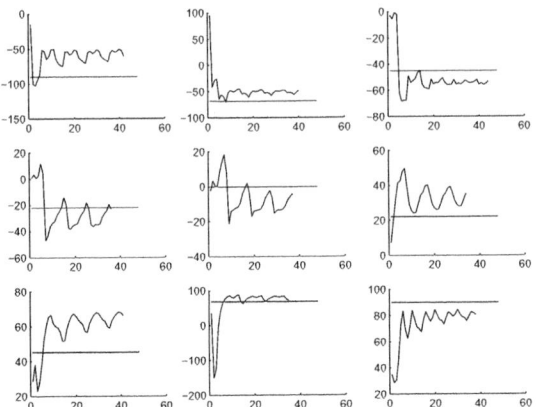

Fig. 6. Results for experiment IV. Each of the diagrams shows the angle calculated from the outputs of the best individual on one fitness case from the test set. The angle is calculated after each execution of the individual in the loop except for the first. The test set was identical to the training set. In each of the nine diagrams the horizontal line show the correct answer.

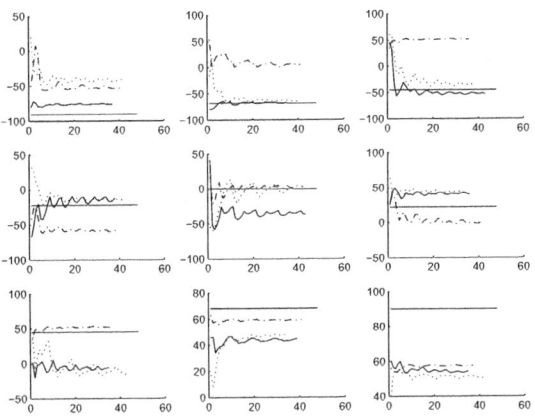

Fig. 7. Results for experiment IV. Each of the diagrams shows the angle calculated from the outputs of the best individual on three fitness cases from the test set. The three fitness cases correspond from sounds from the same direction but from different distances. The distances where 1 m (solid), 2 m (dotted) and 3 m (dashdot). The angle is calculated after each execution of the individual in the loop except for the first. In each diagram, the test set where identical to the training set. In each of the nine diagrams the horizontal line shows the correct answer.

4   Discussion

We first remark that caution should be observed when comparing the results from the different experiments, since the different training sets were recorded with different outer ears. The results of [4] show that if the shape of the human pinna is changed enough, our ability to localize sounds is dramatically reduced.

Let us start with experiment V. The fact that the fitnesses for the best individual evaluated on the training and test sets are almost equally small (see table 1), means that the genetic program generalizes to all sounds in the training set. This is also the case for the other experiments except for experiment II, as can be seen in figures 2 to 5.

In table 1 one can see that the evolved program in experiment V does not generalize to a human speaking voice (Commands) when trained on sawtooth waves. This is not surprising since it is hard for the genetic program even to localize a sawtooth wave sound coming from different directions and distances, see figure 7. This figure also shows that the genetic program does not generalize that well to sounds coming from different distances. The difficulty is probably due to echoes from walls and furniture in the room.

Training on a human voice was also difficult, see figure 3. The best evolved program has a fitness of $41°$. The fitness can be compared to that obtained from a program that always gives the constant output value $\varphi = 0°$. Then the fitness is $58.1°$, which shows that some learning has occurred.

In figure 6 one can see that the angle from one individual varies periodically when evaluated on one fitness case, while converging to the right answer. The period is the same as that of the sawtooth wave the individual is evaluted on. This indicates that the genetic program uses the IID in some way. This periodic pattern was less clear when the genetic programs were evolved on sound coming from different distances, see figure 7. The reason for this may be that the genetic programs must learn to cope with echoes, which makes it more difficult to use the IID.

5   Summary

We have shown that it is possible to evolve a small machine code program that can interpret an unprocessed stream of sampled stereo sound without insertion of domain knowledge or other structure. The evolved program can localize direction of a sound with varying success. On artificial sound the individuals have similar performance to humans measured as angular resolution, but much research remains for this performance to hold under more realistic settings and conditions.

Acknowledgements

This research was supported by NUTEK, the Swedish Board for Technological Development.

References

1. Irie R.E.; *Robust Sound Localization: An Application of an Auditory Perception System for a Humanoid Robot*, thesis, MIT, Cambridge 1993.
2. Blauert J.; *Spatial Hearing, the Psychophysics of Human Sound Localization*, MIT Press, 1997.
3. Banzhaf W., et.al.; *Genetic Programming*, dpunkt.verlag and Morgan Kaufmann Publishers, Heidenberg and San Fransico, 1998.
4. Hofman P.M., et.al., Relearning sound localization with new ears, *Nature neuroscience*, 5, pp. 417-421, 1998.

On the Scalability of Genetic Algorithms to Very Large-Scale Feature Selection

Andreas Moser[1] and M. Narasimha Murty[2]

[1] German Research Center for Artificial Intelligence GmbH
67608 Kaiserslautern, Germany
moser@dfki.de
[2] Department of Computer Science and Automation
Indian Institute of Science
Bangalore - 560 012, India
mnm@csa.iisc.ernet.in

Abstract. Feature Selection is a very promising optimisation strategy for Pattern Recognition systems. But, as an NP-complete task, it is extremely difficult to carry out. Past studies therefore were rather limited in either the cardinality of the feature space or the number of patterns utilised to assess the feature subset performance.

This study examines the scalability of Distributed Genetic Algorithms to very large-scale Feature Selection. As domain of application, a classification system for Optical Characters is chosen. The system is tailored to classify hand-written digits, involving 768 binary features. Due to the vastness of the investigated problem, this study forms a step into new realms in Feature Selection for classification.

We present a set of customisations of GAs that provide for an application of known concepts to Feature Selection problems of practical interest. Some limitations of GAs in the domain of Feature Selection are unrevealed and improvements are suggested. A widely used strategy to accelerate the optimisation process, Training Set Sampling, was observed to fail in this domain of application.

Experiments on unseen validation data suggest that Distributed GAs are capable of reducing the problem complexity significantly. The results show that the classification accuracy can be maintained while reducing the feature space cardinality by about 50%. Genetic Algorithms are demonstrated to scale well to very large-scale problems in Feature Selection.

1 Introduction

The recent development in Information Technology has brought about a tremendous flow of information. Companies, organisations and individuals are literally getting drowned in a flood of measurements.

Pattern Recognition, and in particular pattern classification, offers promising means towards a more refined way of dealing with information. In order to improve the process of classification, Feature Selection may be applied to increase the throughput of pertinent information. It does not only reduce the amount of

S. Cagnoni et al. (Eds.): EvoWorkshops 2000, LNCS 1803, pp. 77–86, 2000.

data taken into account for an Information Processing task, but may also improve the overall process accuracy.

However promising these observations may be, there are serious challenges to meet: Feature Selection is known to be an NP-complete problem. This fact makes it extremely difficult to examine large-scale, real-world domains.

Since the early studies of Feature Selection methods in the seventies, the notion of what "large-scale" actually means has changed. The first experiments dealt with rather *small* problems of some tens of features. With the increase of computational power, studies of *large-scale* tasks (100 - 500 features) became possible. Today, domains with more than 500 features can be said to be truly challenging; subsequently, such problems shall be called *very large-scale*.

There are basically two ways of carrying out a Feature Selection process: *Filter methods* and *Wrapper methods* [1]. The former class of approaches relies on general statistical properties of the problem domain and abstracts from particular classifier models. Wrapper methods on the other hand optimise the performance of a concrete classifier. Whereas Filter methods typically deliver more general results, Wrapper methods are considered to yield a better classification performance for specific tasks [2].

Thus, when aiming at a high classifier accuracy, Wrapper methods seem to be the more promising candidates. But the vast assessment time for classifier performance let such approaches appear infeasible for real-world applications at the first glance. However, advances in the realm of Distributed Artificial Intelligence provide ways to apply Wrapper methods to challenging real-world problems.

The following pages will describe a successful application of Distributed Genetic Algorithms - a Wrapper method - to a very large-scale problem in Feature Selection for an OCR system. First, the general customisation of GAs for Feature Selection is outlined. It is followed by a description of our tailoring of GAs for very large-scale Feature Selection. An experimental section demonstrates the usefulness of the setup. Finally, the results are summarised and possible extensions are suggested.

2   Genetic Algorithms to Select Features

GAs have proved to be capable of dealing with NP-complete problems in various fields. Siedlecky and Slansky suggested their application to Feature Selection for classification [3]. Other Wrapper (and Filter) methods have been applied in various studies [1]. But especially when being faced with very demanding large-scale tasks, GAs appear to be the best choice due to their inherent parallelism and nominal demand for background knowledge.

2.1   Using Simple GAs

In their milestone paper, Siedlecky and Slansky used a Simple GA to generate promising feature subsets. The aim of their study was to reduce the feature space cardinality and at the same time to keep the predictive accuracy acceptably high.

To apply GAs, the two researchers "cover" the patterns with a binary mask through which the classifier perceives the samples - the feature subset. A '1' indicates the presence and a '0' the absence of a feature. The chromosome is obtained by truncating the rows of the feature mask. Siedlecky and Slansky's model has been successfully adopted and refined by many other researchers.

A commonly applied fitness function for a feature subset p consists of a combination of classifier accuracy and subset complexity:

$$f(p) = acc(p) - pf * \frac{\mathcal{F}(p)}{|p|}$$

In this formula, acc denotes an accuracy estimate for the instantiated classifier, and $\mathcal{F}(p)$ is a measure for the complexity of the feature set - usually the number of utilised features. Furthermore, $|p|$ yields the feature space cardinality, and pf is a punishment factor to weigh the multiple objectives of the fitness function. The number of features used by a subset is intended to lead the algorithm to regions of small complexity.

2.2   Going Ahead: Distributed GAs

The setup as described above works pretty well for feature spaces of comparatively small cardinality. But when being confronted with very large-scale domains, the vast assessment time for feature subsets prohibits this approach.

Punch et al employed Distributed GAs for Feature Extraction in order to study higher-order problems [4][1]. They used a Micro Grain Distributed GA to accelerate the evaluation process, yielding up to linear speedups.

Using such conventional Micro Grain GAs, a feature subset is rated by sending it to a remote evaluator. The subset's predictive accuracy is estimated there by running a complete classifier. The results are then returned to a master node and used to guide the GA.

Recently, Distributed *Vertical* GAs were suggested in order to accelerate the evaluation process for classifier optimisations in further [5]. The use of this evaluation scheme facilitates for a study of very large scale domains in Pattern Recognition. This model can be beneficially applied to Feature Selection.

2.3   Past Studies

From today's perspective, past studies applying Genetic Algorithms to Feature Selection were limited in various ways:

  − The majority of researchers investigated domains of rather small complexity, that is below 100 features [3, 6–8].

---

[1] Though Punch et al. call their contribution *"Further Research on Feature Selection and Classification Using Genetic Algorithms"*, they in fact carry out a *Feature Extraction* process.

- Studies working on large scale and very large scale domains relied on a rather small number of training data [9–11], ([12]). Over-fitting tendencies thus were quite likely. Validation was usually not reported.
- Comparative studies of Genetic Algorithms and other methods were either based on small-scale problems [8, 13, 14], or the circumstances for GAs do not appear to be beneficial [12].

The subsequent sections will describe means to study very large-scale domains in Feature Selection while avoiding the above limitations.

3   Towards Very Large Scale Domains

When applying GAs to domains of large complexity in Feature Selection, the major challenge is the overwhelming time complexity of the process.

3.1   Threads of Time

Though the time requirements for the fitness evaluation can be dramatically decreased by virtue of Distributed (Vertical) GAs, the duration of the assessment process remains the threatening limitation of the search process. The GA thus has to invent good solutions in fewer cycles as compared to other fields of application. The crucial tradeoff is to keep the evaluation process as extensive as required - and as concise as possible.

3.2   Impacts on System Design

In our study, several measurements were taken in order to cope up with these requirements. Most of these measurements have been reported before; it is their careful combination that facilitates for a successful applications of GAs to very large-scale Feature Selection.

- *Dynamic Operators* [15] provide an efficient means to enforce convergence of the search process. This is necessary as the time requirements to assess the individuals prevent the GA from converging in a "natural" way.
  After a maturing phase of about one-hundred generations, the selection pressure was increased in our setup in order to lead the algorithm to a - necessarily local - optimum. This was achieved by reducing the mutation rate and by increasing crossover rate and information interchange between individuals. *Section 1* describes further details on the parameter settings.
- In order to enhance the initialisation process, the random generator was manipulated according to an explicit *Initialisation Bias*. In its usual instantiation, the initial population of the GA is obtained by tossing a fair coin for every individual and every feature. Individuals in very large-scale applications will reflect this choice in that they will be settled around the 50% level of complexity. The algorithm being forced to converge fast would thus have a poor initial coverage of the search space.

Using an Initialisation Bias, the probability for a '1' to occur in the initial population is chosen according to

$$\beta_{lin}(p_i) = \frac{b_{max} - b_{min}}{popSize} * i + b_{min}, or$$

$$\beta_{exp}(p_i) = (\frac{b_{max}}{b_{min}})^{(\frac{i}{popSize})^n} * b_{min}$$

In these equations, the Initialisation Bias ranges from b_{min} to b_{max} - either *lin*early or *exp*onentially. *popSize* denotes the number of individuals in the population and i the number of individual p_i $(1 \leq i \leq popSize)$; β_{exp} can be adjusted using the free parameter n.

As the complexity of the individuals can be expected to reflect this bias, the search space coverage may be enlarged by this measurement.

- Apart from the standard mutation operator, *asynchronous mutation* was applied. Dropping the assumption of equal mutation probabilities p_{01} and p_{10} (p_{ij} denoting the probability of 'i' to be changed into 'j') facilitates for an exploration of a broader range of the search space.
- Extending the set of usually applied fitness function as introduced in the former section, *polynomial punishment functions* were applied:

$$f_{pol}(p) = acc(p) - pf_1 * (\frac{\mathcal{F}(p)}{|p|})^2 - pf_2 * \frac{\mathcal{F}(p)}{|p|}$$

The hope was that this measurement would encourage the algorithm to explore regions of small complexity preferably.
- To avoid Positional Bias, *Binomial Crossover* [16] was used: The individuals are combined by tossing a biased coin for every gene. Besides this measurement, AND and OR mappings were tested to encourage the production of individuals in lower and upper ranges of complexity of the search space.
- Also suggested in [16] and widely used in Feature Selection studies, Training Set Sampling (TSS) was tried in order to accelerate the performance assessment. The idea is to randomly select a subset of the classifier training data before estimating the classification accuracy of the subsets. As due to the re sampling the fitness measure changes in the coarse of the search process, individuals inherited from former generations have to be re-evaluated. In the literature, TSS was reported to speed up fitness evaluation without introducing too much of noise into the search process

4   Experiments

To validate the usefulness of the customisations described above, tests were conducted, tackling a very large scale task in Feature Selection. The time-consuming experiments were conducted at the German Research Center for Artificial Intelligence, Kaiserslautern, Germany.

4.1   The Setup

In our study we investigated a Feature Selection problem for an OCR system. Our classifier was tailored to categorise handwritten digits.

The Dataset The learning data consisted of 10000 binary patterns with 24×32 pixels, divided into ten classes of equal cardinality. The data set has been used several times before in the OCR community, most recently by Prakash and Murty [17] and Saradhi [18].

In order to investigate the usefulness of GAs in this very large-scale domain, the learning data was randomly split into a 50% training set, a 30% test set, and a 20% validation set at the beginning of each run. The samples were normalised to equal size in a preprocessing step, using a straightforward scaling process which preserved the Aspect Ratio of the patterns.

The Classifier To rate a set of features, the test set was categorised according to the training set using a One-Nearest-Neighbour Classifier. Several studies proved this model to be a robust, efficient and well-performing approach for OCR-applications [19]. Boolean Distance using the XOR operator facilitated for a bit-parallel mapping of the patterns. Though considerable effort had been spent to accelerate the classification process, the time required to assess a single feature subset amounted to one-hundred-fifty seconds on a SUN ULTRA workstation. To illustrate the impacts of this figure: Given the population to consist of one-hundred individuals and the feature space cardinality to be 768, as in the problem investigated here, an exhaustive search using a single machine would require about 10^{226} years!

During the optimisation process, the validation set was fully kept aside. It was used to judge the generality of the results after the GA terminated. Due to the vast resource requirements, comparative studies with other Feature Selection methods were not conducted. However, the careful experimental setup with a separate validation set facilitates for an objective measurement of the method's performance.

The GA A Distributed Vertical GA was run on thirty SUN workstations of type SPARC-10, SPARC-20 and ULTRA. Its adaptive load balancing provided for an efficient use of the heterogeneous multi-user environment. For further details about the concept of Vertical GAs the reader is referred to [5] and [20].

A population of one-hundred individuals was maintained. To select the parents of the new generation, a combined deterministic and random strategy was used: The best individuals were copied directly into the next generation, and the remaining places in the mating pool were filled by using a fortune wheel based on scaled fitness values. This scaling was done such that the differences in classification accuracy were amplified. A minimal fitness value was added to permit for a selection of individuals with low predictive accuracy. The following equation shows how the fitness value $f(p)$ of an individual p is scaled, given $\bar{f}$ to be the average fitness and f_{max} the maximal fitness in the population:

$$f^{new}(p) = \begin{cases} f^{old}(p) - (2 * \bar{f} - f_{max}) + minFitness, \text{ if } f^{new} \geq 0 \\ minFitness, \hspace{3.5cm} \text{otherwise} \end{cases}$$

Table 1 summarises the most important parameter settings in the coarse of a typical run. *Age* denotes the number of generations passed, and crossBias indicates the probability for the binomial crossover to choose a gene from parent one for child one.

Age	Parameter	Value	Effect
0 - 50	minFitness	0.4	moderate selection pressure
(exploration)	mutationRate	2-3%	focus on exploration
	crossoverRate	75%	moderate chromosome interaction
	crossBias	90%	low gene interaction
50 - 75	minFitness	0.2	increased selection pressure
(growing)	mutationRate	1%	exploration and combination
	crossoverRate	80%	
	crossBias	75%	moderate gene interaction
75 - 100	minFitness	0.2	
(maturing)	mutationRate	0.5%	focus on maturing
	crossoverRate	90%	increased chromosome interaction
	crossBias	60%	increased gene interaction
100 -	minFitness	0	high selection pressure
(convergence)	mutationRate	0%	enforce convergence
	crossoverRate	100%	high chromosome interaction
	crossBias	50%	high gene interaction

Table 1. Dynamic operators

4.2   Results

Using Training Set Sampling Training Set Sampling (TSS) has been used by many researchers in the field of Feature Selection to cope up with the overwhelming time requirements. This technique has been as well tried for the purposes of this study. The training sets have been re-sampled at the 10% and 30% level. *Figure* 1 shows the evolution of the best fitness values on the test data, that is the data used in the optimisation process itself. The training set was re-sampled in 30% ratios. Note that due to this re-sampling the fitness function varies from generation to generation.

As can be seen in the figure, no progress in terms of the performance criterion on the test data is obtained. Measurements making use of the validation data turned out to be even more disastrous. The noise introduced to the search process showed to be too vast for this high-dimensional dataset in conjunction with the big number of training instances. Thus, TSS has not been furtherly considered in this study.

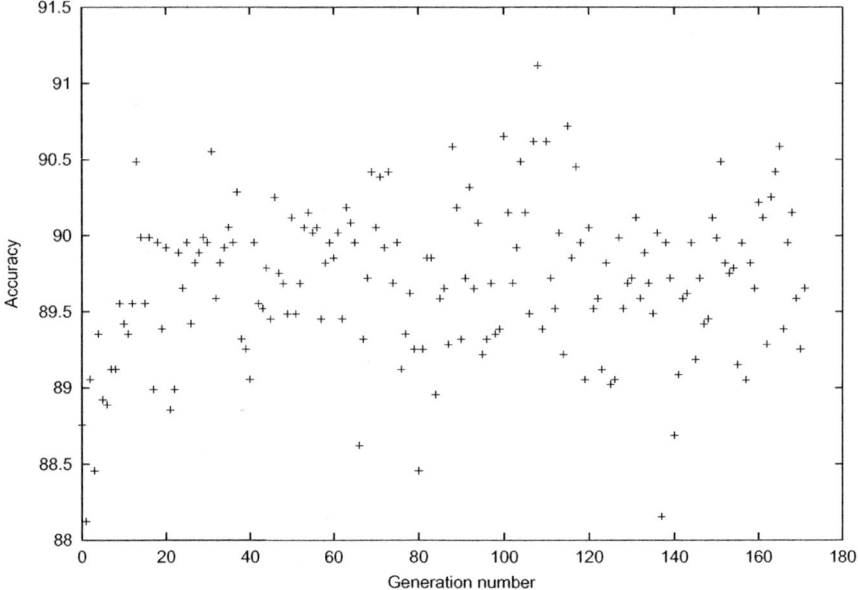

Fig. 1. Best fitness values using Training Set Sampling

Search Space Coverage Using the standard operators, the GA was observed to trace individuals near the 50% level of complexity preferably. This may be blamed to the peculiarities of the problem domain, such as high non-monotonousity of the fitness function, high insensitivity of the NN Classifier to changes in feature space cardinality, the limited maturing phase and sharp performance peaks in ranges of low complexity.

Among the measurements suggested in the former section, asynchronous mutation turned out to be the most effective means to broaden the examined region. The incorporation of a punishment factor into the fitness function alone was not capable of guiding the algorithm to other regions of interest. The use of *AND* mappings in supplement of some of the crossover operations was observed to lead to premature convergence: The algorithm got stuck in suboptimal regions with predictive accuracy below the full feature set.

Scalability Further experiments using the full training set for accuracy estimation were conducted. The results show that GAs scale well to domains of large complexity in Feature Selection: By means of dynamic operators, the GA converged within about two-hundred generations. The number of utilised features could be reduced by about 50% while preserving the predictive accuracy of the classifier.

Figure 2 shows the development of the performance on the unseen validation patterns: After reaching a certain degree of maturity, the calculated feature subsets deliver classification accuracies comparable to or even better than the full feature set. This remarkable result could be reproduced reliably in the consecutive runs.

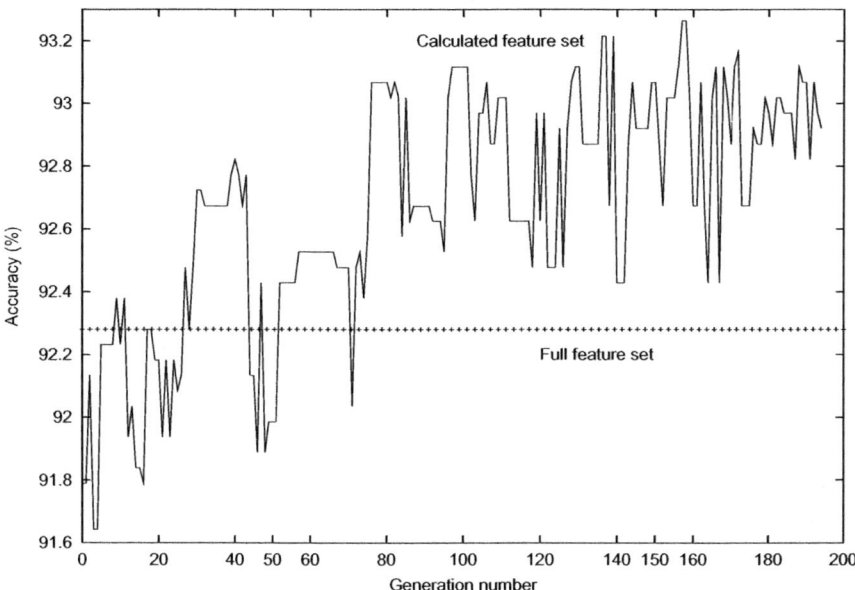

Fig. 2. Validation performance

5   Summary and Outlook

This report aimed at investigating the usefulness of Genetic Algorithms for very large-scale Feature Selection. Several partially already reported measurements were combined in order to cope with the complex problem.

Some limitations of GAs for very large-scale Feature Selection problems were observed and means to overcome them were suggested. Experiments showed that GAs scale well to domains of large complexity in Feature Selection.

A widely applied approximation method, Training Set Sampling, was observed to fail grossly in our application. Following the basic idea of this approach, more suitable customisations should be tried in future studies. Instead of randomly re-sampling the training set, a Prototype Selection process may be carried out simultaneously to Feature Selection. Saradhi applied this idea successfully to small-scale Pattern Recognition problems [18]. This combination appears to be promising for very large-scale tasks, as well.

Future work will aim at a study of further Feature Selection methods in very large-scale domains. The problem of Feature Creation will be addressed in order to enhance the search process in further.

References

1. M. Dash and H. Liu. Feature selection for classification. *Intelligent Data Analysis*, 1997.
2. G. John, R. Kohavi, and K. Pfleger. Irrelevant features and the subset selection problem. *Proceedings of the International Conference on Machine Learning*, 11, 1994.
3. W. Siedlecki and J. Sklansky. A note on genetic algorithms for large-scale feature selection. *Pattern Recognition Letters*, 10:335–347, 1988.
4. Punch, Goodman, Pei, Lai Chia-Shun, P. Hovland, and R. Enbody. Further research on feature selection and classification using genetic algorithms. *Proceedings of the 5th International Conference of Genetic Algorithms*, 1993.
5. A. Moser. A distributed vertical genetic algorithm for feature selection. *Fifth International Conference on Document Analysis and Recognition, Open Research Forum*, 1999.
6. D. Flotzinger. Feature selection by genetic algorithms. *IIG Report Series*, 369, 1993.
7. M. Prakash and M. N. Murty. Feature selection to improce classification accuracy using a genetic algorithm. *Journal of the Indian Institute of Science*, 1997.
8. A. K. Jain and D. Zongker. Feature selection: Evaluation, application and small sample performance. *IEEE Transactions on Pattern Analysis and Machine Intelligence*, 19(2), 1997.
9. J. E. Smith, T. C. Fogarty, and I.R. Johnson. Genetic feature selection for clustering and classification. *Proceedings of the IEEE Colloquium on Genetic Algorithms in Image Processing & Vision; IEEE Digest 1994/193*, 1994.
10. C. Guerra-Salcedo and D. Whitley. Genetic search for feature selection: A comparison between CHC and GENESIS. *Proceedings of the Symposium on Genetic Algorithms*, 1998.
11. J. Yang and V. Honavar. Feature subset selection using a genetic algorithm. *Feature Extraction, Construction and Selection - A Data Mining Perspective*, 1998.
12. F. J. Ferri, P. Pudil, M. Hatef, and J. Kittler. Comparative study of techniques for large-scale feature reduction. *Pattern Recognition in Practice IV*, 1994.
13. I. F. Imam and H. Vafaie. An emprical comparison between global and greedy-like search for feature selection. *Proceedings of the Florida AI Research Symposium*, 1994.
14. E. I. Chang and R. P. Lippmann. Using genetic algorithms to improve pattern classification performance. *Advances in Neural Information Processing*, 3, 1990.
15. D. Beasley, D. R. Bull, and R. R. Martin. An overview of genetic algorithms; part 2: Research topics. *University Computing*, 15(4):58–69, 1993.
16. F. Z. Brill, D. E. Brown, and W. N. Martin. Fast genetic selection of features for neural network classifiers. *IEEE Transactions of Neural Networks*, 3(2), 1992.
17. M. Prakash and M. N. Murty. Growing subspace pattern recognition methods and their neural-network models. *IEEE Transactions on Neural Networks*, 8(1), 1997.
18. V. V. Saradhi. *Pattern Representation and Prototype Selection in Classification*. Master Thesis, Department of Computer Science and Automation, Indian Institute of Science, Bangalore, 1999.
19. L. Holmstroem, P. Koistinen, and E. Oja. Neural and statistical classifiers - taxonomy and two case studies. *IEEE Transactions on Neural Networks*, 8(1), 1997.
20. A. Moser. Distributed genetic algorithms for feature selection. *Diploma Thesis, University of Kaiserslautern, Germany*, 1999.

Combining Evolutionary, Connectionist, and Fuzzy Classification Algorithms for Shape Analysis

Paul L. Rosin
Department of Computer Science
Cardiff University
UK
P.L.Rosin@cs.cf.ac.uk
and
Henry O. Nyongesa
School of Computing
Sheffield Hallam University
UK
H.Nyongesa@shu.ac.uk

Abstract. This paper presents an investigation into the classification of a difficult data set containing large intra-class variability but low inter-class variability. Standard classifiers are weak and fail to achieve satisfactory results however, it is proposed that a combination of such weak classifiers can improve overall performance. The paper also introduces a novel evolutionary approach to fuzzy rule generation for classification problems.

1 Introduction

This paper describes a series of experiments in tackling a difficult classification problem. The data consists of various beans and seeds, examples of which are shown in figure 1. Although some of the objects are larger than others (e.g. almonds compared to lentils) we are interested in classifying them based on their shape alone without using information about their size. This corresponds to the situation where the distance between the objects and the camera is not fixed, and so their apparent imaged sizes would vary. The difficulty of the task lies in the relatively small inter-class difference in shape and the high intra-class differences. In other words, all the objects look similar, appearing roughly elliptical. Although the shapes of some objects (e.g. almonds) are fairly consistent others vary considerably (e.g. corn kernels).

The basis for classifying the objects will be a set of shape properties measured from their silhouettes. Since size information is to be discarded the properties need to be invariant to scaling. Likewise, invariance to position and orientation

S. Cagnoni et al. (Eds.): EvoWorkshops 2000, LNCS 1803, pp. 87–96, 2000.

changes is necessary. Furthermore, it may be useful to include invariance to additional transformations of the shape. For instance, if the determining shape factor of a class is its similarity to an ellipse then the aspect ratio may be irrelevant. The computer vision literature provides a variety of shape measures [1]. A selection of these, in combination with some new shape properties developed by Rosin [2], have been applied to generate a set of 17 measurements of each sample. They can be divided into subgroups according to their properties and/or algorithms:

- *moments #1* – four attributes invariant to rotation, translation, and scaling (invariant under similarity transformations).
- *moments #2* – three attributes invariant to rotation, translation, scaling, and skew (invariant under affine transformations).
- *standard* – four standard attributes – eccentricity, circularity, compactness, and convexity (invariant under similarity transformations).
- *geometric #1* – three measurements of ellipticity, rectangularity, and triangularity (invariant under affine transformations, similarity transformations and stretching along the axes, and affine transformations respectively).
- *geometric #2* – three alternative measurements of ellipticity, rectangularity, and triangularity.

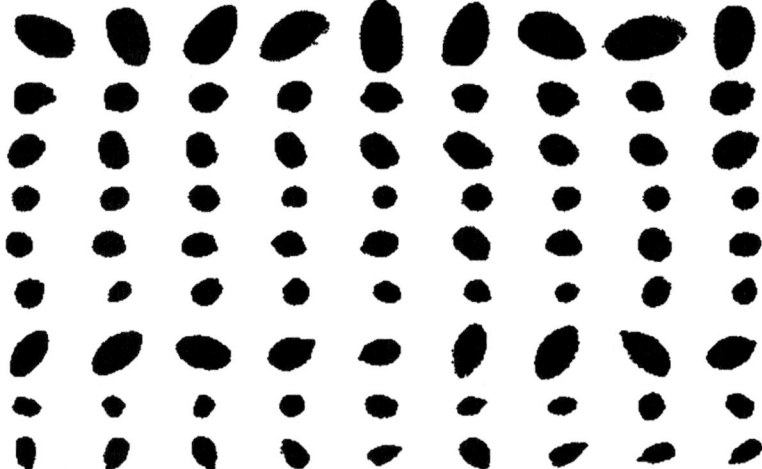

Fig. 1. Examples of data; rows contain: (1) almonds, (2) chickpeas, (3) coffee beans, (4) lentils, (5) peanuts, (6) corn kernels, (7) pumpkin seeds, (8) raisins, (9) sunflower seeds.

In this paper we have investigated four different classification techniques, and combined them in an attempt to improve overall classification. Two implementations of decision trees are used. The first method is the well known C4.5

machine learning method developed by Quinlan [3]. C4.5 induces trees that partition feature space into equivalence classes using axis-parallel hyperplanes (e.g. in 2D this would consist of horizontal and vertical lines). The second approach is OC1 [4] which is a generalisation in that, rather than checking the value of a single attribute at each node, it tests a linear combination of attributes. Feature space is consequently partitioned by oblique hyperplanes. The third method is an ensemble of neural networks that are trained on different features patterns, including one network that incorporates all patterns. Finally, we have also evolved fuzzy rules using evolutionary programming.

2   A Brief Overview of Connectionist, Decision Tree and Fuzzy Classification

Neural networks are connectionist systems that are widely used for regression and function approximation. They comprise input nodes, which are used to provide training data to the network and output nodes used to represent the dependent variables. Input-Output mapping is achieved by adjusting connection weights between the input and output nodes, but more usually, through an intermediate layer of nodes. This characteristic can be modified for classification problems by specifying desirable network outputs to be binary values. Neural networks are reliably used in classification of complex data.

Fuzzy classification is a rule-based approach in which IF-THEN rules are used to categorise data. The rules relate generalised or imprecise groupings of input data, and the decision of a given rule represents a degree of belonging to a given output class. This type of classifier is particularly useful when it is necessary to provide interpretability to the classification system in the form of linguistic rules. However, the process of creating classification rules is often difficult and time-consuming. Several studies have attempted to cope with this problem using learning algorithms, and in particular the use of neural networks.

Decision trees are a well established technique based on structuring the data, according to information theory, into mutually exclusive crisp regions. Classification is generally performed by starting at the tree's root node and checking the value of a single attribute, and depending on its values the appropriate link is followed. This process is repeated at succeeding nodes until a leaf is reached and a classification is assigned. Generating the optimal decision tree is generally NP-hard, and therefore a sub-optimal tree is induced instead, using greedy or randomised hill climbing algorithms for instance. The resulting tree is often pruned; subtrees are replaced by leaf nodes if this reduces the expected error rates. This results in potentially smaller and more accurate trees.

3   Evolutionary Fuzzy Classification

Evolutionary techniques have not been previously applied to fuzzy classification problems. The closest related work studies use genetic algorithms to optimise fuzzy classification rules and their membership functions. One disadvantage with this approach is that it is invariably necessary to pre-specify the structure of the rules, which often results in sub-optimal classification. In this paper, we have proposed a new technique in which fuzzy classification rules of arbitrary size and structure can be generated using genetic programming. This is desirable for complex problems, with large numbers of input features, for which it is not feasible to formulate the structure of rules manually. Furthermore, with such large numbers of features it is usually the case that certain features are not significant in classification of different output classes. Hence, in this case, we can say that genetic programming is used for unconstrained rule discovery and optimisation.

Genetic programming is an evolutionary technique in principle to Holland's genetic algorithms. The main differences are, (1) the structure of a genetic program is a tree, (2) the nodes of the trees are functions (or terminals), which enables the trees to interpreted as programs and (3) the size of each tree in a population is variable, unlike most genetic algorithms where all individuals are the same size. Otherwise, standard operators applicable to genetic algorithms are used in genetic programming.

In this study, the normalised input space was partioned into three fuzzy membership functions, *negative, zero* and *positive* [5]. Let N, Z, P be the fuzzy membership functions. We assumed simple rule constructs comprised of two inputs and one output. The non-terminal nodes of the GP trees represent these simple rules, which are combined to form the complex classification rule. Each simple rule is evaluated by matching its inputs against the fuzzy antecedents and the output is obtained using an AND operator, namely MIN. Thus, a node expressed as ZP(x,y) is interpreted as:

$$\text{IF } (x = Z \text{ AND } y = P) \text{ THEN MIN}(\mu_Z\mid_x, \mu_P\mid_y)$$

where $\mu_Z\mid_x$ is the degree of belonging of x to the fuzzy membership function Z. This type of rule construct is preferable to direct combination of the input parameters because it assists in interpretability of the classification system. There are nine different fuzzy rules which can be formed from the combination of the three membership functions.

The study used GP source code (lilgp) developed at Michigan State University [6]. This was used to evolve a set of complex fuzzy classification rules, one for each output class of data. The function set is comprised of the simple rules described above, while the terminal set comprised random constants and common arithmetic operators. The fitness of the trees on their evaluated outputs were determined against targets of 1.0 for the correct data class and 0.0 otherwise. As an example, the following is a small portion of a complex rule:

$$\text{(ZP (NP a b) (ZN d c))}$$

It can be interpreted as:

$$\text{IF d is Z AND c is N THEN temp1} = \text{MIN}(\mu_Z\mid_d, \mu_N\mid_c)$$
$$\text{IF a is N AND b is P THEN temp2} = \text{MIN}(\mu_N\mid_f, \mu_P\mid_b)$$

IF temp1 is Z AND temp2 is P THEN
out= $\text{MIN}(\mu_Z \mid_{\text{temp1}}, \mu_P \mid_{\text{temp2}})$

4   Voting Schemes

Because neural networks are trained on limited sample sets of representative data there are always significant errors in the generalisation of complex functions. One way to overcome this problem is to train multiple neural networks on independent data sets and then use voting schemes to determine an overall classification [7]. The two popular schemes are commonly known as ensemble and modular networks. In ensembles or committees redundant networks are trained to carry out the same task, with voting schemes being applied to determine an overall classification. On the other hand, it is pointless to train identical neural networks and consideration is thus often given to using different topologies, data sets or training algorithms. In modular networks the classification problem is decomposed into subtasks. For example, neural networks are trained to respond to one class or a group of classes.

The output of a classification network can be used to indicate the degree to which the input features are matched to the different classes. Therefore a simple approach to network combination is to sum the activation levels of corresponding output nodes. A refinement to this scheme is to scale the output levels within a network such that they sum to one. This allows the contributions across networks to be more comparable.

An alternative approach is based on the confusion matrix, a table containing entries c_{ij}, which indicate the frequency that data samples from class i were labeled as class j. Such a table is useful for analysing the performance of a classifier. Our approach is based on the classification accuracies for each class given by

$$\frac{a_{ii}}{\sum_j a_{ij}}$$

The contributions of each classifier are weighted by these expected accuracies. An example of a confusion matrix is shown in Table 1. It can be seen that the classifier is capable of consistently correctly classifying all instances of class 1, but only 75% of class2.

5   Experiments

The experiments compared the performance of four classifiers on the shape classification problem. The training and testing data both consist of 130 samples each containing seventeen continuous attributes of shape. Following standard practice the data was normalised prior to presentation to the neural and fuzzy classifiers, whereas this was not necessary for the decision trees.

The first set of experiments concentrated on evaluating each individual classifier. Examples of their outputs are shown in Figures 2 to 4. Their performances

	1	2	3	4	5	6	7	8	9	% Accuracy
1	11	0	0	0	0	0	0	0	0	100.0
2	0	9	3	0	0	0	0	0	0	75.0
3	0	2	15	1	0	0	0	0	0	83.3
4	1	0	3	12	0	0	0	1	0	70.6
5	0	2	3	3	0	1	1	0	1	0.0
6	0	3	1	1	0	3	2	1	1	25.0
7	1	1	0	0	0	0	16	0	1	84.2
8	0	1	1	2	3	3	1	1	1	7.7
9	0	2	1	0	1	0	0	0	13	76.5

Table 1. Confusion matrix

are shown in Table 2. The bottom three entries correspond to alternative means to deriving a classification based on neural output node activation levels. These are standard winner-take-all (WTA), and using the two decision trees (OC1 and C4.5). It is of interest to note that the simple winner-take-all performs better than the more complex decision trees in combining the outputs of the neural networks.

The second set of experiments compared various voting schemes applied to neural network ensembles. The neural networks were trained on the five subgroups of shape properties, and an additional neural network was trained on the combined seventeen properties. The voting schemes either select only the maximum activation level within a network (winner-take-all), or alternatively all the activations levels are used. The levels may first be scaled, and weighted by the confusion matrix, and are then summed over the networks. The result in Table 3 shows that better performance is achieved when the network ensemble includes the network trained with all properties. Furthermore, the use of the confusion matrix generally improves performance. However, the modular approaches provide little improvement over the basic single network clasification result.

The final set of experiments investigated hybridisation of three classifiers, namely, neural network, fuzzy-GP and C4.5, using the confusion matrix method. This does not rely on potentially incommensurate likelihood values that may be produced by different classifiers. We see from Table 4 that further gains in performance have been achieved, indicating that the neural network and decision tree provide useful complimentary information.

6   Conclusions

This paper has presented a comparison of different classification techniques on a difficult shape analysis problem. Similar to other reports in the literature [8, 9] it has been shown there is no significant differences between the individual techniques on our classification problem. However, we have shown that improvements can be achieved through different combinations of these techniques [10].

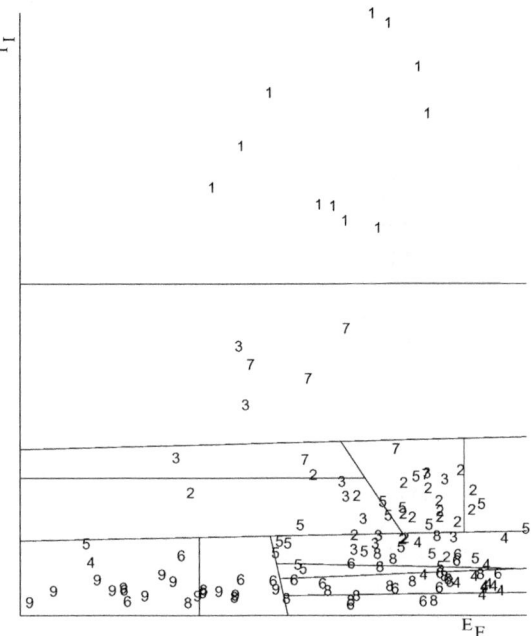

Fig. 2. An example of the partitioning of feature space for the bean classification task by the OC1 decision tree using only two properties. The horizontal axis property is elliptiocity, and the vertical axis property is triangularity.

METHOD	ACCURACY %
C4.5	53
OC1	54
Fuzzy-GP	44
NN + WTA	57
NN + C4.5	52
NN + OC1	51

Table 2. Accuracy for single classifiers

```
Rule 4:
     property 16 <= 0.225411
-> class 6

Rule 7:
     property 1 > 63.7441
     property 1 <= 65.4694
     property 4 > 0.161205
     property 4 <= 0.179735
     property 11 <= 0.967232
     property 13 > 0.710997
     property 15 <= 0.061688
     property 16 > 0.225411
     property 16 <= 0.329981
-> class 8

Rule 8:
     property 1 > 65.4694
     property 4 <= 0.179735
     property 15 <= 0.061688
-> class 6
```

Fig. 3. A typical set of rules generated by C4.5

```
(ZP (NP b
        (PN (- (PN a 0.83860)
               (NP (ZP (- (- (ZN d c)
                             (NZ -0.02504 d))
                          (PZ a f)) h)
                   (PZ e h))) g))
    (ZP a
        (ZP (- (ZN d c)
               (NP d
                   (ZN d c)))
            (ZP (NP f b)
                (ZN d c)))))
```

Fig. 4. A portion of a complex fuzzy-GP rule in prefix notation. Upper case letters denote fuzzy rules, and lower case letters denote ionput shape property values.

METHOD	ACCURACY %	
	Subgroups	Subgroups + Everything
WTA, Sum of activations	39	46
WTA, Sum of activations with scaling	40	46
WTA, Sum of CM	55	58
WTA, Sum of activations, weighted by CM	44	55
WTA, Sum of activations with scaling, weighted by CM	49	56
Sum of activations	42	49
Sum of activations, with scaling	43	51
Sum of activations, weighted by CM	50	56
Sum of activations, with scaling, weighted by CM	51	58

Table 3. Accuracy for NN voting schemes

METHODS	ACCURACY %
NN + C4.5	63
NN + Fuzzy-GP	58
Fuzzy-GP + C4.5	57
NN + C4.5 + FGP	61

Table 4. Hybrid Classifiers

A new approach to generate fuzzy clasification rules using genetic programming, on other hand, has not demonstrated satisfactory result. This may be due to the limited partitioning of the fuzzy input space into only three membership functions implemented in the study and further work will investigate the potential of this method.

References

1. M. Sonka, V. Hlavac, and R. Boyle. *Image Processing, Analysis, and Machine Vision*. Chapman and Hall, 1993.
2. P.L. Rosin. Measuring shape: Ellipticity, rectangularity, and triangularity. *submitted for publication*.
3. J.R. Quinlan. *C4.5: Programs for Machine Learning*. Morgan Kaufmann, 1993.
4. S.K. Murthy, S. Kasif, and S. Salzberg. System for induction of oblique decision trees. *Journal of Artificial Intelligence Research*, 2:1–33, 1994.
5. D. A. Linkens and H. O. Nyongesa. Genetic algorithms for fuzzy control - part i. *IEE Proc - Control Theory Appl.*, 142:161–176, 1995.
6. D. Zongker and B. Punch. *http://isl.cps.msu.edu/GA/software/lil-gp*. 1996.
7. A.J.C. Sharkey, editor. *Combining Artificial Neural Networks*. Springer-Verlag, 1999.

8. H. Chen, P. Buntin, L. She, S. Sutjahjo, C. Sommer, and D. Neely. Expert prediction, symbolic learning, and neural networks: An experiment on greyhound racing. *IEEE Expert*, 9:21–27, 1994.

9. M. Mulholland, D.B. Hibbert, P.R. Haddad, and P. Parslov. A comparison of classification in artificial intelligence, induction versus neural networks. *Chemometrics and Intelligent Laboratory Systems*, 30:117–128, 1995.

10. L.Xu, A. Krzyzak, and C. Y. Suen. Methods of combining multiple classifiers and their application to handwriting recognition. *IEEE Trans. Systems Man and Cybernetics*, 22:418–435, 1992.

Experimental Determination of Drosophila Embryonic Coordinates by Genetic Algorithms, the Simplex Method, and Their Hybrid

Alexander V. Spirov[1], Dmitry L. Timakin[2], John Reinitz[3], and David Kosman[3]

[1] The Sechenov Institute of Evolutionary Physiology and Biochemistry, 44 Thorez Ave., St. Petersburg, 194223, Russia,
[2] Dept. of Automation and Control Systems, Polytechnic University, 29 Polytechnic St, St. Petersburg, 194064, Russia
[3] Dept. of Biochemistry and Molecular Biology, Box 1020 Mt. Sinai Medical School, One Gustave L. Levy Place, New York, NY 10029 USA

Abstract. Modern large-scale "functional genomics" projects are inconceivable without the automated processing and computer-aided analysis of images. The project we are engaged in is aimed at the construction of heuristic models of segment determination in the fruit fly *Drosophila melanogaster*. The current emphasis in our work is the automated transformation of gene expression data in confocally scanned images into an electronic database of expression. We have developed and tested programs which use genetic algorithms for the elastic deformation of such images. In addition, genetic algorithms and the simplex method, both separately and in concert, were used for experimental determination of *Drosophila* embryonic curvilinear coordinates. Comparative tests demonstrate that the hybrid approach performs best. The intrinsic curvilinear coordinates of the embryo found by our optimization procedures appear to be well approximated by lines of isoconcentration of a known morphogen, Bicoid.

1 Introduction

1.1 Computer-Aided Analysis of Biological Images.

The ongoing revolution in molecular genetics has progressed from the large scale automated characterization of genomic sequence to the characterization of the biological function of the genome. These investigations mark the beginning of the era of 'functional genomics' [4]. A key feature of genomic scale approaches is the automated treatment of large amounts of data. Both current and future work in the field is impossible without the automated processing and computer-aided analysis of images in connection with updating interactive electronic image databases [8].

A key aspect of such processing involves the segmentation of individual images and the registration of serial images. Many problems involving the recognition, classification, segmentation and registration of images can be formulated as optimization problems. These optimization problems are typically difficult,

S. Cagnoni et al. (Eds.): EvoWorkshops 2000, LNCS 1803, pp. 97–106, 2000.

involving multiple minima and a complex search space topology. Contemporary approaches based on evolutionary computations are a promising avenue for the solution of such problems (EvoIASP99).

Here we describe a new method for the determination of intrinsic biological coordinates in embryos of the fruit fly *Drosophila melanogaster* by means of genetic algorithms (GAs). GAs, the simplex method, and a hybrid of both were applied to the problem, and we find that hybrid methods perform the best. Our results indicate that these coordinates may be determined by a morphogenetic gradient of the protein *Bicoid*, a result of some biological interest.

1.2   Stripe Straightening: Search of Intrinsic Coordinates of Early Embryo.

Early in the development the fruit fly embryo is shaped roughly like a hollow prolate ellipsoid, composed of a shell of nuclei which are not separated by cell membranes. Deviations from the elipsoidal shape reveal the future polarity of the animal's body: The more pointed end on the long axis makes anterior (head) structures, and the rounder end posterior (tail) structures. From a lateral (side) perspective, one long edge of the embryo is flat and will will make dorsal ("back") structures, while the other long edge is rounded and makes ventral ("underside") structures. In this paper we follow the standard biological convention and show embryos with anterior to the left and (if a lateral view) dorsal up (Fig. 1).

Fig. 1 shows that so called pair-rule stripes (early markers of the future segmental pattern [1]) are not parallel and straight, but have a crescent-like form. The curvature of the stripes is highest at the termini, and minimal at the central part. Each stripe specifies an anterior-posterior (A-P) location, and these stripes can be regarded as contours in an intrinsic coordinate system that is being created by the embryo itself. Another set of embryonic determinants exists for the dorso-ventral (D-V) axis. If the image is smoothly transformed such that the curvilinear coordinates are plotted orthogonally, the stripes appear straight, so the determination of these coordinates can be viewed as a "stripe straightening" procedure. Our task is to understand and characterize this curvilinear coordinate system as it relates to the A-P axis.

Two coupled objectives of this study are:

1. To characterize the intrinsic embryonic curvilinear coordinates.
2. To use carefully characterized and tested computational procedures for the purpose of automatically processing large numbers of images.

2   Methods and Approaches

The work reported here is part of a large scale project to construct a model of segment determination in the fruit fly *D. melanogaster* based on coarse-grained chemical kinetic equations [7]. The acquisition and mapping of gene expression data at a heretofore unprecedented level of precision is an integral part of this

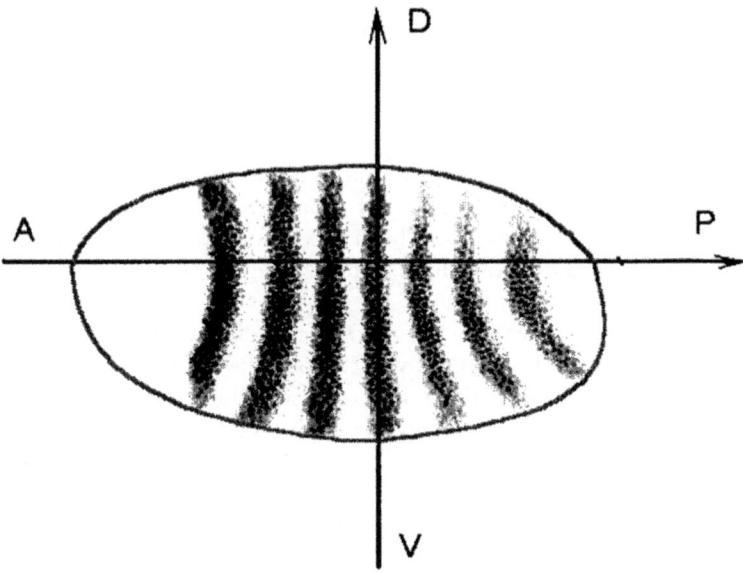

Fig. 1. Image of early (blastoderm stage) fly embryo with crescent-like stripes, in Cartesian physical coordinates. This is a confocally scanned image of an embryo stained by indirect fluorescence (immunostaining with polyclonal antisera against the *EVEN-SKIPPED* segmentation protein). Each small dot is an individual nucleus.

project. The current emphasis in our work is on the automated transformation of gene expression data in confocally scanned images into an electronic database of expression.

2.1 Images of Drosophila Genes Expression.

Transformations of embryonic coordinates begin with data expressed in terms of the average fluorescence level (proportional to gene expression level) of each nucleus, where segmentation proteins exert their biological function. This data was obtained as follows.

Antibodies for 14 protein products of segmentation genes were raised and over 500 images were prepared and scanned [2]. These images were computationally treated by means of the *Khoros* package [6]. Embryos were rotated and cropped automatically such that the physical long axis of the embryo was parallel with the x axis and the short physical axis with the y axis.

Next, the images were segmented [3, Kosman et al. in preparation]. About 2000 segmented and identified nuclei are obtained from each image. Each nucleus is labeled numerically, and the x and y coordinates of its centroid are found, together with the average fluorescence level over that nucleus. The segmented data takes the form of tables in ASCII text format. The result is the conversion of an image to a set of numerical data which is then suitable for further processing.

2.2   Stripe straightening algorithm.

In Fig. 1 the crescent-like pair-rule stripes of an embryo in near saggital projection are shown. We assume that the center of a pair-rule stripe follows a curve of constant A-P position. The origin of the image coordinate system is at the top left, with image coordinates for width w increasing to the right and height h increasing down.

Our goal here is to find the true A-P and D-V coordinates on the image. We approximate the true coordinate system by a Taylor series as follows. We denote the true A-P coordinate by $\hat{x}$ and the true D-V coordinate by $\hat{y}$. We pick the origin of $(\hat{x}, \hat{y})$ and the origin of new image coordinates (x, y) so that they are the same and as close as possible.

We note that there is an A-P position at which a stripe is exactly vertical on its whole length. The center of that stripe defines $\hat{x} = 0$, which is the $\hat{y}$-axis. Each pair-rule stripe other than the one at $\hat{x} = 0$ is curved, and we imagine the $\hat{x}$ axis to intersect each of the stripes at the point where it is exactly vertical. Now we pick new image coordinates x and y such that they have the same origin and orientation as the $(\hat{x}, \hat{y})$ coordinates, that is

$$x = w - w_0 y = -h - h_0 \tag{1}$$

We now turn our attention to $\hat{x}$. For now, we can assume that $\hat{y} = y$. Even if we don't do that, two important things will be true about the relationship between (x, y) and $(\hat{x}, \hat{y})$: (1) The y and $\hat{y}$ axes are coextensive, and (2) The loci $\hat{y} = \text{const}$ are orthogonal to the y and $\hat{y}$ axes as they cross $y = \hat{y} = 0$. Both of these important points follow from the existence of the vertical stripe. We would like to write $\hat{x}$ in terms of x and y, so that

$$\hat{x} = f(x, y). \tag{2}$$

We expand in a Taylor series to third order around the origin. That gives

$$
\begin{aligned}
\hat{x} = \quad & f(0,0) \quad + \quad \frac{\partial f}{\partial x}\big\|_{x=y=0}\, x \quad + \frac{\partial f}{\partial y}\big\|_{x=y=0}\, y \\
& + (1/2)\frac{\partial^2 f}{\partial x^2}\big\|_{x=y=0}\, x^2 + (1/2)\frac{\partial^2 f}{\partial x \partial y}\big\|_{x=y=0}\, xy + (1/2)\frac{\partial^2 f}{\partial y^2}\big\|_{x=y=0}\, x^2 \\
& + (1/6)\frac{\partial^3 f}{\partial x^3}\big\|_{x=y=0}\, x^3 + (1/6)\frac{\partial^3 f}{\partial x^2 \partial y}\big\|_{x=y=0}\, x^2 y + (1/6)\frac{\partial^3 f}{\partial x \partial y^2}\big\|_{x=y=0}\, xy^2 \\
& + (1/6)\frac{\partial^3 f}{\partial y^3}\big\|_{x=y=0}\, y^3.
\end{aligned}
\tag{3}
$$

Now consider the terms and what they mean. $f(0,0) = 0$ by definition. We picked $(\hat{x}, \hat{y})$ such that at the origin $\frac{\partial f}{\partial x} = 1$ and $\frac{\partial f}{\partial y} = 0$. For pure y terms we can say more than that. The fact that the y and the $\hat{y}$ axes are coextensive means that $f(0, y) = 0 \quad \forall y$, and so $\frac{\partial^2 f}{\partial y^2} = \frac{\partial^3 f}{\partial y^3} = 0$ as well. Thus far we have shown that five of the ten terms of the Taylor expansion vanish.

The unit vector $\mathbf{e}_{\hat{x}}$ in the $\hat{x}$ direction is proportional to $\frac{\partial f}{\partial x}$, so $\frac{\partial^2 f}{\partial x^2}$ measures the change in length of $\mathbf{e}_{\hat{x}}$ as we move along the x axis. This means that $\frac{\partial^2 f}{\partial x^2} = 0$. Now consider $\frac{\partial^2 f}{\partial x \partial y}$. This term can be thought of as the rate of change in size of the unit vector $\|\mathbf{e}_{\hat{x}}\| = \frac{\partial f}{\partial x}$ along the y-axis. Along the y-axis where $x = \hat{x} = 0$, $\frac{\partial f}{\partial x} = 1 \quad \forall y$, so that derivatives of this quantity with respect to y vanish, and hence this term of the series vanishes. This has eliminated all but three terms from the series, so now we write the first order model of image transformation as

$$\hat{x} = x + Axy^2 + Bx^2y + Cx^3. \tag{4}$$

All of these terms have a clear interpretation. The xy^2 term is the main one: it gives quadratic D-V curvature that increases with distance from the x-axis. The x^2y term gives residual D-V asymmetry and the x^3 term gives residual A-P asymmetry. Lastly, if one expresses the above equation in terms of w and h, expansion will bring back lower order terms in h and w when expanding

$$\hat{x} = w - w_0 + A(w - w_0)(-h - h_0)^2 + B(w - w_0)^2(-h - h_0) + C(w - w_0)^3 \tag{5}$$

in terms of w and h.

We tested this 1-st order model and found that in more then half of cases it is insufficient for straightening stripes. We expanded the model empirically, with the result that an empirical extension of the 1-st order model is given by

$$\hat{x} = A(w - w_0)(-h - h_0)^2 + B(w - w_0)^2(-h - h_0) + \\ C(w - w_0)^2(-h - h_0)^2 + D(w - w_0)(-h - h_0)^3 \tag{6}$$

We can treat of these additional fourth order members as follows: Cx^2y^2 is a correction term for parabolic splay, while Dxy^3 serves to correct D-V asymmetry. In general, the situation is typical of a polynomial approximation problem—there is one polynomial that is best but there are a number of distinct ones that can approximate it very well.

Preliminary calculations have shown that the best outcome is achieved with an independent deformation of the anterior and posterior half of an embryo. In summary it requires the determination of 8 parameters of a deformation plus an evaluation of values w_0, h_0^1, h_0^2.

2.3   Genetic Algorithms Technique, Simplex Method, and Their Hybrid.

GA Search. The optimization problem of finding the coefficient values for proper elastic transformations was initially implemented with GAs. We have reduced the problem to the determination of factors A, B, C and D of equation 6.

We use the following cost function. Each embryo's image under consideration was subdivided into a series of longitudinal strips. Then each strip is subdivided into boxes and the mean value of the product (*EVEN-SKIPPED* protein) is calculated for each box. Each row of means gives the local profile of *even-skipped* gene expression along each strip. The cost function is computed by comparing each profile and summing the squares of differences between the strips. The task of the GA is to minimize this cost function.

Following the classical GA algorithm, the program generates a population of floating-point chromosomes. Initial chromosomes are randomly generated. After that the program evaluates every chromosome as described above; then, according to the truncation strategy, the average score is calculated. Copies of chromosomes with above average scores replace all chromosomes with a score less than average.

On the next step a predetermined proportion of the chromosome population undergoes mutation, so that one of the coefficients gets a small increment. This cycle is repeated: all chromosomes are consecutively evaluated, the average score is calculated and the winners' offspring substitutes for the losers in the process of reproduction, until an accepted level of stripe straightening is achieved.

Simplex Search. We also solved the optimization problem by the downhill simplex method in multidimensions of Nelder and Mead [5]. The method requires only function evaluation, not derivatives. This is an important speed advantage over gradient methods, since calculation of the gradients requires many more evaluations of the cost function.

A simplex is the geometric figure in N dimensions of $N + 1$ vertices and all their interconnecting line segments. The Nelder-Mead method starts with such a set of $N + 1$ points defining an initial simplex. The downhill simplex method operates by moving the point of the simplex where the function is largest through the opposite face of the simplex to a lower point, and so on until it reaches the vicinity of an extremum.

Hybrid Procedure. Initial experience indicated that that the simplex method is fast but does not give high quality answers, while GAs give excellent answers but are slow. We noted that both multiple simplex runs and GA search perform numerous evaluations for many random points in search space. If we use small increments as mutations we will perform practically the same search by using GAs or the simplex method. If so, we could use a set of chromosomes from the GA technique as a starting simplex for Nelder-Mead optimization. In the hybrid algorithm, we implement a simple evolution strategy with floating-point chromosomes with small mutational increments. Selection and reproduction are performed as described above. In addition, from the very beginning the program links pointers to mutant offspring so as to achieve complete lists of $N+1$ pointers on $N + 1$ relatives.

These "clans of mutants" are ready for simplex procedure. Following the completion of the first list of $N + 1$ pointers the program starts to perform not only mutation, selection, reproduction procedures, but also the simplex procedure for the lowest scoring members of "complete" clans. The more clans achieve completion the more species undergo simplex procedure. In summary, GAs must provide search of global optima together with local ones, while simplex provides fast downhill moving.

3   Results and Discussion

3.1   Search Space Features for Stripe Straightening Problem.

The above described task of image elastic deformations turned out to be a difficult numerical problem. This is caused first of all by the unusual geometry of search space. Fig. 2 gives a picture of its features through crossections of the search space for one typical embryo under consideration.

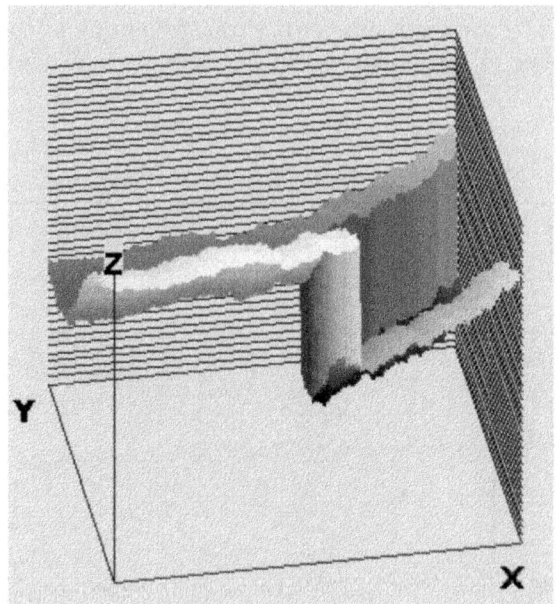

Fig. 2. Search space features for one of crossections $[A + D]$ for typical image. This is surface plot where vertical (Z) axis is evaluation one, while X and Y are A and D coefficients of expression 6.

As we can see this cross section includes two grooves, one of which is deeper than another. The sharp rectangular walls in Fig.2 are caused by penalty conditions. The omission of the penalties gives a smoother surface with one groove, which corresponds to the deeper groove (not shown). In turn, penalties are absolutely needed to avoid highly nonlinear folding of an image instead of smooth deformations.

The bottom of both grooves have several local minima. As a result the simplex search gives in serial searches tens of such local extrema. GA search is more effective and it finds the best local extremum on the bottom of the deepest groove. However to jump from the shallow to the deeper groove is still a difficult task for GA search as well. To overcome this we use large population sizes or a series of runs to achieve the best solution.

3.2   The Results of Stripe Straightening.

After completion of the stripe straightening procedure with 11 coefficients (w_0 and two sets of A, B, C, D and h_0^1, h_0^2) for about two hundred images from the stages when all seven crescent stripes are visible we could compare found coefficient sets. These coefficient sets show considerable diversity, so that we fail to elucidate a general formula of appropriate elastic deformations to achieve satisfactory stripe straightening. However, the resulting transformation of coordinates are very similar for most of the images. Typical example is shown in Fig.3.

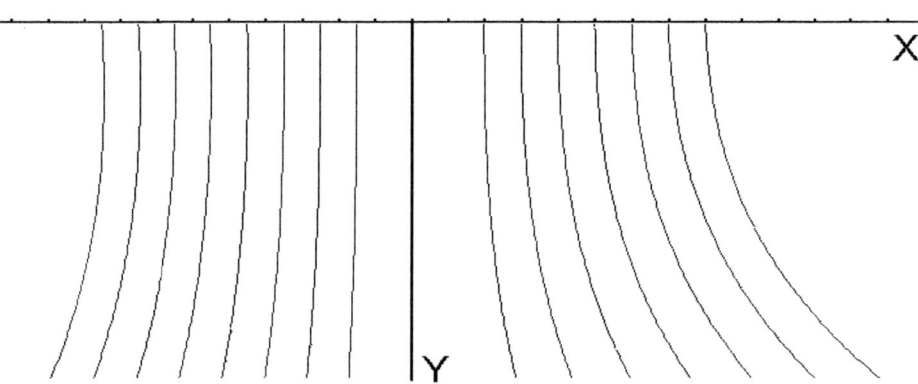

Fig. 3. Typical example of curvilinear coordinates found by our optimization procedures.

On the contrary, comparison of coordinate curves for anterior and posterior halves of embryos reveal small but quite evident differences (Cf. Fig.1 and Fig.3).

A biological subject of interest is the source of the pair-rule stripes' curvature. It is known that in *Drosophila* segmentation the maternally expressed protein *BICOID* forms an anterior-posterior morphogenetic gradient in the egg which controls all following segmentation events [1]. It is interesting that contour lines of a 2D-concentration map of the *BICOID* gradient closely coincide with curvilinear coordinates determined by our method. The full biological implications of this observation will be reported elsewhere.

3.3   Effectivity and Cost of GAs, Simplex and Hybrid Techniques.

In a table 1 the results of the comparative tests on elastic transformation of a typical image by means of the simplex-method, GAs, and GAs with simplex (the hybrid technique) are presented. To find parameters of optimization for a simplex and GAs giving the most effective optimization, careful tuning of the limits in variation of a mutational increment and the range of variability of the initial population was carried out. The hybrid method was tested at the same values of parameters, as GA technique. The result of testing was compared according to the time required for calculation and according to the standard deviation of the result. On each tested procedure 100 independent runs were carried out. Inspection of the table reveals that the simplex-method is fastest, but also the least precise. The GA technique is the most precise, but also requires 10 times as much computing. Our results indicate that the hybrid technique is approximately twice faster than GAs at the same accuracy.

Table 1. Comparison of effectiveness of three approaches on calculation time (in a summarized amount of evaluations) and on a divergence (in a standard deviation values)

Method	Time (in evaluations)	Standard Deviation
Simplex Method	25000	4738.646
GA Technique	301000	753.191
Hybrid Technique	151000	761.530

The problems we encountered with our optimization task are:

1. An abundance of local minima very close to the global minimum. The simplex can get stuck in very small-scale holes, even with starting conditions exceptionally close to a known solution. We need to allow the optimizer to move from a position that is very close to the global minimum towards the global minimum. It seems that this last stage is the difficult part for the Nealder-Mead simplex method. It is possible to produce good scores with a number of distinct end points in polynomial parameter space, suggesting that the problem is probably over-specified.
2. This is the polynomial approximation problem. There will be one polynomial that is the best for the stripe straightening but there are a number of distinct ones that can approximate it very well.

As to the first item, the successful approach to a solution of this problem is to employ hybrid techniques. Genetic algorithms alone are usually slow in optimization problems, since they are too coarse-grained to obtain a solution quickly. On the other hand, downhill algorithms are usually fast (in terms of processor cycles), if they are close to the solution, but tend to get stuck in local minima. Combining both kind of algorithms manages to avoid local minima, and finds solutions accurately.

4   Conclusions

In the task of optimization of parameters of elastic transformation a simplex-method is fastest, but also the least precise, giving the greatest divergence. The GA technique is the most precise, but also requires at least 10 times more time. The hybrid technique is twice faster than GAs at the same accuracy.

The intrinsic curvilinear coordinates of an embryo found by our procedures of optimization appears to be approximated by contour lines of a map of a gradient of the morphogen *bicoid*. It is in the good agreement with known ideas about a governing role of this gradient in consequent processes of segmentation of an early fly embryo.

5   Acknowledgements

This work is supported by INTAS, grant No 97-30950; Russian Foundation for Basic Researches, grant No 96-04-49349; USA National Institutes of Health, grant RO1-RR07801; and CRDF, grant No RB0-685. A.S. wishes to thank Timothy Bowler for stimulating discussions and King-Wai Chu for help with programming.

References

1. Akam,M.: The molecular basis for metameric pattern in the Drosophila embryo. Development **101**(1987) 1–22.
2. Kosman,D. and Reinitz,J.: Rapid preparation of a panel of polyclonal antibodies to Drosophila segmentation proteins. Development, Genes, and Evolution **208** (1998) 290–294.
3. Kosman,D., Reinitz,J. and Sharp D.H.: 1997. Automated assay of gene expression at cellular resolution. In Altman,R., Dunker,K., Hunter,L. and Klein,T. editors, Proceedings of the 1998 Pacific Symposium on Biocomputing, pages 6–17, Singapore: World Scientific Press.
4. Lander, E.S.: The new genomics: Global view of biology. Science **274** (1996) 536.
5. Press,W.H., Flannery,B.P., Teukolsky,S.A. and Vetterling,W.T.: 1988. Numerical Recipes in C: The Art of Scientific Computing. Cambridge: Cambridge University Press.
6. Rasure J. and Young M.: An open environment for image processing software development. In: 1992 SPIE/ISET Symposium on Electronic Imaging, V.1659 of SPIE Processings, SPIE, 1992.
7. Reinitz,J., Kosman,D., Vanario-Alonso,C.E. Sharp,D.: Stripe forming architecture of the gap gene system. Developmental Genetics **23** (1998) 11–27.
8. Sanchez,C., Lachaize,C., Janody,F., et al.: Grasping at molecular interactions and genetic networks in Drosophila melanogaster using FlyNets, an Internet database. Nucleic Acids Research **27** (1999) 89–94.

A Genetic Algorithm with Local Search for Solving Job Problems

L W Cai[1], Q H Wu[2], and Z Z Yong[1]

[1] Department of Electronic Engineering
Shenzhen University, Shenzhen, P. R. China
[2] Department of Electrical Engineering and Electronics
The University of Liverpool, Liverpool, L69 3GJ, U.K.
q.h.wu@liv.ac.uk

Abstract. This paper presents a genetic algorithm specially designed for job shop problems. The algorithm has a simple coding scheme and new crossover and mutation operators. A simple local search scheme is incorporated in the algorithm leading to a combined genetic algorithm(CGA). It is evaluated in three famous Muth and Thompson problems (i.e. MT6×6, MT10×10, MT20×5). The simulation study shows that this algorithm possesses high efficiency and is able to find out the optimal solutions for the job shop problems.

Keywords: Combined genetic algorithm, local search, job shop scheduling.

1 Introduction

The machine scheduling problem can be defined as a triple$\{\mathcal{M}, \mathcal{J}, \mathcal{C}\}$, where $\mathcal{M}$ represents the machine environment, $\mathcal{J}$ contains the jobs involved and $\mathcal{C}$ is an optimality criterion chosen according to the objective of scheduling. A classic job shop problem (JSP) [1] [2] consists of M machines on which N jobs are processed. Each machine can process only one job at a time, the processing of a job on a machine is called an operation. Each job has a specified processing order through all the machines with the corresponding processing time on each machine. This order is called machine sequence. The scheduling problem is to find out the best operation sequences on all machines in order to minimize the makespan. In this case, the makespan implies the criterion to be optimized. It is well known that the job shop scheduling problem is one of the most difficult NP problem. The classic approaches to JSPs are the branch and bound methods(BAB) [3] [4]. The performance of the existing algorithms for JSPs is far from satisfactory.

Over the last two decades, genetic algorithms (GAs) [5] have become more and more popular in the field of optimization. The GAs are a class of stochastic search algorithms that start with a population of randomly generated chromosomes and evolve towards better new chromosomes by applying genetic operators (crossover, mutation and inversion, etc). Two most distinguished characteristics of GAs are: (1) Only the fitness of chromosomes need to be calculated. Therefore the GAs are suitable for such optimization problems as JSPs which have no derivative information provided; (2) The GAs search from a population of solutions rather than a single point. They can find out the global optimal solution of optimization problems with multimodality efficiently. Development of an

S. Cagnoni et al. (Eds.): EvoWorkshops 2000, LNCS 1803, pp. 107-116, 2000.

efficient coding sheme and design of appropriate GA operators have received a great attention in the GA applications [6] ∼ [9].

Recently, the GAs have been applied to solve JSPs [10] ∼ [12]. Local search has been tried to work alongside with GA for JSPs [13], but the algorithm is rather complicated.

This paper presents a GA with a simple coding scheme and new crossover and mutation operators which are introduced to solve JSPs. A simple local search scheme is incorporated in the GA to improve the computation efficiency. In the CGA, the offsprings obtained by crossover and mutation are not used in the next generation directly but used as a seed of subsequent local search. The local search finds out the local optimal chromosomes in the neighborhood of the offsprings, then these chromosomes are used in the next generation. The paper is organized as follows. Section 2 describes the coding scheme. Construction of the new crossover and mutation operators is discussed in section 3. Section 4 introduces the CGA with local search. Section 6 shows the simulation results. Section 5 concludes the paper.

2   The coding scheme

2.1   Representation

In a classic job shop problem, there are N jobs to be processed on M machines. Each job must be processed by every machine. Thus there are a total of $N \times M$ operations. In the coding scheme, each chromosome is a string of $N \times M$ genes. Each gene represents an operation:

$$[gene_1 \ldots gene_i \ldots gene_{N \times M}]$$

In all chromosomes, the machine sequence for every job is strictly held. This constraint is ensured by the population initialization and GA operators.

2.2   Initialization

The operations in a chromosome should not be ordered arbitrarily. For every job, there is a constraint applied from the machine sequence. For the scheduling problem, we have a machine sequence matrix shown in Figure 1. The ith row of the matrix represents the machine sequence of the ith job:

$$S_i = [\, m_{i1} \quad m_{i2} \quad \ldots \quad m_{iM} \,]$$

In order to make a feasible schedule, the constraints for all jobs must be satisfied. The initialization of chromosomes is to take the elements of the machine sequence matrix one at a time in the following way:

(1) Select a non-empty row S_i ($i \in N$) randomly, $m_{i1} \neq \phi$, ϕ denotes an empty element.
(2) Let $gene_j = i_{m_{i1}}, m_{ik} = m_{i(k+1)} (k = 1, ..., M - 1), m_{iM} = \phi, j = j + 1$.
(3) Repeat (1) and (2) until $m_{ij} = \phi, \forall i \in N, j \in M$.

$$\begin{bmatrix} m_{11} & m_{12} & \cdots & m_{1M} \\ m_{21} & m_{22} & \cdots & m_{2M} \\ \vdots & \vdots & \ddots & \vdots \\ m_{N1} & m_{N2} & \cdots & m_{NM} \end{bmatrix}$$

Fig. 1. The machine sequence matrix

2.3 Evaluation

As stated above, a chromosome is a string of $N \times M$ operations. The job sequence of a machine can be obtained easily by scanning a chromosome from left to right to find out the job order for the same machine. Repeating the scanning for every machine, we have a job sequence matrix, given in Figure 2, which corresponds to a chromosome. The ith row of the matrix represents the job sequence of the ith machine:

$$P_i = \begin{bmatrix} j_{i1} & j_{i2} & \cdots & j_{iN} \end{bmatrix}$$

From the machine sequence matrix and the job sequence matrix, an active schedule can be achieved in the following way:

(1) Search a non-empty row P_i $(i \in M)$, if $m_{j_{i1}1} = i$, job j_{i1} will be processed as soon as possible after the ith machine completes the former operation.
(2) Let $j_{ik} = j_{i(k+1)}(k = 1, ..., N - 1), j_{iN} = \phi$; Let $m_{j_{i1}k} = m_{j_{i1}(k+1)}(k = 1, ..., M - 1), m_{j_{i1}M} = \phi$.
(3) Repeat (1) and (2) until $j_{lk} = \phi, \forall l \in M, k \in N$; $m_{ij} = \phi, \forall i \in N, j \in M$.

The fitness of a chromosome is the makespan of its corresponding active schedule.

$$\begin{bmatrix} j_{11} & j_{12} & \cdots & j_{1N} \\ j_{21} & j_{22} & \cdots & j_{2N} \\ \vdots & \vdots & \ddots & \vdots \\ j_{M1} & j_{M2} & \cdots & j_{MN} \end{bmatrix}$$

Fig. 2. The job sequence matrix

3 Operators

3.1 Crossover

The crossover is an important operator in GAs. The performance of GAs depends largely on the construction of crossover operator. By using this operator, a pair of chromosomes (parents) produces a pair of new chromosomes (offsprings) through exchanging and recombining schematic information. For JSPs, the difficulty of constructing the crossover operator is that the machine sequence of every job must be strictly held in newly generated chromosomes.

The construction of crossover operator for JSPs is as follows: At first, two parents are selected; then, $k(1 \leq k \leq N - 1)$ jobs are selected randomly. These k selected jobs correspond to $k \times M$ operations in each parent. A string is produced from the first parent by picking out these operations with their positions unchanged while filling other positions with holes (Hs). Rotating the string forward or backward l positions. The l positions are determined randomly, but in doing so, the relative order of the selected operations and the distance between every two selected operations should be maintained. Filling the holes (Hs) with the unselected operations of the second parent one by one while the relative order of these unselected operations is maintained. Now one offspring has been produced. Similarly, the other offspring can be produced with the selected operations of the second parent and the unselected operations of the first parent.

For example, we have the following machine sequence matrix:

$$\begin{bmatrix} 1 & 2 & 3 & 4 \\ 3 & 1 & 4 & 2 \\ 1 & 3 & 2 & 4 \\ 3 & 2 & 4 & 1 \end{bmatrix}$$

The following two chromosomes are randomly selected to crossover:

$$parent1 : 2_3\ 2_1\ 3_1\ 4_3\ 3_3\ 1_1\ 4_2\ 3_2\ 1_2\ 1_3\ 2_4\ 2_2\ 1_4\ 4_4\ 3_4\ 4_1$$

$$parent2 : 1_1\ 1_2\ 4_3\ 3_1\ 1_3\ 3_3\ 2_3\ 4_2\ 4_4\ 2_1\ 1_4\ 2_4\ 2_2\ 3_2\ 4_1\ 3_4$$

Now, select jobs randomly, for example $job1$ and $job3$ are selected as follows From $parent1$ we have

$$string1 : H\ H\ 3_1\ H\ 3_3\ 1_1\ H\ 3_2\ 1_2\ 1_3\ H\ H\ 1_4\ H\ 3_4\ H$$

Unselected operations of $parent1$ are as follows

$$2_3\ 2_1\ 4_3\ 4_2\ 2_4\ 2_2\ 4_4\ 4_1$$

From $parent2$, we have

$$string2 : 1_1\ 1_2\ H\ 3_1\ 1_3\ 3_3\ H\ H\ H\ H\ 1_4\ H\ H\ 3_2\ H\ 3_4$$

Unselected operations of $parent2$ are as follows

$$4_3\ 2_3\ 4_2\ 4_4\ 2_1\ 2_4\ 2_2\ 4_1$$

Then rotate $string1$ randomly. For example, one position forward, leave the last position as a hole, then we have

$$string3 : H\ 3_1\ H\ 3_3\ 1_1\ H\ 3_2\ 1_2\ 1_3\ H\ H\ 1_4\ H\ 3_4\ H\ H$$

Rotate $string2$ randomly. In this case, zero position is the only choice , then we have

$$string4 : 1_1\ 1_2\ H\ 3_1\ 1_3\ 3_3\ H\ H\ H\ H\ 1_4\ H\ H\ 3_2\ H\ 3_4$$

Fill in the holes of *string3* with the unselected operations of *parent2*, we have one offspring

$$offspr1 \ : \ 4_3 \ 3_1 \ 2_3 \ 3_3 \ 1_1 \ 4_2 \ 3_2 \ 1_2 \ 1_3 \ 4_4 \ 2_1 \ 1_4 \ 2_4 \ 3_4 \ 2_2 \ 4_1$$

Fill in the holes of *string4* with the unselected operations of *parent1*, we have another offspring

$$offspr2 \ : \ 1_1 \ 1_2 \ 2_3 \ 3_1 \ 1_3 \ 3_3 \ 2_1 \ 4_3 \ 4_2 \ 2_4 \ 1_4 \ 2_2 \ 4_4 \ 3_2 \ 4_1 \ 3_4$$

3.2   Mutation

Mutation is a random walk through the solution space, although it plays a secondary role in GAs, it is an insurance policy against premature loss of important information, a well constructed mutation operation can help a GA to get a better solution more easily and more fast.

Relocation is the key of the mutation operator. First, M operations with the same job are selected randomly; then, their positions are all relocated randomly at the same time in such a way that the order of these selected operations and the distance between every two selected operations are strictly held. The order of those unselected operations is also held.

For example, the following chromosome is to be mutated

$$oldstri \ : \ 2_3 \ 2_1 \ 3_1 \ 4_3 \ 3_3 \ 1_1 \ 4_2 \ 3_2 \ 1_2 \ 1_3 \ 2_4 \ 2_2 \ 1_4 \ 4_4 \ 3_4 \ 4_1$$

Operations with job 1 are selected, then we have the following string

$$string5 \ : \ H \ H \ H \ H \ H \ 1_1 \ H \ H \ 1_2 \ 1_3 \ H \ H \ 1_4 \ H \ H \ H$$

Unselected operations are as follows:

$$2_3 \ 2_1 \ 3_1 \ 4_3 \ 3_3 \ 4_2 \ 3_2 \ 2_4 \ 2_2 \ 4_4 \ 3_4 \ 4_1$$

Now, relocate the first selected operation(1_1) randomly, for example, at position 3 (in this case, its new position can be chosen from 1 to 9), and relocate other selected operations accordingly, we have

$$string6 \ : \ H \ H \ 1_1 \ H \ H \ 1_2 \ 1_3 \ H \ H \ 1_4 \ H \ H \ H \ H \ H \ H$$

Fill in the holes with the unselected operations in their previous order, the new chromosome is now produced

$$newstri \ : \ 2_3 \ 2_1 \ 1_1 \ 3_1 \ 4_3 \ 1_2 \ 1_3 \ 3_3 \ 4_2 \ 1_4 \ 3_2 \ 2_4 \ 2_2 \ 4_4 \ 3_4 \ 4_1$$

3.3   Reproduction

Reproduction is basically a process in which individual chromosomes are copied according to their fitness values. The classic reproduction method is the biased roulette wheel method. This method is simple and does well in many practical problems, but the simulation study shows that it is not efficient for job shop scheduling problems, because it forms a very large searching space. The reproduction applied in the proposed GA introduces new chromosomes and reinforces the principle of best survival, which are addressed as follows:

(1) New chromosomes

Some chromosomes are selected randomly and replaced by new randomly generated chromosomes. By doing so, new information will be brought in to prevent premature. Simulation shows that this is very useful in solving job shop scheduling problems. From the simulation study, it is found that two or three new chromosomes used in every generation would provide better performance.

(2) Best survival

 – The best chromosome of a generation is kept to evolve into the next generation.
 – In crossover, two best chromosomes are selected from the two parents and the two offsprings to survive into the next generation.
 – In mutation, the best from the old chromosome and the new chromosome is selected to survive into the next generation.

4   Genetic algorithm with local search for JSPs

4.1   Local search

Local search(LS) is a method to find the best solution from a neighborhood region. A solution x represents a point in the search space, and its neighborhood $N(x)$ is defined as a set of feasible solutions capable of being reached from x by exactly one step transition. For the min $f(x)$ problem, the local search method is described as follows:

One-step local search(OSLS)

(1) Select point x.
(2) Determine the neighborhood $N(x)$ of x and evaluate $f(x)$, $x \in N(x)$.
(3) Find x^* from $\{x, N(x)\}$ according to: $f(x^*) = \min\limits_{z \in \{x, N(x)\}} f(z)$.

Multi-step local search(MSLS)

(1) Select a starting point $x_i, i = 0$.
(2) Determine the neighborhood $N(x_i)$ of x_i and evaluate $f(x_i)$, $x_i \in N(x_i)$.
(3) Find x^* from $\{x_i, N(x_i)\}$ according to: $f(x^*) = \min\limits_{z \in \{x_i, N(x_i)\}} f(z)$.
(4) Let $i = i + 1, x_i = x^*$, repeat (2) to (4) until the termination condition is met.

Figure 3 illustrates the multi-step local search method.

4.2   GA/LS for JSPs

In our algorithms, the neighborhood of a chromosome is defined as the set of all feasible chromosomes produced by changing the positions of two neighboring genes. So the determination of the neighborhood of a chromosome is rather easy. The local search is incorporated in the GA. The GA/LS for JSPs is described as follows:

(1) Initialize the population.
(2) Make crossover on selected chromosomes, do local search on the offsprings.
(3) Make mutation on selected chromosomes, do local search on the offsprings.
(4) Reproduce a new population.
(5) Repeat (2) to (4) until the termination condition is met.

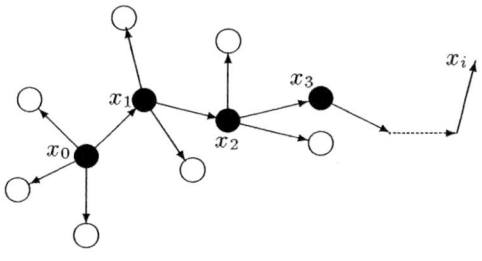

Fig. 3. Multi-step local search

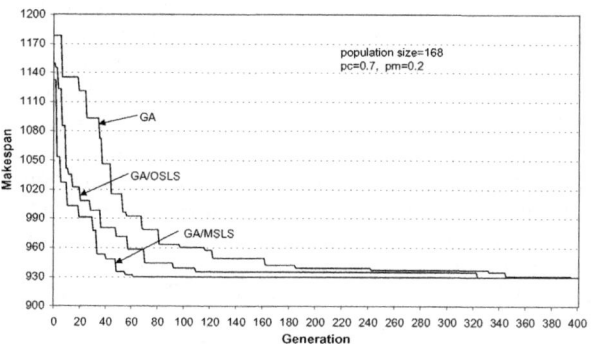

Fig. 4. The best convergence curves for MT10×10

Table 1. Minimum makespans for job shop problems

Algorithm	MT6×6	MT10×10	MT20×5
Barker1985(BAB)	55	960	1303
Carlier1989(BAB)	55	930	1165
Nakano1991(GA)	55	965	1215
Federico1995(GA)	55	946	1178
This paper(CGA)	55	930	1165

5   Conclusion

6   Simulation results

The above genetic algorithm has been implemented in C^{++} to solve three well-known JSP benchmarks [14]:

- MT6×6(6 jobs, 6 machines)
- MT10×10(10 jobs, 10 machines)
- MT20×5(20 jobs, 5 machines)

The minimum makespans achieved by this algorithm for the three benchmarks (MT6×6, MT10×10, MT20×5) are 55, 930 and 1165 respectively. For every benchmark, more than one optimal schedules with the same minimum makespan have been found out. Figure 4 shows the convergence of the CGA, in which the size of the population was chosen as 168, and the crossover and mutation probabilities were given as 0.7 and 0.2 respectively. It can be seen from Figure 4 that the CGA provides a much better convergence performance and the OSLS and MSLS offer different contributions to the CGA. Table 1 compares the minimum makespans obtained by this algorithm and other existing algorithms. The result of CGA is the same as Carlier's BAB and better than other algorithms. But it is well known that the large amount of computation of BAB for JSPs is a notorious problem. Three optimal schedules for the three JSP benchmarks are presented in Table 2, Table 3 and Table 4 respectively. In the tables, 'start' and 'end' represent the starting time and the end time of a job on a machine.

A combined genetic algorithm(CGA) with a simple coding scheme and specially designed crossover and mutation operators has been proposed to solve job shop scheduling problems. A simple local search scheme is incorporated in the algorithm. The optimal schedules for all the three JSP benchmarks [14] can be achieved by the proposed CGA. The simulation results show that the new crossover and mutation operators are efficient. The reproduction involving the 'new chromosomes' has been proved to be useful. It speeds up the searching process and guarantees the optimal results to be found. The algorithm obtains a better convergence by the introduction of the incorporated local search scheme. The shortcoming of the simple coding scheme is that there are more than one chromosome applied for a same schedule. This actually results in an increase of the searching space.

7   Acknowledgement

Mr. L W Cai would like to thank China Government and The University of Liverpool for supporting him to work at The University of Liverpool to undertake this work.

References

1. Coffman, E. G., et al., Computer and Job-Shop Scheduling Theory, U.S.A., John Wiley & Sons, 1976.
2. French, S., Sequencing and Scheduling: An Introduction to The Mathematics of The Job-Shop, England, Ellis Horwood Ltd., 1982.

3. Barker, J. R. and McMahon, G. B., 'Scheduling the general job shop', Management Science, 1985, **31** (5), 594-598.
4. Carlier, J. and Pinson, E., 'An algorithm for solving the job-shop problem', Management Science, 1989, **35** (2), 164-176.
5. Goldberg, D. E., Genetic Algorithms in Search, Optimization, and Machine Learning, Addison-Wesley Publishing Company, 1989.
6. Starkweather, T., *et al.*, 'A comparison of genetic sequencing operators', Proceedings of The Fourth International Conference on Genetic Algorithms, SAN DIEGO, 1991, pp.69-76 41.
7. Bagchi, S., *et al.*, 'Exploring problem-specific recombination operators for job shop scheduling', Proceedings of The Fourth International Conference on Genetic Algorithms, SAN DIEGO, 1991, pp.10-17.
8. Cao, Y.J., Wu, Q.H., 'Mechanical design optimization by mixed-variable evolutionary programming', Proc. IEEE International Conference on Evolutionary Computation, 1997, Indianapolis, USA, pp.443-446.
9. Wu, Q.H., Cao, Y.J., 'Stochastic optimization of control parameters in genetic algorithms', Proc. IEEE International Conference on Evolutionary Computation, 1997, Indianapolis, USA., pp77-80.
10. Nakano, R. and Yamada, T., 'Conventional genetic algorithm for job shop problems', Proceedings of The Fourth International Conference on Genetic Algorithms, SAN DIEGO, 1991, pp.474-479.
11. Federico Della Croce, *et al.*, 'A genetic algorithm for the job shop problem', Computers & Operations Research, 1995, **22** (1), 15-24.
12. Shi, G., 'A genetic algorithm applied to a classic job-shop scheduling problem', International Journal of Systems Science, 1997, **28** (1), 25-32.
13. Yamada, T. and Nakano, R., 'A genetic algorithm with multi-step crossover for job shop scheduling problems', First International Conference on Genetic Algorithms in Engineering Systems: Innovations and Applications: GALESIA,1st,Sheffield, 1995, pp.146-151.
14. Muth, J. F. and Thompson, G. L., Industrial scheduling, Prentice-Hall, Englewood Cliffs, New Jersey, 1963.

Table 2. An optimal schedule for MT6×6

Job	machine 1		machine 2		machine 3		machine 4		machine 5		machine 6	
	start	end	start	end	start	end	start	end	start	end	start	end
1	1	4	25	31	0	1	31	38	49	55	41	44
2	40	50	0	8	8	13	50	54	13	23	27	37
3	21	30	31	32	1	6	6	10	38	45	10	18
4	16	21	11	16	22	27	27	30	30	38	44	53
5	50	53	22	25	13	22	54	55	25	30	37	41
6	30	40	8	11	49	50	11	14	45	49	18	27

Table 3. An optimal schedule for MT10×10

	machine 1		machine 2		machine 3		machine 4		machine 5	
Job	start	end	start	end	start	end	start	end	start	end
1	119	148	445	523	523	532	532	568	568	617
2	76	119	637	665	224	314	568	637	355	430
3	408	493	308	399	532	606	493	532	887	920
4	185	256	0	81	84	179	637	735	256	355
5	256	262	286	308	179	193	370	396	430	499
6	361	408	84	86	0	84	138	233	499	505
7	148	185	86	132	375	388	233	294	698	753
8	275	361	399	445	193	224	813	892	519	551
9	0	76	217	286	421	506	294	370	668	694
10	262	275	132	217	314	375	735	787	787	877

	machine 6		machine 7		machine 8		machine 9		machine 10	
Job	start	end	start	end	start	end	start	end	start	end
1	617	628	645	707	721	777	777	821	821	842
2	753	799	707	753	813	885	895	925	430	441
3	699	709	753	842	709	721	609	699	842	887
4	799	842	355	364	416	501	364	416	766	788
5	308	369	842	895	530	579	499	520	593	665
6	86	138	420	485	505	530	233	281	281	353
7	421	442	388	420	668	698	520	609	442	474
8	445	519	557	645	777	813	699	718	718	766
9	370	421	517	557	579	668	821	895	506	517
10	628	675	375	382	885	930	416	480	517	593

Table 4. An optimal schedule for MT20×5

	machine1		machine2		machine3		machine4		machine5	
Job	start	end	start	end	start	end	start	end	start	end
1	518	547	731	740	822	871	945	1007	1025	1069
2	372	415	441	516	671	717	531	600	722	794
3	479	518	350	441	581	671	801	846	671	683
4	634	705	526	607	1043	1128	1143	1165	794	803
5	37	63	14	36	0	14	63	84	84	156
6	907	954	768	820	333	417	1007	1013	859	907
7	802	863	685	731	871	903	1013	1045	1131	1161
8	954	986	894	940	766	797	1045	1064	1095	1131
9	547	623	988	1028	903	988	725	801	1069	1095
10	415	479	152	237	272	333	484	531	541	631
11	791	802	607	685	1128	1149	689	725	803	859
12	623	634	740	768	491	581	899	945	995	1025
13	287	372	516	526	417	491	600	689	689	722
14	177	276	298	350	50	145	386	484	498	541
15	171	177	237	298	717	766	846	899	355	424
16	76	171	0	2	797	822	173	245	245	310
17	0	37	50	71	37	50	84	173	173	228
18	705	791	820	894	995	1043	1064	1143	907	995
19	276	287	83	152	221	272	297	386	424	498
20	63	76	76	83	145	221	245	297	310	355

Distributed Learning Control of Traffic Signals

Y. J. Cao, N. Ireson, L. Bull and R. Miles

Intelligent Computer Systems Centre
Faculty of Computer Studies and Mathematics
University of the West of England, Bristol, BS16 1QY, UK

Abstract. This paper presents a distributed learning control strategy for traffic signals. The strategy uses a fully distributed architecture in which there is effectively only one (low) level of control. Such strategy is aimed at incorporating computational intelligence techniques into the control system to increase the response time of the controller. The idea is implemented by employing learning classifier systems and TCP/IP based communication server, which supports the communication service in the control system. Simulation results in a simplified traffic network show that the control strategy can determine useful control rules within the dynamic traffic environment, and thus improve the traffic conditions.

1   Introduction

Traffic control in large cities is a difficult and non-trivial optimization problem. Most of the existing automated urban traffic control systems, such as TRANSYT[1], SCATS[2], LVA[3] and SCOOT[4], have a centralized structure, i.e. information gathering and processing, as well as control computations, are carried out in a centralized manner, in which case efficiency is decreased due to the large volume and the heterogeneous character of information [5]. To achieve global optimality, hierarchical control algorithms are generally employed. However, these algorithms have a slow speed of reaction and it has been recognized that incorporating some computational intelligence into lower levels can remove some burdens of algorithm calculation and decision making from higher levels [6]. Recently, there is a growing body of work concerned with the use of evolutionary computing techniques for the control of traffic signals. Montana and Czerwinski [7] proposed a mechanism to control the whole network of junctions using genetic programming [8]. They evolved mobile "creatures" represented as rooted trees which return true or false, based on whether or not the creature wished to alter the traffic signal it has just examined. Cao et al has developed an intelligent local traffic junction controller using learning classifier systems and fuzzy logic [9] and showed that the local controller can determine useful junction control rules within the dynamic environment. Mikami and Kakazu used

S. Cagnoni et al. (Eds.): EvoWorkshops 2000, LNCS 1803, pp. 117–126, 2000.

a combination of local learning by a stochastic reinforcement learning method with a global search via a genetic algorithm. The reinforcement learning was intended to optimize the traffic flow around each crossroad, while the genetic algorithm was intended to introduce a global optimization criterion to each of the local learning processes [12]. Escazut and Fogarty proposed an approach to generate a rule for each junction using classifier systems in biologically inspired configurations [13].

This paper is devoted to developing a distributed learning control strategy for traffic signals. A fully distributed architecture has been developed in which each subsystem is solely responsible for one aspect of the system and where a coherent global control plan emerges from the interactions of the subsystems; no hierarchical structure is included. Such an approach is aimed at increasing the speed of response of the local controller to changes in the environment. To do this, we have developed an agent-alike controller, which is implemented by employing learning classifier systems [14, 15] and TCP/IP based communication server, supporting the communication in the control system. Simulation results in a simplified traffic network show that the control strategy can determine useful control rules within the dynamic traffic environment, and therefore improve the traffic conditions.

2   An Agent-alike Controller

Optimization of a group of traffic signals over an area is a large and multi-agent type real-time planning problem without precise reference model given. To do this planning, each signal should learn not only to acquire its control plans individually through reinforcement learning but also to cooperate with each other. This requires communication between the agents. If each signal simultaneously communicates with each other and controls its phases according to the change of the global traffic flow, the total volume of the area will be well optimised. However, to provide the efficient communication is a difficult task, caused by the inefficient accounting of interactions between subsystems (in decentralized case) and the complex communication structure (in the hierarchical case).

In this work, we developed an agent-alike controller, consisting of a learning classifier system and a communication server, as shown in Figure 1. Rule-based controller as classifier systems lie midway between neural network and symbolic processing systems that can combine the benefits of both. The limitation of specifying a single classifier system is that while it may work for a simple controller, the method does not scale up to complex control systems. Work done so far has addressed this problem by dividing a complex system into its simplest, physical sub-systems, specifying rule-based controller for each of these and thus creating a multi-agent system.

For traffic control problem, we associate an agent to each junction of the traffic network. So, the whole control strategy developed in this work contains a number of distributed, communicating agents, where each agent has a classifier

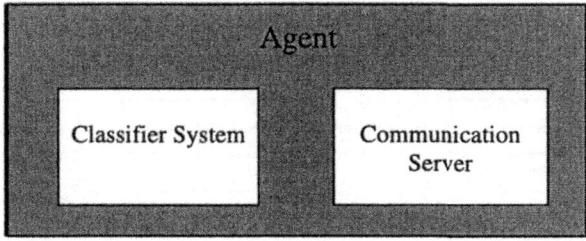

Figure 1: Structure of the agent-alike controller

system (providing the control strategy) and a communication server which is used to connect the agent to the user interface, the application and to other agents.

The two elements of the agent, classifier system and communication server, are separate since; as messages are passed around the agent network, the communication server acts independently of the classifier system to route the message to its neighbours. Another reason for keeping the communication server distinct from the classifier system is that communication is likely to be implementation specific (even in the test applications). Thus it is necessary when specifying the communication server to consider the general requirements of setting up and maintaining the communication in a distributed learning system rather than those in a specific software and hardware implementation.

2.1   Classifier systems

A classifier system is a learning system in which a set (population) of condition-action rules called *classifiers* compete to control the system and gain credit based on the system's receipt of reinforcement from the environment. A classifier's cumulative credit, termed *strength*, determines its influence in the control competition and in an evolutionary process using a genetic algorithm in which new, plausibly better, classifiers are generated from strong existing ones, and weak classifiers are discarded.

A classifier c is a condition-action pair

$$c =< condition >:< action >,$$

with the interpretation of the following decision rule: if a current observed state matches the condition, then execute the action. The condition is a string of characters from the ternary alphabet { 0, 1, # }, where # acts as a wildcard allowing generalization. The action is represented by a binary string and both conditions and actions are initialized randomly. The real-valued *strength* of a classifier is estimated in terms of rewards obtained according to a payoff function. Action selection is implemented by a competition mechanism, where a strength

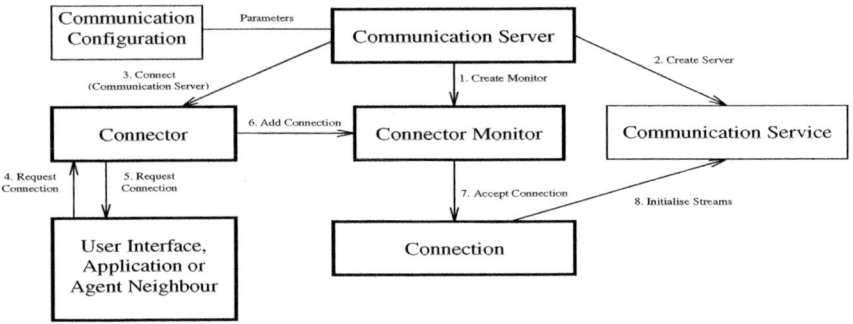

Figure 2: Initialisation structure of communication server

proportionate selection method is usually used. To modify classifier strengths, the given *credit assignment* algorithm is used, e.g. the Bucket brigade [14].

To create new classifiers a standard GA is applied, with three basic genetic operators: selection, crossover and mutation. The GA is invoked periodically and each time it replaces low strength classifiers with the offspring of the selected fitter ones (the reader is referred to [14] for full details).

2.2   Communication Server

The communication server provides the service for each agent connecting to the user interface, application and other agents. All these channels might involve two-way communication. During the initialisation the agents open a communication channel and await a connection message. The channel is tested to ensure the communication is setup correctly as although the configuration parameters have been previously checked for consistency, the parameters may be inconsistent with the physical communication process, also this process might be faulty.

Although the term socket is used in the specification as the medium to connect communication channels in implementation other methods can be used, such as calls to remote objects, when using RMI or DCOM. The basis of the communication initialisation and run-time processes are not affected. The creation of the communication object and binding in a remote registry (on a given hostname and port) replaces the creation of a server socket and calls to the remote object replace read and write calls to the sockets.

Note that it is possible for the communication server to create separate processes to listen on the communication channel for messages, this allows the agent to be reactive to external messages.

The communication with neighbours requires a single channel for incoming messages, and separate channels from sending to each neighbor (except if the messages are broadcast on sent via a proxy). The initialisation of the communication server, shown in Figure 2, involves the following steps:

1. The Communication Server object creates the specific Communication Services (Application, User Interface or Neighbourhood) as specified by the configuration.

2. The Communication Server object create a monitor which maintains the list of current connections

3. The Communication Server object passes the Communication Service object and connection configuration information to the Connector object which, for connection with the User Interface and Application and incoming channel from the neighbouring agents, opens a Server Socket on the specified port and waits for a request to connect. For the outgoing channel to the neighbouring agents the Connector object intermittently requests a connection to the neighbours specified port..

4. The User Interface, Application or Neighbouring Agent sends a request to connect.

5. The request to connect is accepted by the neighbour's server socket.

6. The Connector sends the Communication Service object and open socket to the Communication Monitor.

7. The Communication Monitor object tests the communication channel, if the test succeeds the Communication Service is passed to the Connection object, otherwise the socket is closed and the failure reported.

8. The Connection object starts the thread to handle the connection and passes the input and output streams to the Communication Service object.

3   How to Control Traffic Signals

To control a traffic network, we associate an agent to each junction of the traffic network. The agents are initialised according to the traffic network configuration and user-specified parameters. For the simulated 2×2 traffic network, shown in Figure 3, four agents, i.e., agents I, II, III, and IV, associating with junctions I, II, III, and IV, are need to provide comprehensive control of the network. Agent I has the neighbouring agents II and III, and agent II has the neighbouring agents I and IV, etc. The communication server in each agent provides the control actions of its neighbouring agents, and these information is used to construct control rules for its junction.

The classifier system employed is a version of Wilson's "zeroth-level" system (ZCS) [15]. ZCS is a Michigan-style classifier system, without internal memory. In order to avoid the genetic algorithm manipulating unnecessarily long rules, we extend the binary string representation in ZCS to a more general representation, which uses 0 to L ($L < 10$) for each variable (bit) position instead of the binary code. This reduces the string length significantly and appears to benefit multiple variable problems. For these hybrid strings, mutation in the GA is performed by changing an allele to a randomly determined number between 0 and L other than itself [16].

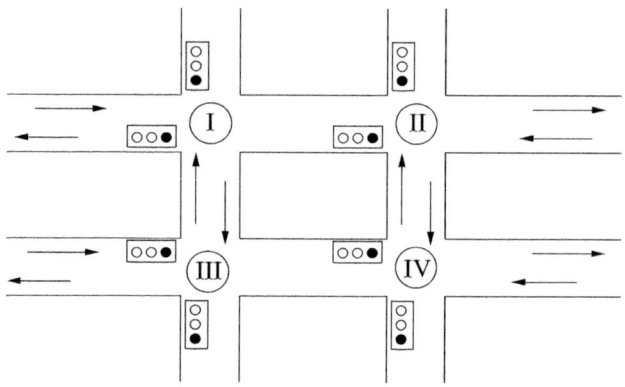

Figure 3: The simulated traffic environment

3.1   Individuals

The classifiers have the representation shown in Figure 4. The condition part of each classifier consists of six bits, which reflects the scalar level of queue length from each direction and the previous actions of the neighbouring agents. In this application, the scalar level of the queue length is set to 4, which ranges from 0 to 3, corresponding to the four linguistic variables, {zero, small, medium, large }. The action part indicates the required state of the signal. For instance, for junction I, the rule 130201:1 says that if the queue from directions east and west are small (1) and zero (0), but the queue from directions south and north are large (3) and medium (2), and the previous neighbourhood junction controllers' actions are vertically red (0) (junction II) and green (1) (junction III), then the traffic light stays green vertically (1) for a fixed period of time.

3.2   Evaluation of actions

We assume that the junction controller can observe the performance around it, let the evaluated performance be P. Traffic volume sensors are set at each of the intersections. They are able to count the numbers of the cars that come from all directions, pass through the intersection and stop at the intersection. In this study, the evaluation function we use to reward the individuals is the average queue at the specific junction. Let q_i denote the queue length from direction i at the intersection ($i = 1, 2, 3, 4$), then the evaluation function is: $f = \frac{1}{4}\sum_{i=1}^{4} q_i$. We thus attempted to minimize this measure. Let us identify the k-th cycle by a subscript k, then f_k for the cycle k is calculated by observing the sensor from the beginning of the k-th cycle to the end of this cycle. Thus, the evaluated performance of the action performed at the k-th cycle is computed as $P_k = f_{k-1} - f_k$. Specifically, if $P_k > 0$, the matched classifiers containing the performed action should be rewarded, otherwise penalized.

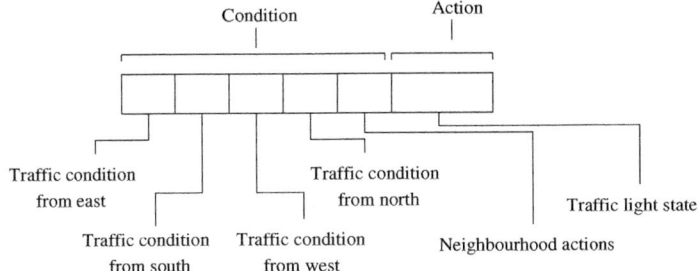

Figure 4: Structure of the classifier system rules

3.3   Reinforcement learning

After the controller has produced an action, the environment judges the output, and accordingly, gives payoff in the following manner:

- Rewards: The objective of the local signal controller is to minimize the average queue length, f_i. We have found the performance-based rewards are helpful in the environments we used in our experiments. The reward function we used was $r = \frac{1}{4} \sum_{i=1}^{4} \left((100 - q_{k,i}^4) + (q_{k,i} - q_{k-1,i})^3 \right)$, where $q_{k,i}$ denotes the queue length of the ith direction at the kth cycle.

- Punishments: We use punishments (i.e., negative rewards). We found the the use of appropriate punishments results in improved performance (in a fixed number of cycles), at least in the environments used in our experiments. We also found that large punishments could lead to instability of the classifiers and slow convergences of the rules. The appropriate punishments should be determined by trial tests.

4   Simulation Results

For the traffic network shown in Figure 3, we developed a simplified traffic simulator, which is similar to the one used in [12]. The simulator is composed of four four-armed junctions and squared roads. Each end of a road is assumed to be connected to external traffic, and cars are assumed to arrive at those roads according to a Poisson distribution. Each intersection has two "complementary" signals: when the horizontal signal is red, the vertical signal is green and *vice versa*. Each of the cars attempts to attain the same maximum speed. When a car passes an intersection, it changes its direction according to the probability associated with that intersection. Specifically, let d_i, $i = 1, 2, 3$, be the next directions for a car, that is, $\{d_i\} = \{$ right, forward, left $\}$. At each of the intersections, the probabilities $\{p_{d_i}\}$ are previously given, where p_{d_i} corresponds to

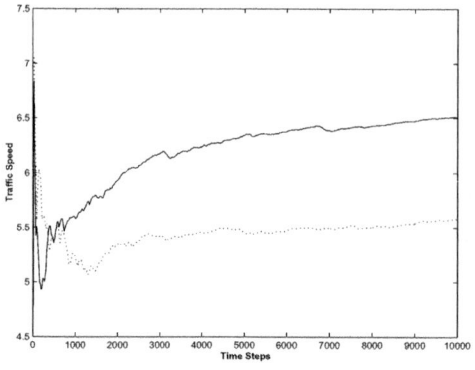

(a) Number of Cars = 30

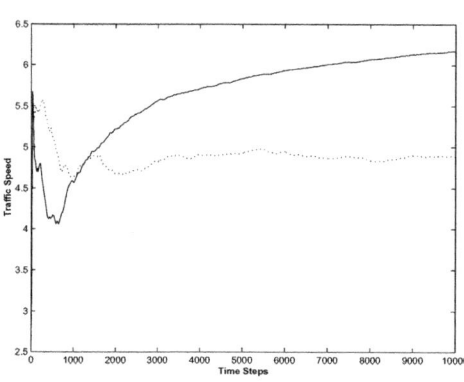

(b) Number of Cars = 60

the probability of selecting an action d_i for the car passing through the intersection. Roads are not endless, thus only a limited number of cars is allowed to be on the road at a given time. If a car reaches the end of the road, then the car is simply removed from the simulation, and another car is generated, entering on a randomly selected road.

For comparison purpose, two types of control strategies are employed: random control strategy and the developed distributed learning control (DLC) strategy. The random control strategy determines the traffic light's state (0 or 1) randomly at 50% of probability; whilst distributed learning control (DLC) strategy determines the traffic light's state according to the action of the winning classifier of the agent. The parameters used for the DLC were as follows:

- Population size: 100
- Mutation probability: 0.05
- Crossover probability: 0.85
- Selection method: Roulette wheel selection

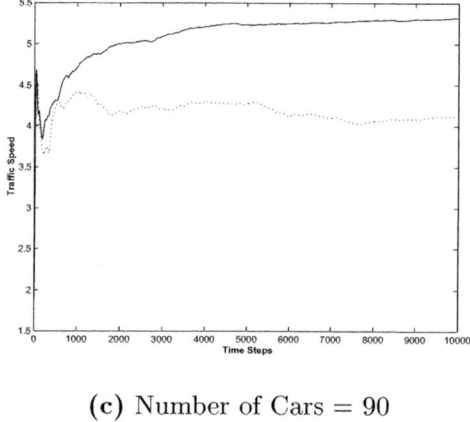

(c) Number of Cars = 90

Figure 5: Comparison performance of the control strategies

As the major task is to test whether the proposed DLC can learn some good rules in the traffic network, experiments were carried out for three different types of traffic conditions. In these simulations, the mean arrival rates for the cars are the same but the number of cars in the area is limited to 30, 60, and 90, corresponding to a sparse, medium, and crowded traffic condition. In all cases, the DLC strategy is found to learn how to reduce the average queue length and improve the traffic speed in the network. For example, Figure 5 shows the average performances of the random control strategy and DLC strategy respectively over 10 runs in all cases, where the solid line represents DLC strategy and the dotted line represents random control strategy. It can be seen that the DLC strategy consistently learns and improves the traffic speed over 10,000 iterations.

5   Conclusion and Future Work

In this paper we have presented a distributed learning control strategy for traffic signals. The simulation results on a simplified traffic environment are encouraging since we have shown that the developed control strategy can learn to coordinate and determine useful control rules within a dynamic environment. This preliminary work needs, of course, a number of extensions. We are currently extending this work in a number of directions, particularly examining ways of improving the learning capability of classifier systems and the performances in much more complicated traffic network.

6   Acknowledgment

This work was carried out as part of the ESPRIT Framework V Vintage project (ESPRIT 25.569).

References

[1] Robertson, D. I.: TRANSYT– A traffic network study tool. Transport and Research Laboratory, Crowthorne, England (1969)

[2] Luk, J. Y., Sims, A. G. and Lowrie, P. R.: SCATS application and field comparison with TRANSYT optimized fixed time system. In Proc. IEE Int. Conf. Road Traffic Signalling, London (1982)

[3] Lowrie, P. R.: The Sydney coordinated adaptive traffic system. In Proc. IEE Int. Conf. Road Traffic Signalling, London (1982)

[4] Hunt, P. B., Robertson, D. I., Bretherton, R. D. and Winston, R. I.: SCOOT–A traffic responsive method of co-ordinating traffic signals. Transport and Research Laboratory, Crowthorne, England (1982)

[5] Scemama, G.: Traffic control practices in urban areas. Ann. Rev. Report of the Natl Res. Inst. on Transport and Safety. Paris, France (1990)

[6] Al-Khalili, A. J.: Urban traffic control – a general approach. IEEE Trans. on Syst. Man and Cyber. **15**, (1985) 260–271

[7] Montana, D. J. and Czerwinski, S.: Evolving control laws for a network of traffic signals. Proc. of 1st Annual Conf. on Genetic Programming, (1996) 333–338

[8] Koza, J. R: Genetic Programming. MIT Press, Cambridge, MA (1992)

[9] Cao, Y. J., Ireson, N. I., Bull, L. and Miles, R.: Design of Traffic Junction Controller Using a Classifier System and Fuzzy Logic. In Computational Intelligence: Theory and Applications, Reusch, B. (ed), Lecture Notes in Computer Sciences, **1625**, Springer Verlag, (1999) 342–353

[10] Cao, Y. J. and Wu, Q. H.: An improved evolutionary programming approach to economic dispatch. International Journal of Engineering Intelligent Systems, **6**, (2), (1998) 187–194

[11] Cao, Y. J. and Wu, Q. H.: Optimisation of control parameters in genetic algorithms: a stochastic approach. International Journal of Systems Science, **20**, (2), (1999) 551–559

[12] Mikami, S. and Kakazu, K.: Genetic reinforcement learning for cooperative traffic signal control. Proceedings of the IEEE World Congress on Computational Intelligence, (1994) 223–229

[13] Escazut, C. and Fogarty, T. C.: Coevolving classifier systems to control traffic signals. In Koza, J. R (ed): Late breaking papers at the Genetic Programming 1997 Conference, Stanford University, (1997) 51–56

[14] Holland, J. H.: Adaptation in Natural and Artificial Systems. MIT Press, Cambridge, MA (1992)

[15] Wilson, S. W.: ZCS: A zeroth level classifier system. Evolutionary Computation, **2**, (1994) 1–18

[16] Cao, Y. J. and Wu, Q. H.: A mixed-variable evolutionary programming for optimisation of mechanical design. International Journal of Engineering Intelligent Systems, **7**, (2), (1999) 77–82

Time Series Prediction by Growing Lateral Delay Neural Networks

Lipton Chan and Yun Li

Centre for Systems and Control, and Department of Electrical and Electronics Engineering,
University of Glasgow, Glasgow G12 8LT, U.K.
Email: L.Chan@elec.gla.ac.uk

Abstract. Time-series prediction and forecasting is much used in engineering, science and economics. Neural networks are often used for this type of problems. However, the design of these networks requires much experience and understanding to obtain useful results. In this paper, an evolutionary computing based innovative technique to grow network architecture is developed to simplify the task of time-series prediction. An efficient training algorithm for this network is also given to take advantage of the network design. This network is not restricted to time-series prediction and can also be used for modelling dynamic systems.

1. Introduction

Dynamic modelling addresses the modelling problem from data of a dynamic system. A dynamic system is a system which has internal states represented in an abstract phase or time space. Its future state and outputs depends on its current state. They can be mathematically described by an initial value problem [1]. An example of dynamic modelling is the modelling of time series data, where predictions has to be made on the future values of the time series based on current values of the series. This type of modelling tries to capture the geometry and geometrical invariants of a dynamic system from past outputs of the system [2]. The use of past outputs, delay co-ordinates, to model dynamic systems can be traced back as far as 1927 to the work of Yule, who used auto-regression (AR) to create a predictive model for sunspot cycles [1], [3].

The most popular method of modelling time-series data today is the statistical method of Box-Jenkins. The Box-Jenkins methodology search for an adequate model from AR, moving average (MA), auto-regression moving average (ARMA), and auto-regression integrated moving average (ARIMA) [5]. This modelling is a three-stage process: identification, estimation, and diagnostics.

The identification involves the use of sample auto correlation functions (SACF) and sample partial auto correlation functions (SPACF) to analysis the linear relationships of the time series with its lagged images. The estimation process involves finding a model, one of AR, MA, ARMA, and ARIMA, with a theoretical ACF and PACF similar to the SACF and SPACF of the time series. The third stage, diagnostics, involves residual analysis, goodness of fit statistics, and cross-validation. The limitation of this method is that human decision and associated errors are inherent

S. Cagnoni et al. (Eds.): EvoWorkshops 2000, LNCS 1803, pp. 127–138, 2000.

in every stage of the process [6]. Also the SACF and SPACF functions measure only linear relationships.

Neural networks have been shown to be universal function approximators [4]. By using delay co-ordinates as the inputs of a neural network, it can be used as a non-linear approximator for a delay differential equation. With an appropriate structure and learning strategy, this delay differential equation can be tuned to have similar behaviour to the dynamic system being modelled.

Neural networks are commonly used for time-series predictions also. Different network architectures have been employed to tackle the prediction of time-series data, for example the multi-layer perceptrons [7], the finite-duration impulse response (FIR) networks [8], and the recurrent networks [9]. Good prediction results have been obtained from such networks. The difficulty with these architectures is that they are static, fixed before the training is begun. That is, the designer needs to decide on the number of delay co-ordinates to use, the number of hidden neurons to have, etc.

In this paper, a novel network architecture and training strategy is proposed for the modelling of dynamic systems which alleviates the designer of much of these decisions. The network structure is not static and changes during the training process which makes use of evolutionary algorithms (EA). A description of the novel network architecture is given in Section 2. The evolutionary technique for growing and training the network is described in Section 3. In Section 4, results of predicting chaotic time series using this technique and network are shown. Section 5 demonstrates the application of time series prediction on real data, and shows a method of improving prediction. Finally, the paper is concluded in Section 6.

2.   Lateral Delay Neural Network

A neural network requires memory to have dynamic behaviour [8]. This memory can be delay elements in the architecture or the use delay co-ordinate inputs. The number of delay co-ordinate inputs relates to the embedding dimension of the system, and the number of hidden neurons is dictated by necessary degrees of freedom [10]. Both of these have to be estimated for static network architectures.

The principle of growing the network structure one hidden neuron at a time has shown to be a fast and efficient method of approximating a function with a neural network [11]. Several such incremental learning algorithms have been implemented [11], [12]. It is desirable to use these incremental approaches to speed the learning process and reduce the dimensionality of optimisation. Yet there are two structural parameters that needs to be estimated, namely, numbers of delays and hidden neurons. The design of the lateral delay neural network (LDNN) combines these two parameters into one for the implementation of incremental learning algorithms. This allows for the increased speed and efficiency of modelling of dynamic systems.

The combination of delay elements and hidden neurons is shown in Fig. 1 of the network architecture. This architecture is incremental with the hidden neurons forming a one-way chain of delayed elements. The simplicity of design is facilitated by not requiring structural decisions. Improving the performance of network is achieved by adding a neuron to the lateral chain. There are very few synaptic weights associated with each neuron, thus the complexity of the weight optimisation is very low.

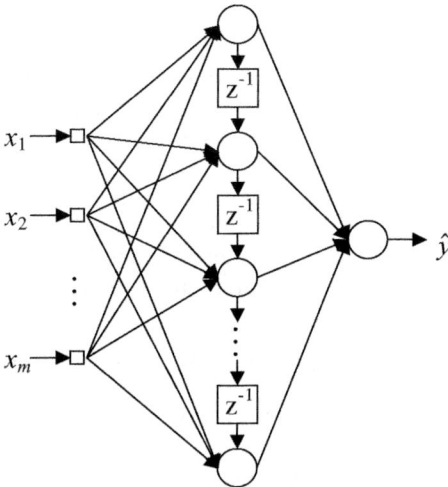

Fig. 1. Architecture of LDNN.

The LDNN can be used in dynamic modelling with its inputs $x \in \Re^m$ being the m system inputs and its output $\hat{y}$ being the predicted system output. In time-series prediction, there is only one input which this the current value of the time-series, and the output is the predicted future value of time-series, i.e., $x = [x_n]$ and $\hat{x}_{n+1} = \hat{y}$, where x_n is the nth data in the series.

3. Evolutionary Network Growing and Training

3.1. EA-only Training

Due to the LDNN architecture's simplicity and incremental design, training algorithms can be devised to take advantage of these properties. EAs can be used to train the network incrementally by first optimising the synaptic weights of the network with only one hidden neuron, i.e., no delay elements yet. The best solution found is then used to "hot-start" subsequent EA optimisations of the network with an added hidden neuron. This can be iterated until the desired accuracy is reached or until over-fitting begins to occur.

Using EA for training in this way can produce good results, but the dimensionality of the optimisation space increases rapidly. If there is a weight associated with each synaptic connection and one for the threshold of each neuron, then the dimensionality involved is $(m+3)n-1$ where m is the number of inputs and n is the number of hidden neurons.

3.2.   Orthogonal Training with EA

The dimensionality of the optimisation can be greatly reduced by keeping the optimised weights of the previous network fixed and only optimising the weights of the newly added neuron. A learning algorithm based on this idea is given for this LDNN here.

Beliczynski gave an incremental algorithm for training one-hidden-layer perceptrons [11]. This algorithm is slightly modified to accommodate the learning of LDNNs and lateral delays; the proofs can be found in his work [11].

Thus, first assume that the time series to be modelled is defined by a finite set of input-output pairs:

$$\{(x_1, f(\boldsymbol{a}_1, x_1)), (x_2, f(\boldsymbol{a}_2, x_2)), \ldots, (x_N, f(\boldsymbol{a}_N, x_N))\} \tag{1}$$

where $f : \Re^{c,1} \to \Re$ is the system output (the next data), x_j is an input (the current data), $x_j \in \Re$, $\boldsymbol{a}_j$ is the internal state of the system, $\boldsymbol{a}_j \in \Re^c$, $j = 1, \ldots, N$, and N is the number of input-output pairs. It follows that for time series,

$$x_{j+1} = f(\boldsymbol{a}_j, x_j) \tag{2}$$

where $j = 1, \ldots, N-1$. Define the vector $\boldsymbol{X}$ of inputs, i.e., the time series data in order, as

$$\boldsymbol{X} = [x_1, x_2, \ldots, x_N]^T \in \Re^N \tag{3}$$

and output vector

$$\begin{aligned} \boldsymbol{F}(\boldsymbol{X}) &= \left[f(\boldsymbol{a}_1, x_1), f(\boldsymbol{a}_2, x_2), \ldots, f(\boldsymbol{a}_N, x_N) \right]^T \in \Re^N \\ &= [x_2, x_3, \ldots, x_{N+1}]^T \end{aligned} \tag{4}$$

Let $g_k(\hat{a}_{j,k}, x_j)$ denote the function of a hidden neuron, where $g_k : \Re^{1,1} \to \Re$, $\hat{a}_{j,k}$ is the delayed input to the hidden neuron, for every neuron $k = 1, \ldots, n$, and time $j = 1, \ldots, N$, where $n \le n_{max}$ and n_{max} is the maximum number of hidden neurons allowed.

According to the structure of the network

$$\hat{a}_{j,k} = \begin{cases} 0 & \text{if } j = 1, \text{or } k = 1 \\ g_{k-1}(\hat{a}_{j-1,k-1}, x_{j-1}) & \text{otherwise} \end{cases} \tag{5}$$

where $k = 1, \ldots, n$, and $j = 1, \ldots, N$. The delays of the network constitutes the internal state of the network, thereby the network state vector $\hat{\boldsymbol{a}}_j$ and the state matrix $\hat{\boldsymbol{A}}_n$ can be defined as follows

$$\hat{\boldsymbol{a}}_j = [\hat{a}_{j,1}, \hat{a}_{j,2}, \ldots, \hat{a}_{j,n}]^T \in \Re^n \tag{6}$$

$$\hat{\boldsymbol{A}}_n = [\hat{\boldsymbol{a}}_1, \hat{\boldsymbol{a}}_2, \ldots, \hat{\boldsymbol{a}}_N]^T \in \Re^{N,n} \tag{7}$$

Define the vector $\boldsymbol{G}_k(\boldsymbol{X})$ which is composed of the outputs of the kth hidden neuron for the whole time series

$$G_k(X) = [g_k(\hat{a}_{1,k}, x_1), g_k(\hat{a}_{2,k}, x_2), \ldots, g_k(\hat{a}_{N,k}, x_N)]^T \in \Re^N$$
$$= [\hat{a}_{2,k+1}, \hat{a}_{3,k+1}, \ldots, \hat{a}_{N,k+1}, g_k(\hat{a}_{N,k}, x_N)]^T \tag{8}$$

and define the matrix

$$H_n(X) = [G_1(X), G_2(X), \ldots, G_n(X)] \in \Re^{N,n} \tag{9}$$

Also let W_n denote the vector of weights of the output neuron

$$W_n = [w_1, w_2, \ldots, w_n]^T \in \Re^n \tag{10}$$

where w_k is the weight of the synaptic connection from neuron k to the output neuron, $k = 1, \ldots, n$.

Thus the network's predictions, $F_n(X)$, of the time series X is

$$F_n(X) = H_n(X)W_n \in \Re^N \tag{11}$$

Now the training error in network prediction can be defined as

$$E_n(X) = \begin{cases} F(X) & \text{for } n = 0 \\ F(X) - F_n(X) & \text{for } 1 \le n \le n_{\max} \end{cases} \tag{12}$$

and the mean squared training error is

$$e_n(X) = \frac{1}{N} \|E_n(X)\|^2 \tag{13}$$

From [11], $\|E_n(X)\|$ is non-increasing, and the maximum rate of decrease in $\|E_n(X)\|$ occurs when the newly-added neuron, g_{n+1}, is chosen such that

$$\sup_{g \in G} \left| E_n(X)^T \frac{G_{n+1}(X)}{\|G_{n+1}(X)\|} \right| \tag{14}$$

is achieved. This is provided that $n_{\max} < N$. In practice, N has to be large for time-series prediction and $n_{\max} \ll N$. The quantity in Equation 14, is the scalar product of the two vectors, and by maximising it, the two vectors are made quasi-parallel. The error is also orthogonal to the output of every hidden neuron in the network.

There are two optimisations to be performed for each new neuron being added. The optimisation of the newly added hidden neuron, g_n, and W_n the weights of the output neuron. These are optimised using EAs. The dimensionality of these search are $m+1$ for g_n and $n+1$ for W_n, where m is the number of inputs and n is the total number of hidden neurons. The whole orthogonal incremental training process is summarised in the flowchart in Fig. 2.

The hidden-layer neuron function, $g_k(\hat{a}_{j,k}, x_j)$, can use different activation functions. Neither of the two training algorithms discussed in this section have many restrictions on the activation functions used for the hidden neurons. For all the time-series predictions made in this work, sigmoid functions have been used as activation functions for the hidden neurons. A linear function is chosen for the output neuron's activation. The use of sigmoid activation functions means that the network output is bounded, thus the time-series being modelled must also be bounded.

The training algorithms in this section are static; the network does not change after the training process. Thus the systems that can be modelled are autonomous systems and the time-series that can be predicted are stationary.

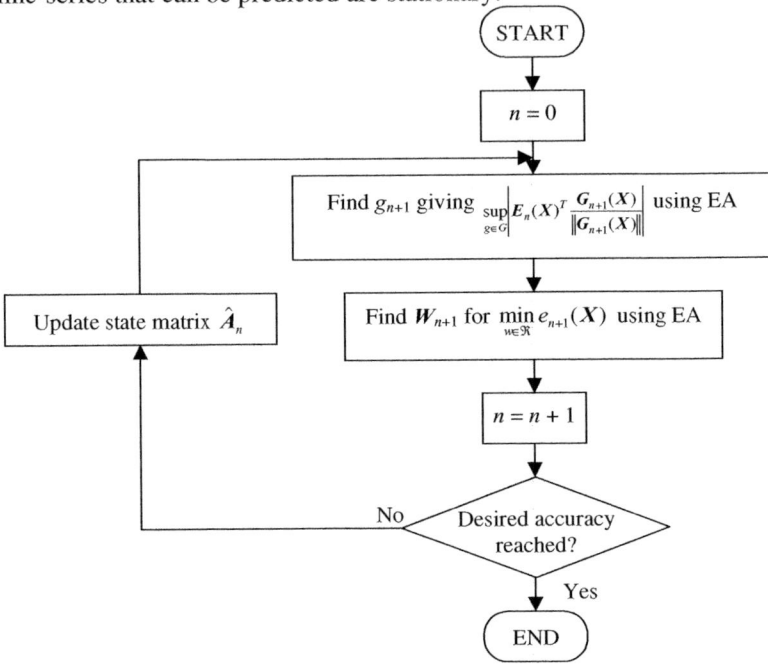

Fig. 2. Flowchart of the orthogonal incremental training process.

The results shown in Fig. 3 are the times taken to train an additional neuron by the two training algorithms as the number of neurons increases. These times are the average times of 5 network trainings for the prediction of the Mackay-Glass chaotic time-series. The additional training time taken by the orthogonal training algorithm, with reduced complexity, stays near constant – rising only slightly as the number of hidden neurons increases. Whereas, with the EA-only training, the extra time taken per neuron increases exponentially with the addition of hidden neurons. Experimentation shows that the EA-only algorithm often produces very good prediction accuracy with very few neurons, while the orthogonal algorithm may need many more neurons to achieve the same level of accuracy. Due to the low computational power required by the orthogonal training algorithm, the rest of the prediction results in this paper are obtained using this training method.

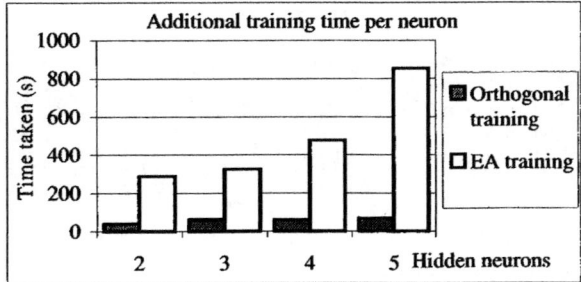

Fig. 3. Comparison of training times taken by the two training algorithms when adding new neurons.

4. Prediction of Chaotic Series

4.1. Logistic Map

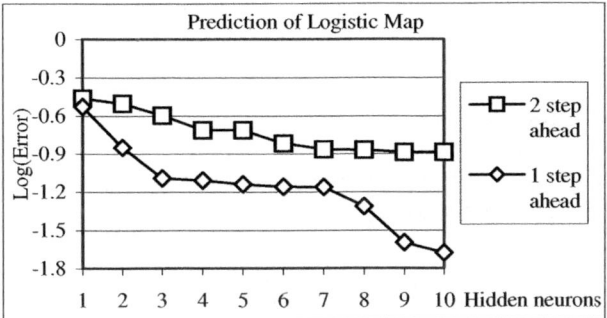

Fig. 4. RMS validation errors of LDNN prediction 1 step and 2 steps into the future of the logistic map.

The logistic map is a chaotic series given by the equation

$$x_{n+1} = 4x_n(1-x_n) \tag{15}$$

Using the orthogonal training algorithm, a LDNN is trained to predict the logistic map 1 step into the future, and another LDNN to predict 2 steps into the future. The resulting validation errors of prediction, of data unseen during training, for up to 10 hidden neurons are plotted in Fig. 4. The use of incremental techniques guarantees that the training error monotonically decreases. The validation errors show the model's generalisation capability. The prediction validation values of the 10 hidden neuron network are compared with the actual values in Fig. 5, showing a close match.

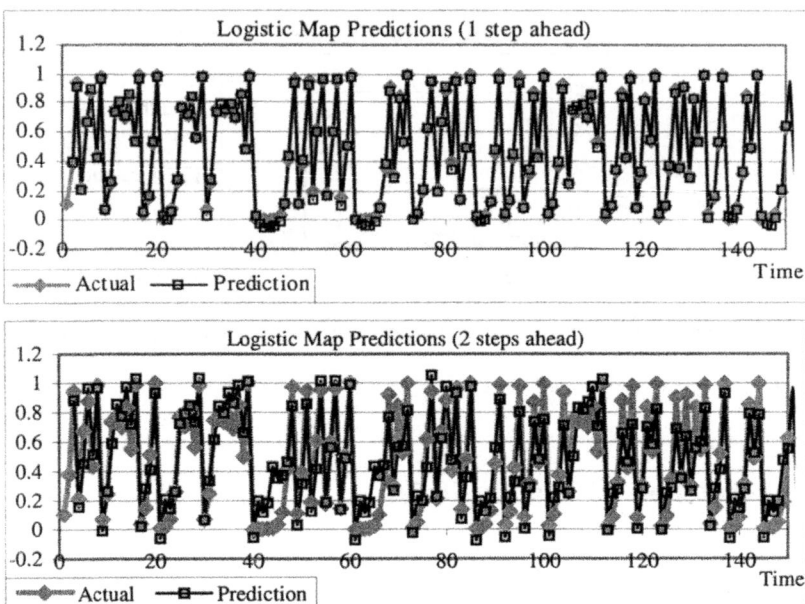

Fig. 5. 1 step and 2 steps ahead logistic map prediction results with 10 hidden neurons.

4.2.   Mackay-Glass Series

The Mackay-Glass differential equation produces chaotic time-series that are quasi-periodic. It is defined as

$$\dot{x} = -bx_t + \frac{ax_{t-\tau}}{1+x_{t-\tau}^{10}} \qquad (16)$$

The decrease of prediction validation errors with the increase of hidden neurons can be seen in Fig. 6 for the prediction of the Mackay-Glass series up to 5 steps into the future. The prediction validation results are shown in relation to the actual data values for 1, 3, and 5 steps ahead prediction in Fig. 7. It can be seen that prediction results are very close to the actual data values, though the predict results get more erratic the further into the further the prediction is.

All these predictions are made with only one input to the LDNN. It may be possible to improve on these results by explicitly embedding other distant delay co-ordinates as inputs to the network, for example x_{t-5}, x_{t-15}, etc.

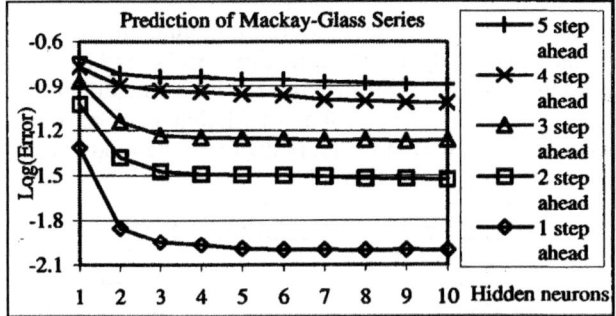

Fig. 6. RMS validation errors of predictions of the Mackay-Glass series at several steps into the future.

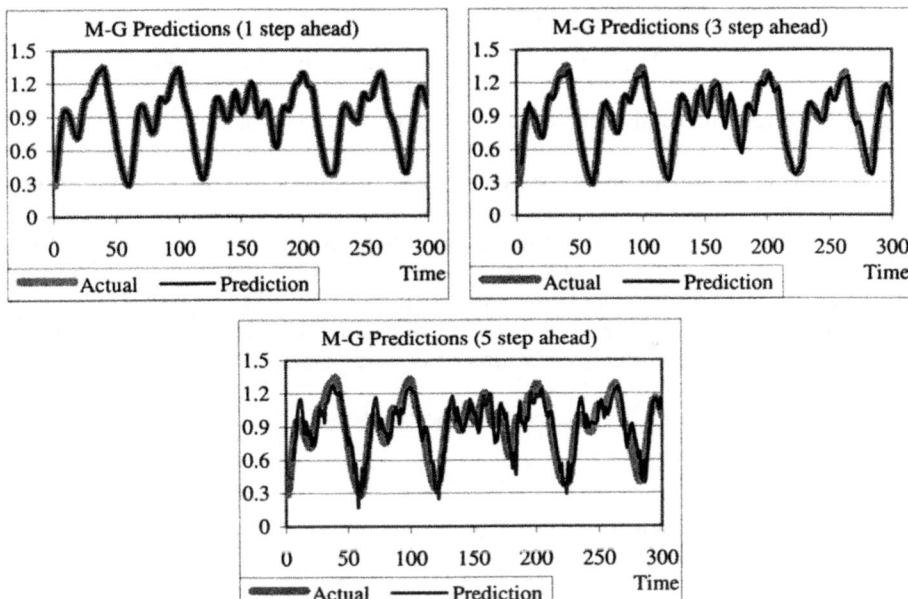

Fig. 7. 1, 3 and 5 steps ahead Mackay Glass prediction results with 10 hidden neurons.

5.   Prediction of Sunspot Numbers

Sunspots numbers are indicators for the level of solar activity in the Sun. The sunspot numbers form a cycle of approximately 11 years. Here, yearly sunspot numbers are used to train a LDNN, demonstrating the use of this network and its ease of application. The data used are yearly numbers from 1851 to 1998 [13].

Fig. 8 shows the RMS errors of prediction validation as the number of hidden neurons increases. Using incremental training guarantees the decrease in RMS training errors, and the analysis of validation errors will show the generalisation ability of the LDNN model. From Fig. 8, the validation errors appear to start continually increasing when the number of hidden neurons reached 6-7. Further addition of hidden neurons will most likely decrease the model's generalisation ability and increase over-fitting to the training data.

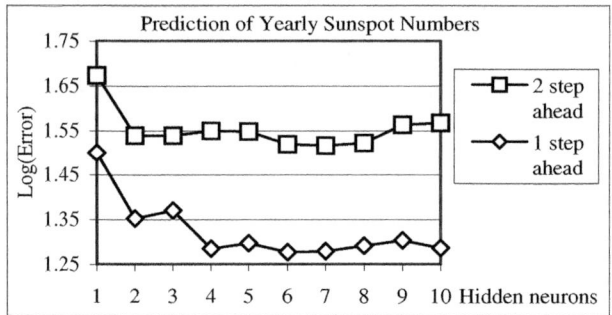

Fig. 8. RMS validation errors for the prediction of yearly sunspot numbers.

To improve on these results, one can choose to use a different activation function for subsequently added hidden neurons on the detection of over-fitting. Alternatively, explicitly embedding delay co-ordinates as inputs to the network can be employed. The latter approach is used in this case and the results of Fig. 8 can be used to suggest which delay co-ordinate to embed. Since over-fitting occurs at about the addition of the 7th hidden neuron and hence the best network outputs depend on only up to the 7th delay co-ordinate, the 8th delay co-ordinate is chosen, i.e.,

$$\hat{x}_{n+1} = f_1(\boldsymbol{a}_n, x_n, x_{n-7}) \tag{17}$$

$$\hat{x}_{n+2} = f_2(\boldsymbol{a}_n, x_n, x_{n-7}) \tag{18}$$

where $\boldsymbol{a}_n$ is the internal state of the model, and x_n is the nth value of the time series.

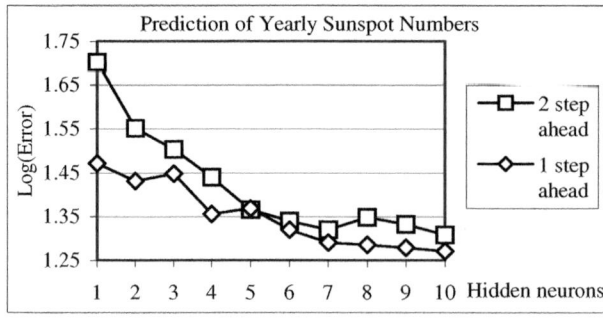

Fig. 9. RMS validation errors with the explicit embedding of an extra delay co-ordinate.

The resulting decrease in RMS validation errors with increasing number of hidden neurons can be seen in Fig. 9 when using an extra delay co-ordinate. Over-fitting does not occur so soon with this arrangement and generalisation is good. The training and validation results for 1 and 2 steps ahead predictions are shown in Fig. 10 and 11 respectively.

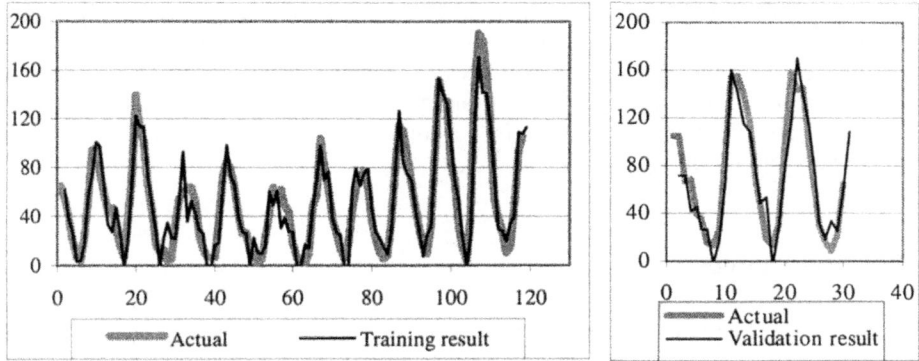

Fig. 10. Training and validation results for 1 step ahead prediction of sunspot numbers.

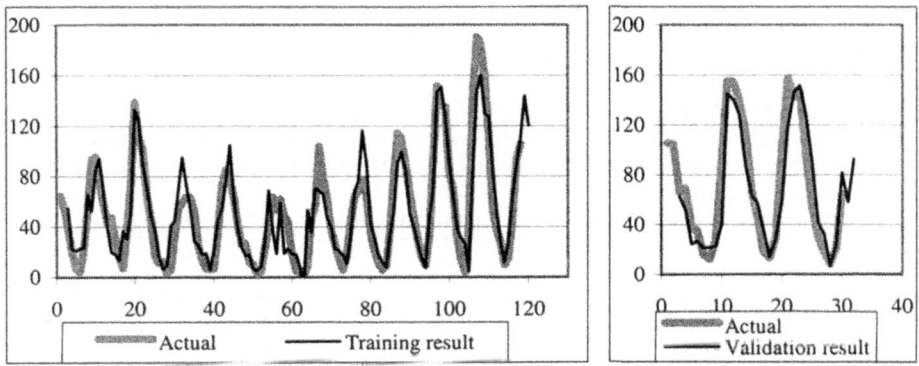

Fig. 11. Training and validation results for 2 step ahead prediction of sunspot numbers.

6. Conclusions

A novel architecture and growing technique are proposed in this paper which simplifies the design process of neural networks for time-series prediction. The evolutionary incremental training algorithm presented for this network architecture shows itself to be a fast and efficient method of approximating the network to a dynamic system. The training algorithm allows for different combinations of activations functions to be used in the network. It is possible that a change in hidden neuron activation function will increase the rate of error reduction, when this rate is converging.

In addition to the prediction of time-series, the network may be used for the modelling of dynamic systems.   Improved results may be obtained with the addition of exogenous variables or explicit embedding of delay co-ordinates. With static training as given in this paper, autonomous systems and stationary time-series can be modelled. This methodology is validated by the application to a real example, namely the prediction of sunspot numbers. Its ease and efficiency is demonstrated along with explicit embedding of delay co-ordinates to improve results. For the modelling of non-autonomous systems and non-stationary time-series, real-time adaptation of this network may be required, which is a subject of ongoing research here in CSC, University of Glasgow.

7.   References

1. Meiss J.: Nonlinear Science FAQ, May 1999, Ver. 1.3.1. Internet FAQ Consortium, http://www.faqs.org/faqs/sci/nonlinear-faq/. Online.
2. Packard N.H., et al.: Geometry from a Time Series. Physical Review Letters, Vol. 45, No. 9. (1980) 712-716
3. Yule G.U.: Philosophical Transactions of the Royal Society of London A, Vol. 226. (1927) 267
4. Hornik K., Stinchcombe M., and White H.: Multilayer Feedforward Networks are Universal Approximators. Neural Networks, Vol. 2. (1989) 359-366
5. Neerchal N.K.: Time Domain, August 1999. Time Series Tutor: An Interactive Introduction to Time Series Analysis, http://math.umbc.edu/~nagaraj/. Online.
6. Elkateb M.M., Solaiman K., and Al-Turki Y.: A comparative study of medium-weather-dependent load forecasting using enhanced artificial/fuzzy neural network and statistical techniques. Neurocomputing, Vol. 23. (1998) 3-13
7. Conway A.J., et al.: A neural network prediction of solar cycle 23. Journal of Geophysical Research, Vol. 103, No. A12. (1998) 29733-29742
8. Haykin S.: Neural Networks. Macmillan. (1994)
9. Matsuoka M., Golea M., and Sakakibara Y.: Columnar Recurrent Neural Network and Time Series Analysis. Fujitsu Scientific & Technical Journal, Vol. 32, No. 2. (1996) 183-191
10. Lowe D., and Hazarika N.: Complexity modelling and stability characterisation for long term iterated time series prediction. IEE Conference Publication, No.440. (1997) 53-58
11. Beliczynski B.: Incremental Approximation by One-Hidden-Layer Neural Networks: Discrete Functions Rapprochement. IEEE International Symposium on Industrial Electronics Vol.1. (1996) 392-397
12. Fritzke B.: Fast learning with incremental RBF networks. Neural Processing Letters 1. (1994) 2-5
13. Sunspot Numbers, October 1999. Solar-Terrestrial Physics Division of the National Geophysical Data Center, http://www.ngdc.noaa.gov/stp/stp.html. Online.

Trajectory Controller Network and Its Design Automation through Evolutionary Computing

Gregory Chong and Yun Li

Centre for Systems & Control, Department of Electronics & Electrical Engineering
University of Glasgow, Glasgow, G12 8LT, UK.

gregccy@elec.gla.ac.uk

Abstract. Classical controllers are highly popular in industrial applications. However, most controllers are tuned manually in a trial and error process though computer simulation. This is particularly difficult when the system to be controlled is nonlinear. To address this problem and help design of industrial controllers for a wider range of operating trajectory, this paper proposes a trajectory controller network (TCN) technique based on linear approximation model (LAM) technique. In a TCN, each controller can be of a simple form, which may be obtained straightforwardly via classical design or evolutionary means. To co-ordinate the overall controller performance, the scheduling of the TCN is evolved through the entire operating envelope. Since plant step response data are often readily available in engineering practice, the design of such TCN is fully automated using an evolutionary algorithm without the need of model identification. This is illustrated and validated through a nonlinear control example.

1 Introduction

A dynamic engineering system is usually nonlinear and complex in practice. Plant dynamics may vary significantly with changes of operating conditions. Therefore, the use of a single nominal linear model under one operating condition, and hence controllers designed out of such a plant model, are often unreliable and inadequate to represent a practical system. The recently developed local controller network techniques [5], have provided some effective solutions to these problems, but they are based on locally linearised models.

To address these problems more completely for a wider range of operating trajectories and to make use of plant step-response data that are often readily available in engineering practice, this paper proposes a trajectory controller network (TCN) technique based on linear approximation model (LAM) technique [2]. Such a LAM network is obtainable directly from plant step-response by fitting nonlinear trajectories between two operating levels. As preliminaries to design, this modelling technique is outlined in Section 2.

In a TCN, each controller can be of a simple form, such as a proportional plus integral plus derivative (PID) controller, which may be obtained straightforwardly via

S. Cagnoni et al. (Eds.): EvoWorkshops 2000, LNCS 1803, pp. 139–146, 2000.

classical design or evolutionary means. To co-ordinate the overall controller performance, the scheduling of the TCN is evolved through the entire operating envelope. This is detailed in Section 3. Section 4 illustrates and validates the TCN technique through a nonlinear control example. Finally, conclusions are drawn in Section 5.

2   Linear Approximation Model for Nonlinear System Modelling

Here, the LAM to approximate a nonlinear plant is illustrated through an example. The plant used for the example is a twin-tank coupled nonlinear hydraulic system that models liquid-level found in chemical and diary plants. The scaled down model can also be found in the laboratory. Based on the Bernoulli's mass-balance and flow equations, the system structure is given by:

$$
\begin{bmatrix} \dot{h}_1 \\ \dot{h}_2 \end{bmatrix} = \begin{bmatrix} -\operatorname{sgn}(h_1 - h_2)\dfrac{c_1 a_1}{A}\sqrt{2g|h_1 - h_2|} \\ \operatorname{sgn}(h_1 - h_2)\dfrac{c_1 a_1}{A}\sqrt{2g|h_1 - h_2|} - \dfrac{c_2 a_2}{A}\sqrt{2g(h_2 - H_0)} \end{bmatrix} + \begin{bmatrix} \dfrac{Q_1}{A} & 0 \\ 0 & 0 \end{bmatrix} \begin{bmatrix} v_i \\ 0 \end{bmatrix}
\tag{1}
$$

The system input is the voltage applied to the pump, v_i, and the system output is the liquid level in tank 2, h_2. The coefficients of the twin tank are tabulated in Table 1. The non-linearity of the plant model is clearly plotted as shown in Fig. 1.

Table 1. Nonlinear system parameters

Height of water in tank 1	h_1 (m)
Height of water in tank 2	h_2 (m)
minimum height of water in tank	$H_0 = 0.03$ m
Cross sectional area of tank 1&2	$A = 0.01$ m^2
Discharge coefficient of orifice 1	$c_1 = 0.53$
Discharge coefficient of orifice 2	$c_2 = 0.63$
Cross sectional area of orifice 1	$a_1 = 0.0000396$ m^2
Cross sectional area of orifice 2	$a_2 = 0.0000386$ m^2
Gravitational constant	$g = 9.81$ m s^{-2}
Per-volt Pump Flow rate	$Q_i = 0.000007$ (m^3 s^{-1} V^{-1})
Flow rate from tank 1 to tank 2	Q_1 (m^3 s^{-1})
Discharge rate	Q_0 (m^3 s^{-1})

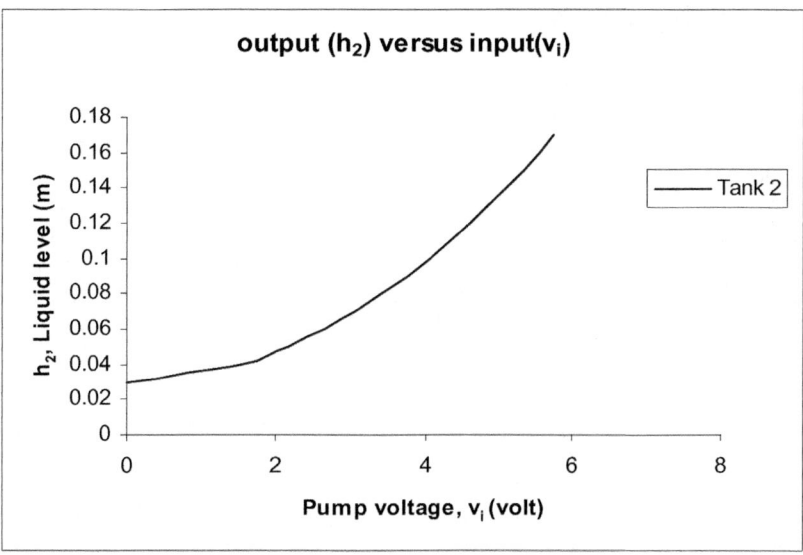

Fig. 1. Non-linearity of the plant. h_2 is the steady state liquid level of tank 2 when the input voltage v_i is applied to the pump

If three operating points on the nonlinear trajectory are used, they can be chosen based on the rate of change or equally divided from the full operating range of the liquid level. The simple division is used because of the trajectory capability of the LAM models. The three operating points are 0.05m, 0.1m and 0.15m. Generated step responses from the LAM at these points are shown in Fig. 2.

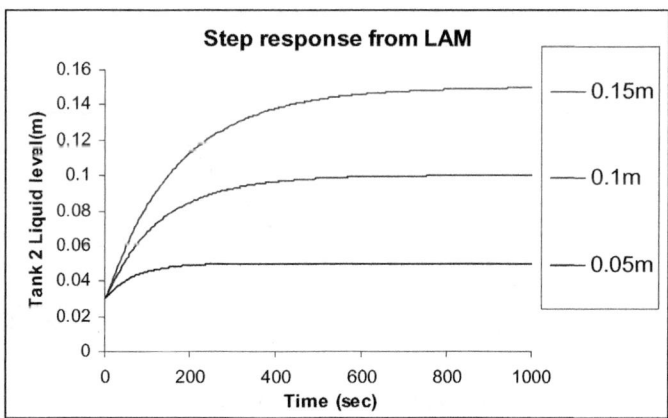

Fig. 2. Step responses at three operating points of a LAM network

3   Evolving a TCN

Many control system design methods are based on linear systems analysis. Here, a simple PID control system is design for a LAM. To apply the TCN technique to nonlinear plants, controllers must be designed for the entire LAM network of the nonlinear system.

This is to provide adequate performance across the operating envelope of the system. Three PID controllers in the TCN are scheduled or switch between them as shown in Fig. 3. During controller operation, a variable indicating operating point is monitored and different controllers (or controller parameters) are activated according to this scheduling variable. In this design, the plant output $y(t)$ is used as scheduling variable to schedule the output of the controllers.

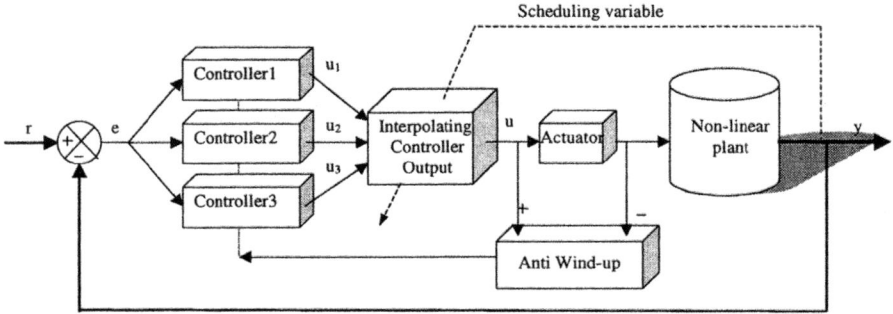

Fig. 3. Multiple controllers based trajectory controller network.

The TCN uses a linear interpolation or weighting schedule as shown in Fig. 4.

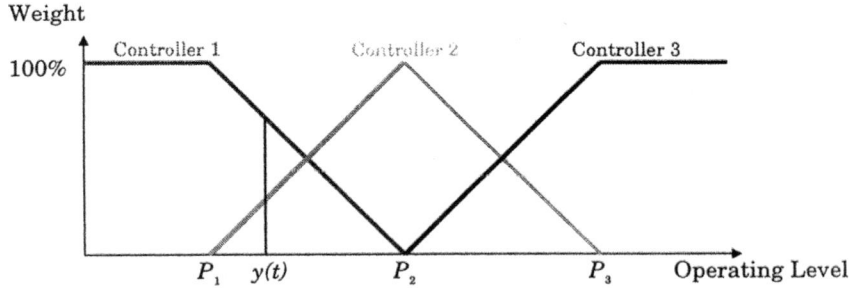

Fig. 4. A simple interpolation schedule in forming a TCN

Therefore, at any output level $y(t)$, the individual controller outputs $u_i(t)$ are interpolated giving a final controlling output $u(t)$ using equation (2), where $P_1=0.05$m, $P_2=0.1$m and $P_3=0.15$m.

$$u(t) = \begin{cases} \dfrac{P_{i+1} - y(t)}{P_{i+1} - P_i} \times u_i(t) + \dfrac{y(t) - P_i}{P_{i+1} - P_i} \times u_{i+1}(t) & \text{if } P_i \leq y(t) \leq P_{i+1} \\[2ex] u_i & \text{if } y(t) < P_i \\[2ex] u_n & \text{if } y(t) > P_n \end{cases}$$

(2)

where $i=1,\ldots,n$-1 and n is the total number of linear controller

Here, interpolation may also applied to the controller parameters K_p, K_i and K_d.

4   Design Example and Validation

4.1 Generating Trajectory Controllers from Step Responses

Individual PID controllers from a step-response trajectory to each of the three operating points are generated from the PIDeasy™ design automation package [4], as shown in Fig. 5.

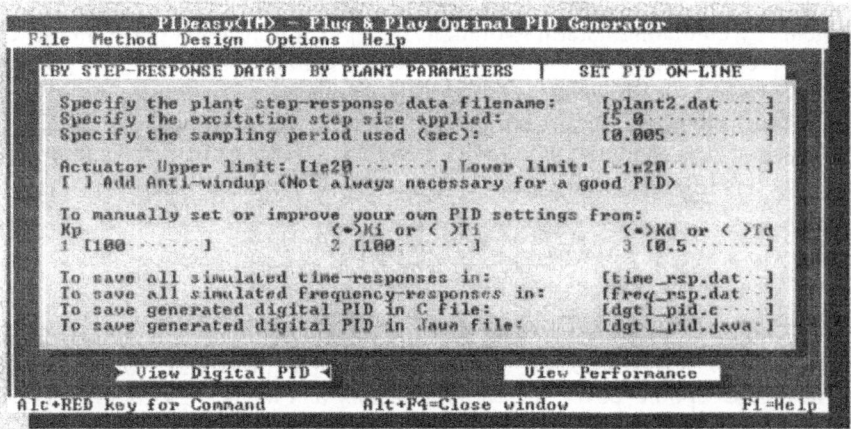

Fig. 5. Direct design from plant response using PIDeasy™

PIDeasy™ analyses step response data and generates an appropriate PID controller from them. At each operating point, fitting the step response generated from a LAM produces fast generation of PID controller. The closed loop responses at these operating points are plotted on the same graph shown in Fig. 6. Note that the fast generation of PID controllers using linear PIDeasy™ technique is tested against the nonlinear plant. This reveals the need of network tuning.

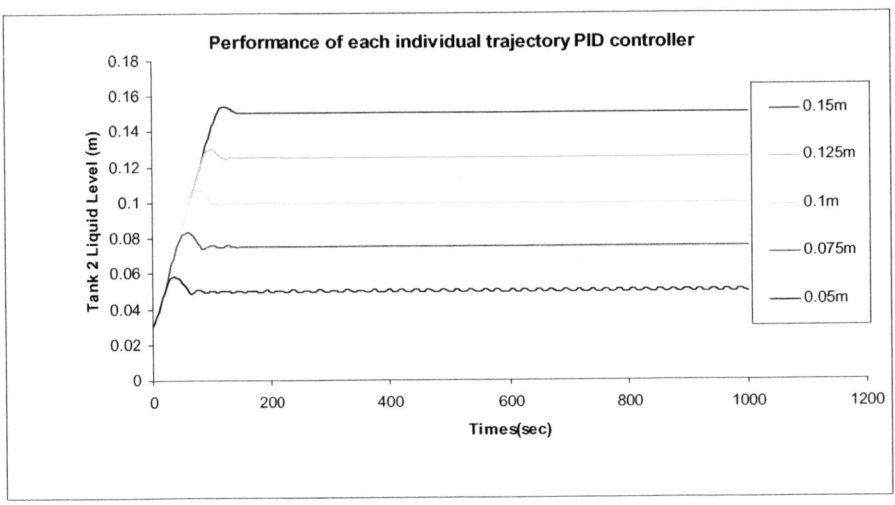

Fig. 6. Performance of each Individual trajectory PID controller

4.2.   Networking Through Evolution

A genetic algorithm (GA) provides globally optimal solutions to engineering design problems by emulating natural evolution. A population of potential solutions evolves using the evolutionary operators of crossover, mutation and selection to approach optimal solutions. One advantage of a GA for optimisation is that the objective or fitness function needs not to be differentiable. Here, the objective function to be minimised is the summation of all errors across the entire TCN at n reference operating points within a given time period m.

$$J = \sum_{ref=1}^{n} \sum_{t=0}^{m} |e(t)|$$

(3)

Here, there are $n=5$ reference levels used to evaluate the error tracking performance as shown in Fig.7. The 5 reference levels are set to 0.05m, 0.075m, 0.1m, 0.125m and 0.15m. These 5 points cover the whole trajectory and including two unseen operating points. Each reference is tested for a period of $m=1000$ sec, where

$$e(t)=|r(t)-y(t)|.$$

(4)

represents the tracking error between the closed-loop output $y(t)$ and the command $r(t)$.

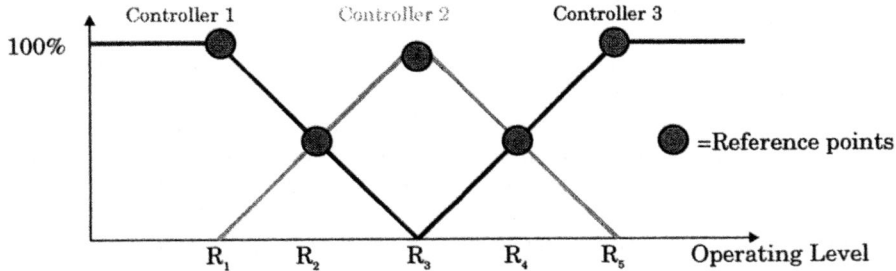

Fig. 7. Evaluation points in the operating envelope.

To evolve TCN based on the fast generated PID controllers out of LAM, all the parameters of the three linear controllers and the scheduling weights, are evolved simultaneously, at the seen and unseen operating points along in operating envelope. The closed loop responses of the finally evolved TCN at the end of 50 generation with a population size of 50 are shown in Fig.8 for all of the tested operating levels depicted in Fig.7. It can be seen that the linear TCN provided an excellent solution to the nonlinear control problem.

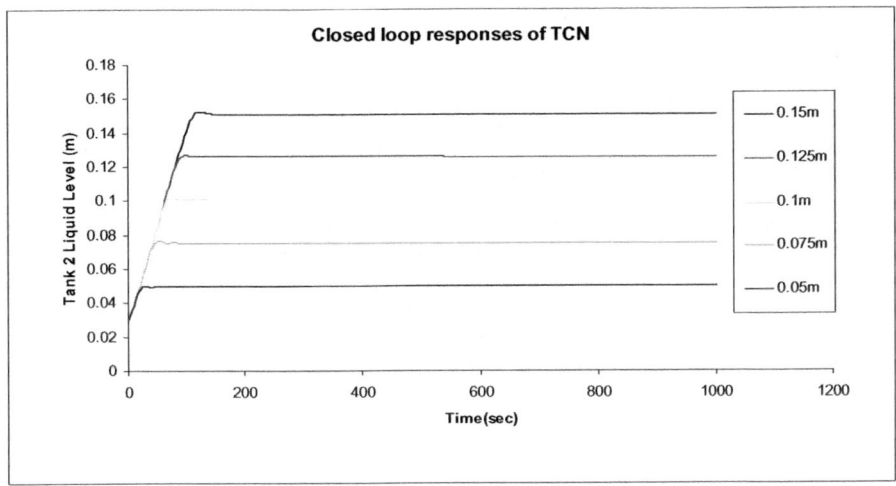

Fig. 8. Closed loop responses of the TCN at operating points including the unseen ones at 0.125m and 0.075m.

5   Discussion and Conclusion

To assist control system design for a wide range of operating envelope for nonlinear plants, this paper has developed a trajectory controller network (TCN) technique based on linear approximation model (LAM) technique. The example shows that the linear TCN used to control a nonlinear system performs well in the entire operating envelope. This offers potential benefits and simplicity for control of nonlinear systems. The results show that the GA based automatic controller network design for nonlinear systems is possible and useful. Such a network is easily designed from sampled response data.

References

1.   G.J. Gray, D.J. Murray Smith, Y. Li, K.C. Sharman, T. Weinbrenner: Nonlinear model structure identification using genetic programming. Control Engineering Practice, Vol.6, No.11. (1998) 1341-1352
2.   Y. Li and K.C. Tan: Linear approximation model network and its formation via evolutionary computation, Academy Proceedings in Engineering Sciences (SADHANA), Indian Academy of Sciences, Invited paper (1999)
3.   D.E. Goldberg: Genetic Algorithm in Search, Optimisation and Machine Learning, Addison-Wesley, Reading (1989)
4.   Y. Li, W. Feng, K.C. Tan, X.K. Zhu, X. Guan and K.H. Ang: PIDeasy™ and automated generation of optimal PID controllers, The Third Asia-Pacific Conference on Measurement and Control, Dunhuang, China, Plenary paper. (1998) 29-33
5.   G.J. Gray, Y. Li, D.J. Murray-Smith and K.C. Sharman: Specification of a control system fitness function using constraints for genetic algorithm based design methods, Proc. First IEE/IEEE Int. Conf. on GA in Eng. Syst.: Innovations and Appl., Sheffield. (1995) 530-535
6.   Y. Fathi: A linear approximation model for the parameter design problem, European Journal Of Operational Research, Vol.97, No.3. (1997) 561-570
7.   Klatt and Engell: Gain-scheduling trajectory control of a continuous stirred tank reactor, Computers & Chemical Engineering, Vol.22, No.4-5. (1998) 491-502
8.   G. Corriga, A. Giua, G. Usai: An implicit gain-scheduling controller for cranes, IEEE Transactions On Control Systems Technology, Vol.6, No.1. (1998) 15-20

Evolutionary Computation and Nonlinear Programming in Multi-model-robust Control Design

Dorothea Kolossa

dorothea.kolossa@
daimlerchrysler.com

Georg Grübel [*]

georg.gruebel@ieee.org

Abstract. An algorithmic parameter tuning methodology for controller design of complex systems is needed. This methodology should offer designers a great degree of flexibility and give insight into the potentials of the controller structure and the consequences of the design decisions that are made. Such a method is proposed here. For an exploratory phase a new pareto-ranked genetic algorithm is proposed to generate an evenly dispersed set of near optimal, global, solutions. By pair-wise preference statements on design alternatives a linear program is set up as a formal means for selecting the solution with best overall designer satisfaction. In a following interactive design phase using nonlinear programming techniques with a priori decisions on allowed quality levels, a best tuning compromise in competing requirements satisfaction is searched for while guaranteeing pareto-optimality. In particular, this two-phase tuning approach allows the designer to balance nominal control performance and multi-model control robustness.

1 Introduction

Control engineering work is mainly occupied with adapting a control system architecture with given control law structure, sensors and actuators, to needs of changed product requirements or new product versions. This is called 'incremental design'. In practice this occurs much more often than starting control system design afresh. The essence of incremental design is adaptation by tuning the control law parameters, partial replacement or augmentation of the control law structure by dynamic compensators, filters, and signal limiters, and tuning the overall structure in concurrence with the basic control law parameters.

Common industrial practice is hardware-in-the-loop, manual, tuning called 'calibrating'. However, since there may be very many parameters to be tuned manual tuning is not efficient neither in required engineering costs nor in exploiting the full potential of the chosen system architecture with respect to multivariate requirements. Therefore, 'virtual product engineering' based on high fidelity system model simulations is more and more becoming the engineering life style of choice. CACSD, i.e., Computer Automated Control System Design, is the discipline to provide the

---

[*] EvoScondi Node, DLR- Institute of Robotics and Mechatronics, Oberpfaffenhofen

S. Cagnoni et al. (Eds.): EvoWorkshops 2000, LNCS 1803, pp. 147–157, 2000.

pertinent technology in control engineering, making use of recent advances in automated symbolical / numerical system dynamics modeling and algorithmic parameter search via nonlinear programming or evolutionary computation. In particular, CACSD supports the design computation loop depicted in Figure 1, cf. [1].

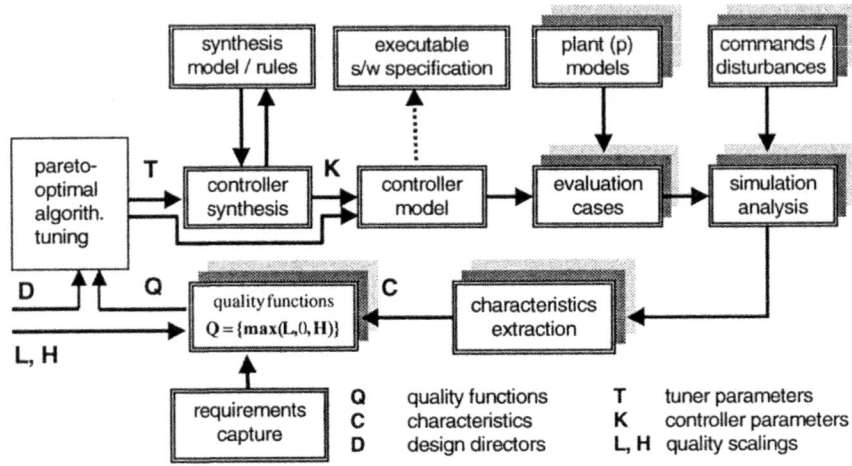

Fig. 1. CACSD generic computation loop to support incremental control design

This CACSD computation loop is generic in that it allows to incorporate any controller structure to be tuned. Tuning parameters T may relate to controller parameters either directly, $K = T$, e.g., in PID control [2], or indirectly via an analytic synthesis method, $K = f(T, \text{synthesis model})$, e.g., in H∞ [3] control, as well as fuzzy control, $K = f(T, \text{fuzzy rules})$, where in the latter case possible tuning parameters are the scaling factors of the membership functions and the weighting factors of the fuzzy control rules [4]. Furthermore, it allows to apply any analysis method, e.g., linear methods in state space and frequency domain as well as non-linear time simulation, to cope with non-commensurable control quality evaluations simultaneously.

To assess requirements satisfaction, design requirements are captured formally by quality functions. The fuzzy-type interval formulation of quality functions, cf. Section 2, allows to deal with quality levels, e.g., levels of 'good', 'acceptable', 'bad', requirement satisfaction. The data obtained by evaluation of all the quality functions feed a tuning algorithm to compute pareto-optimal tuning parameter values. For this kind of data-driven tuning both evolutionary algorithms, cf. Section 3, or non-linear programming algorithms, cf. Section 4, can be used.

Pareto-optimality lends itself not to a unique solution. Therefore, in Section 3.1 a new multi-objective genetic algorithm is proposed, which has the special property that it yields evenly dispersed solutions in or near to the pareto-optimal set, thus making best use of evolutionary computation to produce a rich set of design alternatives. Having many alternatives available to choose from, selection of the best candidate needs to follow a formal approach. This is dealt with in Section 3.2.

Nonlinear programming formulations in Section 4 are used to generate dedicated pareto-optimal design alternatives either to attain optimal *designer satisfaction* or to iterate quantitative compromises in competing *requirements satisfaction*.

This suggests a two-phase tuning procedure to achieve multi-model control robustness, Section 5. In phase one a global multi-objective design for a nominal plant model instantiation is carried out using evolutionary computation, and in phase two interactive nonlinear programming computations are applied to compromise nominal control behavior with off-nominal behavior characterized by a number of off-nominal plant model instantiations. By this tuning off-nominal behavior is to become at least 'acceptable' while nominal behavior is to be kept within the 'good' quality level.

2 Requirements Capture and Satisfaction Assessment

For design assessment design characteristics like system damping, steady state error, gain and phase margins, or maximum control rate, have to be transformed into a quality value which indicates the degree to which requirements are met. Two kinds of mathematical formulations are commonly in use: positive definite 'the-smaller-the-better' functionals of time and frequency responses, e.g., [2], which ought to be minimized, and inequalities on the design characteristics, which ought to be satisfied as constraints, e.g., [5], [3].

Advantages of the two approaches for quality modeling can be combined by the-smaller-the-better interval quality functions, where requirement satisfaction is considered 'good' for one range with function value zero, 'acceptable' in a range with function value not greater than one, and 'bad' outside a limiting range.

Such an interval quality function, $q(c)$, is mathematically defined on the design characteristics, c, by the max-operator (1) with four interval values $b_l < g_l < g_h < b_h$ compliant with 'bad', 'acceptable' and 'good' characteristics values, cf. Figure 2,

$$q(c) = \max\{L(c), 0, H(c)\},$$ (1)
$$L(c) = (c - g_l)/(b_l - g_l), \qquad b_l < g_l$$
$$H(c) = (c - g_h)/(b_h - g_h), \qquad g_l < g_h < b_h.$$

Requirement satisfaction is assessed as the better, the smaller a value this quality function assumes. Furthermore, the max-formulation fits to fuzzy logic AND-operation with max-operator [7] to make overall 'good', 'acceptable', 'bad' system quality statements in the vein of fuzzy logic. It also allows to combine an enumerated set of commensurable quality characteristics c_k, $L_k(c_k)$, $H_k(c_k)$, to form a compound quality function for, e.g., taking account of all eigenvalues concurrently or handling all values $c_k := c(t_k)$ of a discretized time response as an entity in requirements capture for robust tracking thumbprint performance [6]. With $H:=0$, the 'good' interval is open to the right, see the example of Figure 2, with $L:=0$ it is open to the left.

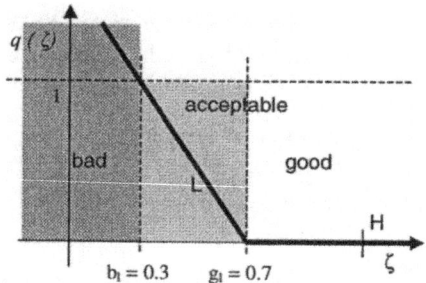

Fig. 2. Quality function for eigenvalue damping, with $\zeta > 0.7$ 'good' and > 0.3 'acceptable'

3 Pareto-Tuning by Genetic Algorithms and Design Preference

Control systems parameter tuning is always a multi-objective problem with competing requirements of control performance versus control effort and robustness. Hence the designer's prime task is to search for a suitable tradeoff while generating feasible solutions. This search ought to be confined to the set of 'best achievable' compromise solutions known as pareto-optimal solutions. Generally, a design alternative a_i is said to be pareto preferred to an alternative a_j if all quality measures q of a_i are better (smaller) than or equal to those of a_j, with at least one being strictly better. Thus, a pareto-optimal, or non-dominated, solution is one where no quality measure can be improved without causing degradation of at least one other quality measure.

To make an informed tradeoff decision, the designer needs a rich set of design alternatives as well as formal methods to support a systematic selection process.

Such a methodology is proposed here. A new pareto-ranked genetic algorithm generates an evenly dispersed set of design alternatives of near optimal solutions, giving the designer a global overview of what can be accomplished with the used controller structure, and by the method of preference-directed design that solution, which results in greatest overall designer satisfaction, is selected.

3.1 A New Genetic Algorithm to Generate an Evenly Dispersed Set of Solutions

A natural way of finding the pareto set by genetic algorithms was proposed by Fonseca and Fleming [8]. This approach measures the fitness of an individual by the number of other individuals that dominate it in the sense of pareto preference. Accordingly, a population is ranked, where the best solutions will be the non-dominated ones. Thus, the non-dominated solutions will always be most likely to be selected, leading to a convergence of the population to the pareto set.

Figure 3 gives an example of how a population would be ranked with this algorithm, where minimization of two quality measures C_1, C_2 is the goal. In this example, there are three non-dominated solutions which are ranked with a zero; all other solutions are dominated and their rank is determined by how many other solutions are better in the pareto sense.

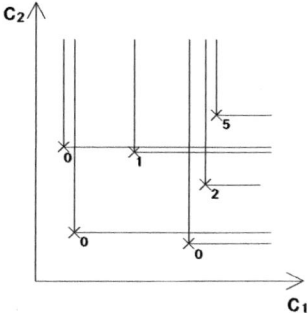

Fig. 3. Pareto ranking according to Fonseca and Fleming [8]

However, due to genetic drift, the population tends to converge to a single point on the pareto surface. Since it is the aim of an exploratory design phase to provide the designer with a rich basis of alternatives for possible tradeoff decisions, finding only one pareto solution is unsatisfactory. Thus, genetic drift should be avoided, which can be accomplished by methods like multiple sub-populations or penalizing overcrowded neighborhoods (fitness sharing) [3].

A variant of fitness sharing is proposed here for the purpose of finding an *evenly dispersed* near-pareto-optimal set of solutions. The idea is to incorporate the proximity of other individuals into the fitness of one individual such that individuals in remote regions of the search space will enjoy an advantage over those in more overcrowded regions. Since some individuals in the crowded regions should remain, so as to not distract the genetic algorithm from promising regions of the search space, penalties for crowding are limited to values less than 1. As the fitness score from pareto ranking is an integer, this ensures that the individuals are always ranked first by pareto domination but among individuals of the same pareto rank, the individuals are forced to spread out evenly.

Having mapped the n criteria values to [0,1] by $c_i - (c_i - c_{i,min})/(c_{i,max} - c_{i,min})$ the penalty to attain an evenly dispersed set is computed by the following formula:

$$p = (1 - \delta) \cdot \sqrt{\sum_{i=1..n} c_{i,current}^2 - c_{i,closest}^2} \cdot 1/\sqrt{n} \ . \tag{2}$$

where δ is chosen small but greater than zero, e.g., $\delta = 0.1$, $c_{i,j}$ signifies the ith criterion value of the jth individual, *'current'* is the individual for which the penalty is calculated and *'closest'* is the individual which is the closest preceding the current one in a list of all individuals sorted by pareto-rank. This list is sorted in ascending order, so that the best individuals come first, and between two individuals of the same rank, the position in the list is decided by random.

Thus, the penalty is always smaller than 1 and the adjusted fitness value is always strictly positive, which is necessary for some genetic search algorithms.

This variant of a multiobjective genetic algorithm has been implemented by means of the Genetic and Evolutionary Algorithm Toolbox [9]. Results show that this ranking procedure assures convergence towards the entire pareto set as opposed to either converging to just a part of it or favoring remote areas of the search space with less than optimal criteria values. The evolutionary optimization is carried out until either a certain number of generations is reached or a liveliness criterion is met. But since no analytical convergence criterion is used there is no guarantee for obtaining exact optimal solutions rather than attaining a near pareto-optimal set.

One example of how this multiobjective genetic algorithm fares with penalties (2) can be seen in Figure 4. The optimization results are shown together with the boundary of the set of all possible solutions. As measure for the degree of spread along the pareto surface the average standard deviation in the obtained pareto value set, computed over 1000 runs and scaled to a maximum of 1, is shown in Table 1.

Table 1. Average standard deviation in computed pareto value set

	Without Penalties	With Penalties
Average Standard Deviation	0.1493	0.2417

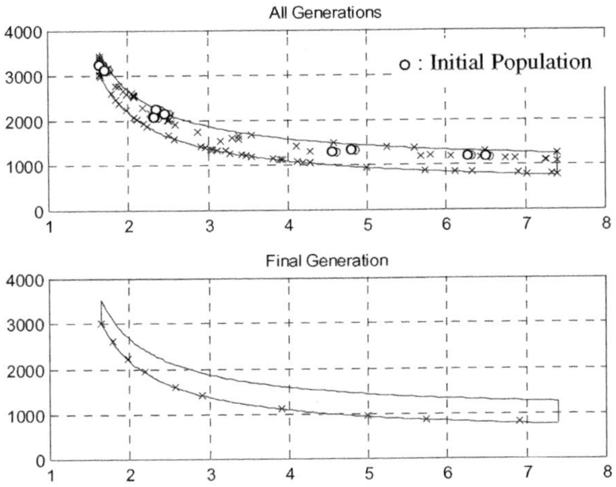

Fig. 4. Example: convergence to evenly dispersed pareto-optimal value set

3.2 Design Selection for Best Designer Satisfaction

After an evenly dispersed set of solutions is found, that solution which gives highest designer satisfaction is to be selected. Commonly, designer satisfaction is measured by a normalized weighted-sum value function $v(q)$ and a design is judged the better the smaller the value $v(q)$:

$$v = \sum w \cdot q\ (T), \text{ with: } \sum_k w_k = 1, \forall k : w_k > 0, q_k \geq 0. \tag{3}$$

The problem, however, is how to attribute a priori the numerical weights w_k in compliance with designer's intentions to formally decide on the 'best' solution with minimal value $v(q)$. The a posteriori approach of preference-directed design selection [10] copes with this problem by so-called imprecise value functions based on a number of pair-wise preference statements to be made by the designer.

If a designer makes a preference statement "Tuning T_i is superior to T_j" then by (3) this implies

$$v(T_i) = \sum_k w_k \cdot q_{ik}(T_i) < \sum_k w_k \cdot q_{jk}(T_j), \tag{4}$$

which can be rewritten as

$$\sum_k w_k (q_{jk}(T_j) - q_{ik}(T_i)) > 0, \quad \sum_k w_k = 1, \forall k : w_k > 0. \tag{5}$$

With known q_{ik}, q_{jk}, a set of preference statements (5) describes a subspace for admissible weights w_k. In [10] this constitutes an 'imprecise value function' to check whether a compatible preference ordering exists among other design alternatives as well. This is formally decided by solvability of the following linear program:

$$\min_w \sum_k w_k (q_{2*l-1,k}(T_{2*l-1}) - q_{2*l,k}(T_{2*l})) \tag{6}$$

s.t.

$$\sum_k w_k (q_{jk}(T_j) - q_{ik}(T_i)) > 0$$

$$\vdots$$

$$\sum_k w_k (q_{2*l-1,k}(T_{2*l-1}) - q_{2*l,k}(T_{2*l})) > 0$$

$$\sum_k w_k = 1, \forall k : w_k > 0,$$

where for $j = (2*l-1)$ and $i = (2*l)$ the quality function values $q_{2*l-1,k}$, $q_{2*l,k}$ are those of a pair of additional inferior and superior design alternatives, respectively.

Based on a few compatible preference statements (5) this formalism allows to partially order the set of design alternatives for selecting 'the best' one. The imprecise value function is a means to prune inferior paths during design space search thus reducing the complexity of the selection process. Having finally selected the best available design alternative, a compatible set of weights can be computed by (6), minimizing the difference of the value function of the chosen alternative to its ideal, i.e. superior, value of designer satisfaction, which is characterized by $q_k = 0$.

4 Pareto-Tuning by Interactive Nonlinear Programming

The genetic algorithm with pareto preferred ranking, as proposed in Section 3.1, fits well to the nature of evolutionary computation since on return to the many function evaluations that evolutionary computation requires it yields a rich set of solutions evenly dispersed in or close to the entire pareto-optimal set. Furthermore, genetic algorithms cope well with a large number of parameters and with a large search space, which makes them likely to find the global instead of a local solution in multimodal problems. Together with formal decision support, as dealt with in Section 3.2, this is well suited for selecting a posteriori a design candidate with best designer satisfaction among a number of global design alternatives.

On the other side, nonlinear programming approaches to attain pareto-optimal solutions are based on an analytical optimality condition, which makes them very efficient to compute just one, a priori dedicated, pareto-optimal solution in the local neighborhood of where the algorithms gets started. Furthermore, the necessary Karush-Kuhn-Tucker optimality conditions yield a numerical convergence condition that allows to attain a pareto optimum with high accuracy. This makes nonlinear programming algorithms suitable for 'fine tuning'. If parameterized in a decision intuitive way, interactive, declarative search to attain a specific, 'best', compromising solution for requirement satisfaction under local design conflicts becomes feasible.

Nonlinear Programming can be used to compute pareto-optimal tuning values T by solving [11] the normalized weighted-sum minimization problem, cf. (3)

$$\min_T \sum_k w_k q_k, \quad s.t. \sum_k w_k = 1, \forall k : w_k > 0, q \geq 0, \tag{7}$$

or the min-max optimization problem

$$\min_T \max_i \{q_i(T)/d_i\}, \quad s.t. \ 0 \leq q_k(T) \leq d_j, \ k = \{i, j\}. \tag{8}$$

Optimization (7) or (8) is a sufficient condition to yield a pareto-optimal solution with parameterized specific properties. There are standard algorithms of nonlinear programming, like SQP, to be used starting with a (global) solution by Section 3.1.

Formulation (7) yields the solution of optimal designer satisfaction with respect to attributed weights as found, e.g., by the a posteriori decision procedure of Section 3.2. A priori selection of weights is not decision intuitive [11] to attain specific properties.

Formulation (8) yields a pareto-optimal solution depending on parameters d. This formulation is well suited for compromising competing requirements by using these parameters iteratively as a priori 'design directors' to balance requirements satisfaction within a feasible solution set. A decision intuitive approach to choose design directors d is now proposed, which is inspired by the performance vector decision systematics due to Kreisselmeier [12].

Initialization Step: Start by a global solution with designer satisfaction according to Section 3 and pareto-optimize this solution by solving the unconstrained min-max problem with $d_i = \{1\}$, $j \in \emptyset$. This yields a balanced solution where all quality functions get pareto-minimal and the value function for designer satisfaction is further improved if the start solution is not yet an optimized one.

By solving the unconstrained min-max problem, the question for the main conflicts in requirements satisfaction can be answered: If $T = T^*$ is a minimizer, then

$$\max\{q_i(T^*)\} = q_{c1}^* = q_{c2}^* = \alpha^* > 0, \tag{9}$$

which means that the values of quality functions $q_{c1}, q_{c2}, \ldots$ belonging to the most competing requirements are equal and that they have the largest value among all quality functions. Moreover, $\alpha \leq 1$ characterizes a feasible ('acceptable') solution, which gives room for compromising the most competing requirements within the set of pareto-optimal alternatives.

Iterative Compromising Steps: Starting with a pareto-optimal solution, satisfaction of competing requirements cannot be improved simultaneously. This means that lowering the value of one quality function, $q_{c1}(T)$, can be achieved only at the expense of a higher value of another, $q_{c2}(T)$, and vice versa. Different compromise solutions q_{c1} versus q_{c2} can be achieved by different choices of the design directors d in an iterative procedure:

With a given pareto-optimal solution $Q^{(v-1)} = \{q_i^{(v-1)}, q_c^{(v-1)}\}$, for the next iteration step decide which of the most conflicting quality functions $\{q_{c1}, q_{c2}\}$ shall be improved, say q_{c1}. This choice may be made dependent on the weights that are associated to these quality functions via the formalism of Section 3. Then, concatenate $k^{(v)} \in \{i, c1\}$, and choose $d_k = 1$ and d_{c2} such that $q_{c2}^{(v-1)} < d_{c2}^v$ (≤ 1). Solving the *constrained* min-max problem,

$$\min_T \max_k \{q_k^{(v)}(T)\} \quad s.t. q_{c2}(T) \leq d_{c2}^{(v)}, \tag{10}$$

then attains the best possible solution in the sense that all quality functions of interest are minimized up to the constraint of the quantified limit of degradation one declares to be acceptable for the main conflicting quality function. Thus in an iterative procedure one can search for a 'best' compromise satisfaction of competing requirements.

Iterative compromising is best carried out in an *interactive mode of working* which needs fast algorithms to execute the CACSD computation loop of Figure 1. In addition it needs visual decision support on various information levels to best grasp design problem complexity. In particular, a graphical user interface [13] with a parallel coordinates display of the many quality functions, used as interactive steering aid to detect compromise conflicts and to choose design directors at runtime, greatly enhances engineering productivity. ANDECS_MOPS is such an environment [14].

5   Pareto-Optimal Multi-model Robustness Tuning

Feedback control suffers from potential stability problems, but if properly designed feedback reduces parameter sensitivity. Therefore design of controllers, which are stability and performance robust with respect to off-nominal operation, is of prime concern. Analytical robust-control theory, like μ-synthesis, relies on analytical stability criteria and pertinent (linear) plant model and uncertainty descriptions. Thus, it is restricted to problems with specific, commensurable, performance measures.

A completely general approach to robust control design is so-called multi-model design as implied by the CACSD tuning loop, cf. Figure 1. It is applicable to any kind of (non-linear) plant models and non-commensurable performance measures since it relies only on the data of the performance measures and not on their analytical description. Structural independence makes this kind of robust control approach applicable to any type of controller, i.e., PID, observer feedback, fuzzy control, etc.

The idea of robust multi-model design is to state the design problem for a nominal plant model instantiation reflecting nominal operation conditions and nominal system parameters within pertinent tolerance bands. Then, the same problem is stated for a number of off-nominal model instantiations reflecting worst case plant behavior, e.g., fast, lightly damped, and slow, over-damped, behavior within the range of assumed operation conditions and parameter uncertainty intervals. The quality functions of all these formulations are concatenated to a single multiobjective problem for which a satisfying pareto-optimal solution is to be found. This approach is highly competitive in comparison to other (analytic) robust control approaches [15].

The solution approaches that are dealt with in Sections 2, 3, 4, are particularly suitable to be combined for this type of multi-model, multi-objective, robust control design in form of a *two-phase design procedure* :

In phase 1, only a nominal model instantiation is considered and interval quality functions formulated according to Section 2 are optimized by the multiobjective genetic algorithm of Section 3.1 to yield a rich set of global, pareto-optimal, design alternatives as basis for designer preference selection, Section 3.2.

In phase 2, the 'best' nominal performance achieved in phase 1 is embedded in 'good' intervals by re-scaling the quality levels as required. This solution is used to start further tuning under the aspect of robustness: The off-nominal design cases are added to the nominal case and interactive nonlinear programming iterations according to Section 4 are simultaneously applied to all design cases to robustify the result of the first, nominal, design phase.

Thus, the user is allowed to make quantitative trade-off decisions concerning nominal versus robust performance. In these decisions off-nominal control behavior, characterized by worst-case plant model instantiations, should become at least 'acceptable', while nominal behavior is to be kept within a 'good' quality level.

6   Conclusion

A parameter tuning methodology to support control design automation is described. It uses a multiobjective genetic algorithm with fitness sharing to find a rich set of global solutions evenly dispersed in or near to the pareto-optimal set, from which a design candidate for best designer satisfaction is formally selected via pair-wise preference statements. Then min-max nonlinear programming is applied for compromise tuning to attain a pareto-optimal solution with best tradeoffs in requirements satisfaction. An on-line interactive mode of working using nonlinear programming in the compromising phase is supported by a systematics for choosing 'design directors' as allowable upper bounds to limit the expense one is willing to pay in making tradeoffs. Together with requirements capture by fuzzy-type interval quality functions, this two-phase approach is well suited to quantitative multi-model-robust control design with non-commensurable performance and robustness requirements.

References

1. Grübel, G.: Perspectives of CACSD: Embedding the Control System Design Process into a Virtual Engineering Environment. Proc. IEEE Int. Symposium on Computer Aided Control System Design, Hapuna-Beach, Hawaii (1999) 297-302
2. Feng, W., Li, Y.: Performance Indices in Evolutionary CACSD Automation with Application to Batch PID Generation. Proc. IEEE Int. Symposium on Computer Aided Control System Design, Hapuna-Beach, Hawaii (1999) 486-491
3. Chipperfield, A.J., Dakev, N.V., Fleming, P.J., Whidborne, J.F.: Multiobjective Robust Control Using Evolutionary Algorithms. Proc. IEE Int. Conf. Industrial Technology (1996) 269-273
4. Joos, H.-D., Schlothane, M., Grübel, G.: Multi-Objective Design of Controllers with Fuzzy Logic. Proc. IEEE/IFAC Joint Symposium on Computer Aided Control System Design, Tucson, AZ (1994) 75-82
5. Zakian, V., Al-Naib, U.: Design of Dynamical and Control Systems by the Method of Inequalities. Proc. Institute of Electrical Engineers, Vol. 120, No. 11. (1973) 1421-1427
6. Tan, K.C., Lee, T.H., Khor, E.F.: Control System Design Automation with Robust Tracking Thumbprint Performance Using a Multi-Objective Evolutionary Algorithm. Proc. IEEE Symposium on Computer-Aided Control System Design, Hapuna-Beach, Hawaii (1999) 498-503
7. Kienitz, K.H.: Controller Design Using Fuzzy Logic – A Case Study. Automatica, Vol. 29, No. 2. (1993) 549-554
8. Fonseca, C.M., Fleming, P.J.: An Overview of Evolutionary Algorithms in Multiobjective Optimization. Evolutionary Computing, Vol. 3, No. 1. (1995) 1-16
9. Pohlheim, H.: Genetic and Evolutionary Algorithm Toolbox for Use with Matlab - Documentation. Technical Report, Technical University Ilmenau, (1996)
10. d'Ambrosio, J.G., Birmingham, W.P.: Preference-directed Design. Artificial Intelligence for Engineering Design, Analysis and Manufacturing, Vol. 9. (1995) 219-230
11. Miettinen, K.M.: Nonlinear Multiobjective Optimization, Kluwer Academic Publishers, (1998)
12. Kreisselmeier, G., Steinhauser, R.: Application of Vector Performance Optimization to Robust Control Loop Design of a Fighter Aircraft. Int. Journal Control, Vol. 37, No. 2. (1983) 251-284.
13. Finsterwalder, R., Joos, H.-D., Varga, A.: A Graphical User Interface for Flight Control Development. Proc. IEEE Symposium on Computer-Aided Control System Design, Hapuna Beach, Hawaii (1999) 439-444
14. Grübel, G., Finsterwalder, R., Gramlich, G., Joos, H.-D., Lewald, S.: ANDECS: A Computation Environment for Control Applications of Optimization. In: Control Applications of Optimization, R. Bulirsch, D. Kraft, eds., Int. Series of Numerical Mathematics, Vol. 115. Birkhäuser Verlag, Basel (1994) 237-254
15. Grübel, G.: Another View on the Design Challenge Achievements. In: Robust Flight Control – A Design Challenge, Magni, J.F., Bennani, S., Terlouw, J.C., Eds., Lecture Notes in Control and Information Sciences 224, Springer Verlag, Berlin Heidelberg New York (1997) 603-609

Benchmarking Cost-Assignment Schemes for Multi-objective Evolutionary Algorithms

Konstantinos Koukoulakis, Dr Yun Li

Department of Electronics and Electrical Engineering, University of Glasgow

Abstract.

Currently there exist various cost-assignment schemes that perform the necessary scalarization of the objective values when applied to a multi-objective optimization problem. Of course, the final decision depends highly on the nature of the problem but given the multiplicity of the schemes combined with the fact that what the user ultimately needs is a single compromise solution it is evident that elaborating the selection of the method is not a trivial task. This paper intends to address this problem by extending the benchmarks of optimality and reach time given in [1] to mutliobjective optimization problems. A number of existing cost-assignment schemes are evaluated using such benchmarks.

1. Introduction

Having in mind the number of existing approaches to cost-assignment one could presume that the next step would be an appropriate choice. The concept of Pareto dominance has proven to be a great aid towards the formulation of the various schemes but further thinking reveals that what the user would like to have is simply a single compromise solution and not all of the solutions that form the Pareto-optimal set. As stated in [2], "although a Pareto-optimal solution should always be a better compromise solution than any solution it dominates, not all Pareto-optimal solutions may constitute acceptable compromise solutions". Therefore, what is needed is a performance index that can be used for the evaluation of the suitability of each scheme in the context of a specific problem.

To address this issue, two benchmarks used presented in Section 2. Section 3 describes a benchmark problem used for the comparison. Section 4 presents the evolutionary algorithm employed along with an outline of the various cost-assignment methods. Comparison is made between a number of evolutionary algorithms in Section 5. Conclusions are drawn in section 6.

2. The Benchmarks

Two benchmarks used for the evaluation of the different approaches have been defined in [1]. In this section they are going to be briefly presented.

S. Cagnoni et al. (Eds.): EvoWorkshops 2000, LNCS 1803, pp. 158-167, 2000.
© Springer-Verlag Berlin Heidelberg 2000

The first benchmark is called 'optimality'. Suppose that we have a test function $f(x): X \rightarrow F$, where $X \subseteq R^n$, $F \subseteq R^m$,

> where R^n represents the search space in n dimensions,
> R^m represents the space of all the possible objective values,
> n is the number of parameters,
> m is the number of the objectives
> $f \in F$ is the collection of the individual objective elements

Also, consider the theoretical objective vector $f_o = \{ f(x) \}$ that contains the objective values that can ultimately be reached. Finally, consider an objective reached as in Eq. 1.

$$f\left(\overset{\wedge}{x_o} \right) = \overset{\wedge}{f_0}, \overset{\wedge}{x_o} \in X \qquad (1)$$

with $\overset{\wedge}{x_o}$, representing a corresponding solution found.

The optimality measures how close an objective reached is to the theoretical objective vector and is calculated using the formula in Eq.2.

$$Optimality\left(\overset{\wedge}{f_o} \right)_\alpha = 1 - \frac{\left\| f_o - \overset{\wedge}{f_o} \right\|_\alpha}{\left\| \bar{f} - \underset{-}{f} \right\|_\alpha} \in [0,1] \qquad (2)$$

where $\bar{f}$ and $\underset{-}{f}$ are the upper and lower bounds of f respectively.

Any norm can be used to evaluate the optimality of an objective and this paper uses the Euclidean metric (a=2) for this purpose.

The above formula needs to be refined when the problem addressed is a non-dominant or non-commensurate one since no such concept as 'overall optimality' can be assessed in a problem of this kind. Since this is the case for this paper, the 'distance to demands' method explained in [3], is used here.

The second benchmark this short study uses is one that measures the convergence of the algorithm and is called 'reach time'. The reach time is defined as the total number of function evaluations performed by the algorithm by which the optimality of the best individual first reaches b.

$$\mathrm{Re}\,ach\_time\big|_b = C^b \qquad (3)$$

For the purposes of the tests, b is set to 0.999, a certainly high value that may not always be reached by the algorithm. Because of that, a single algorithm terminates when either the set optimality threshold is reached or 20n generations of size 20nxm have been evolved. Those termination conditions are identical to the ones used in [1] with the latter one meaning that the algorithm is not supposed to perform worse than an $O(n^2)$ algorithm in terms of computational time.

3. The Problem

A set of two objective functions (Fonsceca and Fleming, 1995) was chosen for the evaluation of the cost-schemes. The functions of Eq.4 and Eq.5 were chosen in an effort to produce as "standard" a Pareto optimal front as possible.

$$f_1(x) = 1 - \exp\left(-\sum_{i=1}^{n}\left(x_i - \frac{1}{\sqrt{n}}\right)^2\right) \quad (4)$$

$$f_2(x) = 1 - \exp\left(-\sum_{i=1}^{n}\left(x_i + \frac{1}{\sqrt{n}}\right)^2\right) \quad (5)$$

Each individual consists of a real-valued vector of n parameters. For the purposes of this paper n was set to the value of 2, with each parameter coded in the interval [-2, 2).

The individuals that form the Pareto-optimal set belong on the line shown in Fig. 1. Functions f1 and f2 are plotted for n=2 in figures 2 and 3 respectively.

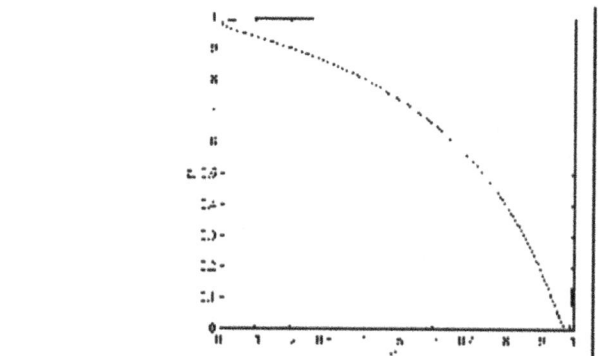

Fig. 1. Pareto optimal front

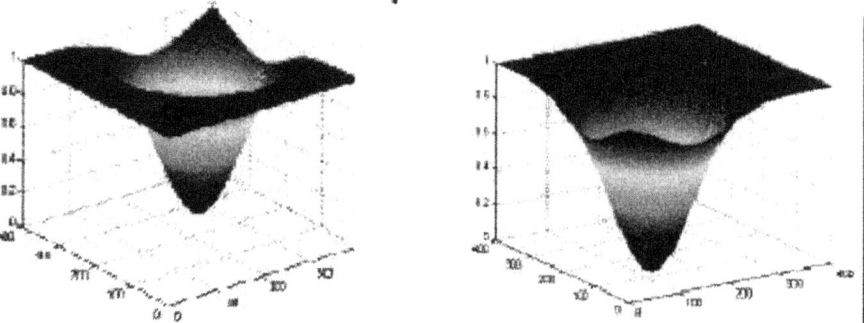

Fig. 2. f1 plotted for n=2   **Fig. 3.** f2 plotted for n=2

4.The Algorithm

4.1 Selection, Crossover and Mutation

The evolutionary algorithm that was used is quite a simple and straightforward one. It uses a binary tournament selection scheme to form the mating pool of the individuals after, of course, the cost assignment procedure has taken place. Each individual in the mating pool then randomly mates with another one using arithmetic crossover since each individual consists of a real-valued vector. Arithmetic crossover, as described in [4], is a canonical intermediate recombination operator, which produces the i-th component of the offspring by averaging it with some weight as defined in Eq.6.

$$x_i^{/} = \alpha x_{1i} + (1 - \alpha)x_{2i} \qquad (6)$$

Next, the offspring are evaluated and then refined by the simulated annealing (SA) technique. The SA positive mutation cycles where conducted using a non-linear Boltzmann learning schedule as the one employed in [5]. For the purposes of mutation, the creep mutation operator (Davis, 1989) was employed. As suggested in [6], entrapment must be alleviated in the case of this operator when used with a bounded small random amount for mutation. As such is the case here, entrapment can be said to have been partially overcome by the probabilistic nature of the SA technique, which maintains a probability of retaining lesser-valued individuals.

After SA, the parents are merged with the offspring the new population is formed with binary tournament selection.

The parameter settings of the algorithm, most of which are suggested in [5], are listed in table 1. The optimality threshold was set to a high value as the tests were intended to prohibit the algorithm to reach it so that a clearer picture of each scheme's behavior could be obtained.

Table 1. Parameter settings

Optimality threshold	0.999
Number of parameters	2
Weight vector	{ 1, 1 }
Priority vector	{ 1, 0=top }
Goal vector	{ 0.0, 0.0 }
Number of generations	160
Population size	320
Arithmetic crossover constant	0.5
Creep mutation probability	0.05
Tournament size	2
Boltzmann constant	5E-06
Initial temperature	1E05
Final temperature	1
Initial annealing factor	0.3
Transient constant	16

4.2 The Cost-Assignment Schemes

Each scheme is successively described below with a minimization problem assumed.

4.2.1 The Weighted Sum Method

According to this approach, all of the objectives are weighted by positive coefficients defined by the user and are added together to obtain the cost.

$$\Phi : \Re^n \to \Re$$

$$f(x) = \sum_{k=1}^{m} w_k f_k(x) \qquad (7)$$

,where $\ddot{O}$ denotes the cost assignment scheme and $\mathbf{x}$ is the parameter vector.

It must be noted that the same weights that are used here are also used to weight the objective vectors prior to the calculation of the norms of the optimality benchmark measure as suggested in [1].

4.2.2 The Minimax Method

This method tries to minimize the maximum weighted difference between the objectives and the goals, with the weights and the goals supplied by the user.

$$\Phi : \Re^n \to \Re$$

$$f(x) = \max_{k=1\ldots m} \frac{f_k(x) - g_k}{w_k} \qquad (8)$$

4.2.3 The Target Vector Method

This approach minimizes the distance of the objective vector from the goal vector using a defined distance measure. Again the user supplies the goals. The Euclidean metric was used as the distance measure in this case.

$$\Phi : \Re^n \to \Re$$

$$f(x) = \left\| [f(x) - g] W^{-1} \right\|_\alpha \qquad (9)$$

4.2.4 The Lexicographic Method

Here, the objectives are assigned distinct priorities and the selection proceeds with the comparison of the individuals with respect to the objective of the highest priority with any ties resolved by a successive comparison with respect to the objective with the second-highest priority, until the lowest priority objective is reached.

$$\Phi : \Re^n \rightarrow \{0,1,\ldots\mu-1\}, \mu = pop\_size$$

$$f(x_i) = \sum_{j=1}^{\mu} l\big(f(x_j)\ell < f(x_i)\big), \mu = pop\_size \qquad \textbf{(10)}$$

where l(condition) evaluates to unity if condition is true and

$$f(x_j)\ell < f(x_i) \Leftrightarrow \exists p \in \{1\ldots m\}:$$
$$\forall k \in \{p,\cdots,m\}, f_k(x_j) \le f_k(x_i) \wedge f_p(x_j) < f_p(x_i) \qquad \textbf{(11)}$$

4.2.5 Pareto Ranking (Goldberg's Approach)

According to the definition in [6] all non-dominated individuals are assigned a cost of one and then removed from contention with the next set of non-dominated individuals assigned a cost of two until the whole population has been ranked.

$$\Phi : \Re^n \rightarrow \{1,\ldots\mu\}, \mu = pop\_size$$

$$f(x_i) = \{ \begin{array}{l} 1 \Leftarrow not\big(f(x_j)p < f(x_i)\big), \forall j \in \{1\ldots\mu\} \\ \phi \Leftarrow not\big(f(x_j)p < f(x_i)\big), \forall j \in \{1\ldots\mu\} \end{array} \qquad \textbf{(12)}$$

$$\backslash \{l : \Phi(f(x_l)) < \phi\}$$

,where the p< condition denotes partial domination of the individual j over the individual i and is true if and only if

$$\forall k \in \{1\ldots m\}$$
$$f_k(x_j) \le f_k(x_i) \wedge \exists k \in \{1\ldots m\} : f_k(x_j) < f_k(x_i) \qquad \textbf{(13)}$$

4.2.6 Pareto Ranking (Fonseca and Fleming's Approach)

Proposed in 1993, this approach ranks an individual according to the number of individuals that dominate him.

$$\Phi : \Re^n \rightarrow \{0,1,\ldots\mu-1\}, \mu = pop\_size$$

$$f(x_i) = \sum_{j=1}^{\mu} l\bigg(f(x_j)p < f(x_i)\bigg), \mu = pop\_size \qquad \textbf{(14)}$$

4.2.7 Pareto Ranking (With Goals and Priorities)

This approach combines the pareto-optimality concept with goal and priority information. Equal priorities may be assigned to different objectives with both the priorities and the goals supplied by the user. Individuals are compared as in the lexicographic method but it is also affected from whether the individuals attain the goals set or not.

$$\Phi : \Re^n \to \{0,1,\dots \mu -1\}, \mu = pop\_size$$
$$f(x_i) = \sum_{j=1}^{\mu} l\left(f(x_j) \underset{g}{\prec} f(x_i) \right) \mu = pop\_size \quad (15)$$

,where

the condition within the brackets denotes preferrability of the j-th individual over the i-th individual and

g is the *preference vector*, a vector that contains the goals of each objective grouped by priority

As for the evaluation of the condition it is deemed too detailed to mention here but is fully described in [2].

5. Comparison Results

For each method, 10 experiments were carried out each with a random initial population, with an experiment terminating either when the optimality threshold has been reached or $400mn^2$ generations have been evolved. A discussion of the results obtained follows.

5.1 Pareto Front Sampling and Diversity

A cost-assignment scheme is considered successful if it has managed to offer a diverse sample of the pareto-optimal front as quickly as possible. With this in mind, a short discussion for each scheme tested follows.

A snapshot of the population in which the most optimal individual was found for each scheme can be seen at figures 4-10. Remember that the goals were set to 0.0 for both objectives.

The weighted sum approach was unable to sample the concave region of the line, focusing entirely on the zero-cost line $f_2 = -f_1$.

With identical, equally weighted goals, the minimax scheme failed to sample the pareto front. Nevertheless, with appropriate goal and weight settings it can prove successful, but surely less successful than the pareto-based approach with goals and priorities, which provides a better sampling in a quick and more efficient manner using only the goal information available.

The target vector scheme has introduced better sampling diversity than the minimax approach in roughly the same time. The diversity is even better than the pareto-based approach with goals and priorities but the latter scheme is significantly quicker in providing its results.

Fonseca and Fleming's approach along with Goldberg's original one has indeed most quickly sampled a very good proportion of the pareto-optimal front with the former being better at that. It must be noted that they have performed better at that than the last scheme without using any information available. This leads us to the conclusion that in the case of unattainable goals both of these schemes can offer a better sample of the front than the last scheme in some applications.

The last approach has performed very well using both the goal and priority information. It is interesting to compare it with the lexicographic cost-assignment scheme, which only uses priorities. It can clearly be seen that the latter scheme has driven the population to the minimisation of f_2, which has a higher priority over f_1. So, it can be said that the lexicographic method needs an aid for better results and the most obvious one is niching combined with mating restriction.

5.2 Optimality

As far as optimality is concerned, as can be seen in Fig. 11, Goldberg's and Fonseca and Fleming's approaches have both quickly given optimal solutions without using the goal information and also have the added bonus of good diversity. Of course with attainable goals, the pareto-based approach with goals and priorities should be the quickest cost-assignment scheme to offer the most optimal solution.

5.3 Reach Time

As no method managed to reach the high optimality threshold of 0.999 (max = 1.0), all of them had a reach time of $400mn^2 = 51200$.

6. Conclusions

The cost-assignment scheme acts as the driving force of the algorithm. Performing the scalarization of the objectives, it is the determining factor of evaluation. The purpose of this paper was to expose the magnitude of its impact on (a) the quality of the sampling of the Pareto-optimal front and (b) on the speed at which this quality is achieved.

The goals were deliberately set unattainable because they were intended to be a means to push the population towards the front. The optimality benchmark itself worked in accord with the goal settings so that it could act as an observer of the algorithm's behaviour rather than a strict evaluator.

It is thus concluded that for a given problem, the sampling of the Pareto-optimal front is generally easier achieved with the Pareto-based cost schemes. As for the rest of the schemes, it is thought that their usefulness can only be experienced with proper tuning of their associated parameters.

As a further study, it would be interesting to test all of the schemes in the context of a harder problem, that is a problem with a more diverse front. Finally, it is believed that this testing should employ a wide range of weights so that the promising aggregating 'target vector' scheme can be examined more closely.

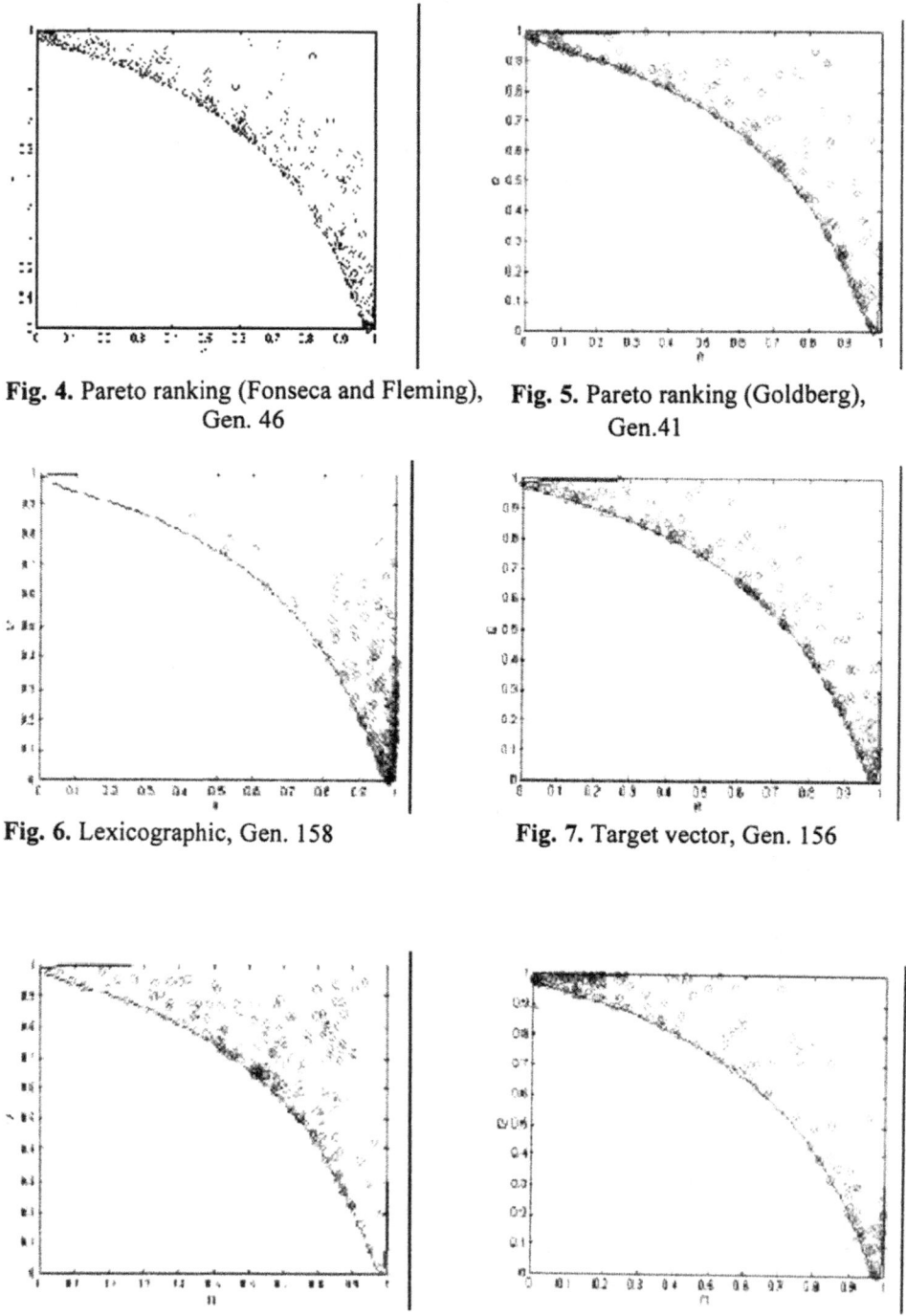

Fig. 4. Pareto ranking (Fonseca and Fleming), Gen. 46

Fig. 5. Pareto ranking (Goldberg), Gen.41

Fig. 6. Lexicographic, Gen. 158

Fig. 7. Target vector, Gen. 156

Fig. 8. Minimax, Gen. 145

Fig. 9. Weighted sum, Gen. 67

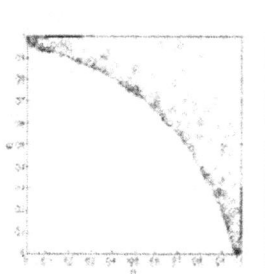

Fig. 10. Pareto ranking (goals and priorities), Gen. 47

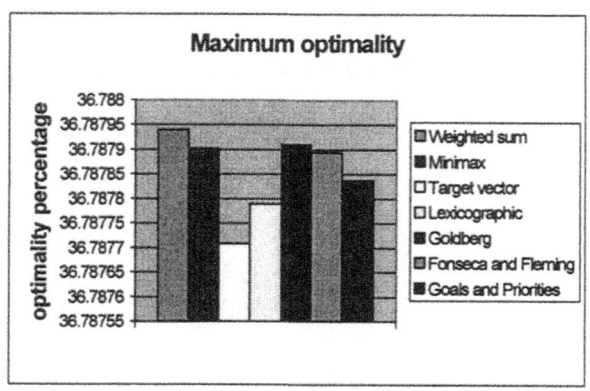

Fig. 11. Maximum optimality

References

1. Benchmarks for testing evolutionary algorithms, The Third Asia-Pacific Conference on Measurement and Control, Dunhuang, China, 31 Aug. - 4 Sept 1998, 134-138. (W. Feng , T. Brune, L. Chan, M. Chowdhury, C.K. Kuek and Y. Li).

2. Back T., Fogel D. B. and Michalewicz Z., *Handbook of Evolutionary Computation* (New York, Oxford: Oxford University Press, Bristol, Philadelphia: Institute Of Physics Publishing, 1997)

3. Michalewicz Z., Genetic Algorithms + Data structures = Evolution programs. (Berlin: Springer-Verlag, 1992)

4. Michalewicz Z., Nazhiyath G. and Michalewicz M, *A note on the usefulness of geometrical crossover for numerical optimization problems,* Proc 5th Ann. Conf. on Evolutionary Programming ed L. J. Fogel, P. J. Angeline and T. Back (Cambridge, MA: MIT Press, 1996)

5. Tan K.C., *Evolutionary methods for Modelling and Control of Linear and Nonlinear Systems,* Ph.D. thesis (Department of Electronics and Electrical Engineering, University of Glasgow, 1997)

6. Goldberg D. E., Genetic algorithms in Search, Optimization and Machine Learning (Reading, MA: Addison-Wesley, 1989)

7. Davis L., *Adapting operator probabilities in genetic algorithms,* Proc 3rd Int. Conf. on GAs (Fairfax, VA, June 1989) ed J. D. Schaffer (San Mateo, CA: Morgan Kaufmann) pp 61-69

8. Fonseca C. M. and Fleming P. J., *Multiobjective genetic algorithms made easy: selection sharing and mating restriction* (First Int. Conf. on GAs in Eng. Systems: Innovations and Applications, Sheffield, UK, 1995) pp 45-52

Automatic Synthesis of Both the Topology and Parameters for a Controller for a Three-Lag Plant with a Five-Second Delay Using Genetic Programming

John R. Koza
Stanford University, Stanford, California
koza@stanford.edu

Martin A. Keane
Econometrics Inc., Chicago, Illinois
makeane@ix.netcom.com

Jessen Yu
Genetic Programming Inc., Los Altos, California
jyu@cs.stanford.edu

William Mydlowec
Genetic Programming Inc., Los Altos, California
myd@cs.stanford.edu

Forrest H Bennett III
Genetic Programming Inc.
(Currently, FX Palo Alto Laboratory, Palo Alto, California)
forrest@evolute.com

Abstract

This paper describes how the process of synthesizing the design of both the topology and the numerical parameter values (tuning) for a controller can be automated by using genetic programming. Genetic programming can be used to automatically make the decisions concerning the total number of signal processing blocks to be employed in a controller, the type of each block, the topological interconnections between the blocks, and the values of all parameters for all blocks requiring parameters. In synthesizing the design of controllers, genetic programming can simultaneously optimize prespecified performance metrics (such as minimizing the time required to bring the plant output to the desired value), satisfy time domain constraints (such as overshoot and disturbance rejection), and satisfy frequency domain constraints. Evolutionary methods have the advantage of not being encumbered by preconceptions that limit its search to well-traveled paths. Genetic programming is applied to an illustrative problem involving the design of a controller for a three-lag plant with a significant (five-second) time delay in the external feedback from the plant to the controller. A delay in the feedback makes the design of an effective controller especially difficult.

S. Cagnoni et al. (Eds.): EvoWorkshops 2000, LNCS 1803, pp. 168–177, 2000.
© Springer-Verlag Berlin Heidelberg 2000

1   Introduction

The process of creating (synthesizing) the design of a controller entails making decisions concerning the total number of processing blocks to be employed in the controller, the type of each signal processing block (e.g., lead, lag, gain, integrator, differentiator, adder, inverter, subtractor, and multiplier), the values of all parameters for all blocks requiring parameters, and the topological interconnections between the signal processing blocks. The latter includes the question of whether or not to employ internal feedback (i.e., feedback inside the controller).

The problem of synthesizing a controller to satisfy prespecified requirements is sometimes solvable by analytic techniques (often oriented toward producing conventional PID controllers). However, as Boyd and Barratt stated in *Linear Controller Design: Limits of Performance* (1991),

"The challenge for controller design is to productively use the enormous computing power available. Many current methods of computer-aided controller design simply automate procedures developed in the 1930's through the 1950's ..."

This paper describes how genetic programming can be used to automatically create both the topology and the numerical parameter values (i.e., the tuning) for a controller directly from a high-level statement of the requirements of the controller. Genetic programming can, if desired, simultaneously optimize prespecified performance metrics (such as minimizing the time required to bring the plant output to the desired value as measured by, say, the integral of the time-weighted absolute error), satisfy time-domain constraints (involving, say, overshoot and disturbance rejection), and satisfy frequency domain constraints. Evolutionary methods have the advantage of not being encumbered by preconceptions that limit their search to well-traveled paths.

Section 2 describes an illustrative problem of controller synthesis. Section 3 provides general background on genetic programming. Section 4 describes how genetic programming is applied to control problems. Section 5 describes the preparatory steps necessary to apply genetic programming to the illustrative control problem. Section 6 presents the results.

2   Illustrative Problem

The illustrative problem entails creation of both the topology and parameter values for a controller for a three-lag plant with a significant (five-second) time delay in the external feedback from the plant output to the controller such that plant output reaches the level of the reference signal in minimal time (as measured by the integral of the time-weighted absolute error), such that the overshoot in response to a step input is less than 2%, and such that the controller is robust in the face of disturbance (added into the controller output). The delay in the feedback makes the design of an effective controller especially difficult (Astrom and Hagglund 1995). The transfer function of the plant is

$$G(s) = \frac{Ke^{-5s}}{(1+\tau s)^3}$$

A controller presented in Astrom and Hagglund 1995 (page 225) delivers credible performance on this problem for values of $K = 1$ and $\tau = 1$.

To make the problem more realistic, we added an additional constraint (satisfied by the controller presented in Astrom and Hagglund 1995) that the input to the plant is limited to the range between -40 and +40 volts. The plant in this paper operates over several different combinations of values for K and τ (whereas the controller designed by Astrom and Hagglund was intended only for $K = 1$ and $\tau = 1$).

3   Background on Genetic Programming

Genetic programming is an automatic technique for generating computer programs to solve, or approximately solve, problems.

Genetic programming (Koza 1992; Koza and Rice 1992) is an extension of the genetic algorithm (Holland 1975). Genetic programming is capable (Koza 1994a, 1994b) of evolving reusable, parametrized, hierarchically-called automatically defined functions (ADFs) so that an overall program consists of a main result-producing branch and one or more reusable and parameterizable automatically defined functions (function-defining branches). In addition, architecture-altering operations (Koza, Bennett, Andre, and Keane 1999; Koza, Bennett, Andre, Keane, and Brave 1999) enable genetic programming to automatically determine the number of automatically defined functions, the number of arguments that each possesses, and the nature of the hierarchical references, if any, among such automatically defined functions.

Genetic programming often creates novel designs because it is a probabilistic process that is not encumbered by the preconceptions that often channel human thinking down familiar paths. For example, genetic programming is capable of synthesizing the design of both the topology and sizing for a wide variety of analog electrical circuits from a high-level statement of the circuit's desired behavior and characteristics (Koza, Bennett, Andre, and Keane 1999; Koza, Bennett, Andre, Keane, and Brave 1999). Five of the evolved analog circuits in that book infringe on previously issued patents while five others deliver the same functionality as previously patented inventions in a novel way.

Additional information on current research in genetic programming can be found in Banzhaf, Nordin, Keller, and Francone 1998; Langdon 1998; Ryan 1999; Kinnear 1994; Angeline and Kinnear 1996; Spector, Langdon, O'Reilly, and Angeline 1999; Koza, Goldberg, Fogel, and Riolo 1996; Koza, Deb, Dorigo, Fogel, Garzon, Iba, and Riolo 1997; Koza, Banzhaf, Chellapilla, Deb, Dorigo, Fogel, Garzon, Goldberg, Iba, and Riolo 1998; Banzhaf, Poli, Schoenauer, and Fogarty 1998; Banzhaf, Daida, Eiben, Garzon, Honavar, Jakiela, and Smith 1999; Poli, Nordin, Langdon, and Fogarty 1999; at web sites such as www.genetic-programming.org; and in the *Genetic Programming and Evolvable Machines* journal (from Kluwer Academic Publishers).

4   Genetic Programming and Control

Controllers can be represented by block diagrams in which the blocks represent signal processing functions, in which external points represent the controller's input(s) and output(s), and in which cycles in the block diagram correspond to internal feedback inside the controller. Genetic programming can be extended to the problem of creating both the topology and parameter values for a controller by establishing a mapping between the program trees used in genetic programming and the block diagrams germane to controllers.

The number of result-producing branches in the to-be-evolved controller equals the number of control variables that are to be passed from the controller to the plant. Each result-producing branch is a composition of the functions and terminals from a repertoire (below) of functions and terminals.

Program trees in the population during the initial random generation (generation 0) consist only of result-producing branch(es). Automatically defined functions are introduced incrementally (and sparingly) into the population on subsequent generations by means of the architecture-altering operations. Each automatically defined function is a composition of the functions and terminals appropriate for control problems, references to existing automatically defined functions, and (possibly) dummy variables (formal parameters) that permit parameterization of the automatically defined function. Automatically defined functions provide a mechanism for internal feedback (recursion) within the to-be-evolved controller. Automatically defined functions also provide a mechanism for reusing useful substructures.

Each branch of each program tree in the initial random population is created in accordance with a constrained syntactic structure. Each genetic operation executed by genetic programming (crossover, mutation, reproduction, or architecture-altering operation) produces offspring that comply with the constrained syntactic structure.

Genetic programming has recently been used to create a controller for a particular two-lag plant and a three-lag plant (Koza, Keane, Yu, Bennett, and Mydlowec 2000). Both of these genetically evolved controllers outperformed the controllers designed by experts in the field of control using the criteria originally specified by the experts.

5   Preparatory Steps

Six major preparatory steps are required before applying genetic programming: (1) determine the architecture of the program trees, (2) identify the terminals, (3) identify the functions, (4) define the fitness measure, (5) choose control parameters for the run, and (6) choose the termination criterion and method of result designation.

5.1   Program Architecture

Since there is one result-producing branch in the program tree for each output from the controller and this problem involves a one-output controller, each program tree has one result-producing branch. Each program tree in the initial random generation (generation 0) has no automatically defined functions. However, in subsequent generations, architecture-altering operations may insert and delete automatically defined functions (up to a maximum of five per program tree).

5.2   Terminal Set

A constrained syntactic structure permits only a single perturbable numerical value to appear as the argument for establishing each numerical parameter value for each signal processing block requiring a parameter value. These numerical values initially range from -5.0 to +5.0. These numerical values are perturbed during the run by a Gaussian mutation operation that operates only on numerical values. Numerical constants are later interpreted on a logarithmic scale so that they represent values in a range of 10 orders of magnitude (Koza, Bennett, Andre, and Keane 1999).

The remaining terminals are time-domain signals. The terminal set, T, for the result-producing branch and any automatically defined functions (except for the perturbable numerical values mentioned above) is

T = {REFERENCE_SIGNAL, CONTROLLER_OUTPUT, PLANT_OUTPUT,
        CONSTANT_0}.

Space does not permit a detailed description of the various terminals used herein (although the meaning of the above terminals should be clear from their names). See Koza, Keane, Yu, Bennett, and Mydlowec 2000 for details.

5.3   Function Set

The functions are signal processing functions that operate on time-domain signals (the terminals in T). The function set, F, for the result-producing branch and any automatically defined functions is

F = {GAIN, INVERTER, LEAD, LAG, LAG2,
        DIFFERENTIAL_INPUT_INTEGRATOR, DIFFERENTIATOR,
        ADD_SIGNAL, SUB_SIGNAL, ADD_3_SIGNAL, DELAY, ADF0, ...,
        ADF4}.

ADF0, ..., ADF4 denote automatically defined functions added during the run by architecture-altering operations.

The functionality of each of the above signal processing functions is suggested by their names and is described in detail in Koza, Keane, Yu, Bennett, and Mydlowec 2000.

5.4   Fitness

Genetic programming is a probabilistic algorithm that searches the space of compositions of the available functions and terminals. The search is guided by a fitness measure. The fitness measure is a mathematical implementation of the high-level requirements of the problem. The fitness measure is couched in terms of "what needs to be done" — not "how to do it."

The fitness measure may incorporate any measurable, observable, or calculable behavior or characteristic or combination of behaviors or characteristics. The fitness measure for most problems of controller design is multi-objective in the sense that there are several different (usually conflicting) requirements for the controller.

The fitness of each individual is determined by executing the program tree (i.e., the result-producing branch and any automatically defined functions that may be invoked) to produce an interconnected sequence of signal processing blocks — that is, a block diagram for the controller. A SPICE netlist is then constructed from the block diagram. The SPICE netlist for the resulting controller is wrapped inside an appropriate set of SPICE commands. The controller is then simulated using our modified version of the SPICE simulator. The 217,000-line SPICE3 simulator (Quarles, Newton, Pederson, and Sangiovanni-Vincentelli 1994) is an industrial-strength simulator. It is run as a submodule within our genetic programming system. The SPICE simulator returns tabular output and other information from which the fitness of the individual is then computed.

The fitness of a controller is measured using 13 elements consisting of 12 time-domain-based elements based on a modified integral of time-weighted absolute error (ITAE) and one time-domain-based element measuring disturbance rejection.

The fitness of an individual controller is the sum (i.e., linear combination) of the detrimental contributions of these 13 elements of the fitness measure. The smaller the sum, the better.

The first 12 elements of the fitness measure evaluate how quickly the controller causes the plant to reach the reference signal and the controller's success in avoiding overshoot. Two reference signals are used. The first reference signal is a step function that rises from 0 to 1 volts at $t = 100$ milliseconds while the second rises from 0 to 1 microvolts at $t = 100$ milliseconds. The two step functions are used to deal with the non-linearity caused by the limiter. Two values of the time constant, τ, are used (namely 0.5 and 1.0). Three values of K are used, namely 0.9, 1.0, and 1.1. Exposing genetic programming to different combinations of values of step size, K, and τ produces a robust controllers and also prevents genetic programming from engaging in pole elimination. For each of these 12 fitness cases, a transient analysis is performed in the time domain using the SPICE simulator. Table 1 shows the elements of the fitness measure in its left-most four columns.

The contribution to fitness for each of these 12 elements of the fitness measure is based on the integral of time-weighted absolute error (ITAE)

$$\int_{t=5}^{36} (t-5)|e(t)|A(e(t))BCdt \, .$$

Because of the built-in five-second time delay, the integration runs from time $t = 5$ seconds to $t = 36$ seconds. Here $e(t)$ is the difference (error) at time t between the delayed plant output and the reference signal. The integral of time-weighted absolute error penalizes differences that occur later more heavily than differences that occur earlier.

We modified the integral of time-weighted absolute error in four ways. First, we used a discrete approximation to the integral by considering 120 300-millisecond time steps between $t = 5$ to $t = 36$ seconds. Second, we multiplied each fitness case by the reciprocal of the amplitude of the reference signals so that both reference signals (1 microvolt and 1 volt) are equally influential. Specifically, B is a factor that is used to normalize the contributions associated with the two step functions. B multiplies the difference $e(t)$ associated with the 1-volt step function by 1 and multiplies the difference $e(t)$ associated with the 1-microvolt step function by 10^6. Third, the integral contains an additional weight, A, that varies with $e(t)$. The function A weights all variation up to 102% of the reference signal by a factor of 1.0, and heavily penalizes overshoots over 2% by a factor 10.0. Fourth, the integral contains a special weight, C, which is 5.0 for the two fitness cases for which $K = 1$ and $\tau = 1$, and 1.0 otherwise.

The 13th element of the fitness measure is based on disturbance rejection. The penalty is computed based on a time-domain analysis for 36.0 seconds. In this analysis, the reference signal is held at a value of 0. A disturbance signal consisting of a unit step is added to the CONTROLLER_OUTPUT at time $t = 0$ and the resulting disturbed signal is provided as input to the plant. The detrimental contribution to fitness is 500/36 times the time required to bring the plant output to within 20 millivolts of the reference signal of 0 volts (i.e., to reduce the effect to within 2% of the 1-volt disturbance signal) assuming that the plant settles to within this range within 36 seconds. If the plant does not settle to within this range within 36 seconds, the detrimental contribution to fitness is 500 plus the absolute value of the plant output in volts times 500. For example, if the effect of the disturbance was never reduced below 1 volts, the detrimental contribution to fitness would be 1000.

A controller that cannot be simulated by SPICE is assigned a high penalty value of fitness (10^8).

5.5   Control Parameters

The population size, M, was 500,000. A maximum size of 150 points (functions and terminals) was established for each result-producing branch and a maximum size of 100 points was established for each automatically defined function. The other parameters for controlling the runs are the default values that we apply to a broad range of problems (Koza, Bennett, Andre, and Keane 1999).

5.6   Termination

The run was manually monitored and manually terminated when the fitness of many successive best-of-generation individuals appeared to have reached a plateau. The single best-so-far individual is harvested and designated as the result of the run.

5.7   Parallel Implementation

This problem was run on a home-built Beowulf-style (Sterling, Salmon, Becker, and Savarese 1999; Bennett, Koza, Shipman, and Stiffelman 1999) parallel cluster computer system consisting of 1,000 350 MHz Pentium II processors (each accompanied by 64 megabytes of RAM). The system has a 350 MHz Pentium II computer as host. The processing nodes are connected with a 100 megabit-per-second Ethernet. The processing nodes and the host use the Linux operating system. The distributed genetic algorithm with unsynchronized generations and semi-isolated subpopulations was used with a subpopulation size of $Q = 500$ at each of $D = 1,000$ demes. Two processors are housed in each of the 500 physical boxes of the system. As each processor (asynchronously) completes a generation, four boatloads of emigrants from each subpopulation (selected probabilistically based on fitness) are dispatched to each of the four toroidally adjacent processors. The migration rate is 2% (but 10% if the toroidally adjacent node is in the same physical box).

6   Results

The best individual in generation 0 has a fitness of 1926.498.

The best-of-run controller emerged in generation 129 (figure 1). This best-of-run controller has a fitness of 522.605. The result-producing branch of this best-of-run individual has 119 points (functions and terminals) and 95, 93, and 70 points, respectively, in its three automatically defined functions. Note that genetic programming employed a 4.8 second delay (comparable to the five-second plant delay) in the transfer function of the evolved pre-filter. This best-of-run controller from generation 129 has a better value of fitness for a step size of 1 volt, an internal gain, K, of 1.0, and a time-constant, τ,of 1.0 (the specific case considered by Astrom and Hagglund 1995).

Figure 2 compares the time-domain response to step input of the best-of-run controller from generation 129 (triangles) with the controller in Astrom and Hagglund 1995 (squares) for a step size of 1 volt, an internal gain, K, of 1.0, and a time-constant, τ,of 1.0.

Figure 3 compares the disturbance rejection of the best-of-run controller from generation 129 (triangles) with the controller in Astrom and Hagglund 1995 (squares) for a step size of 1 volt, an internal gain, K, of 1.0, and a time-constant, τ,of 1.0.

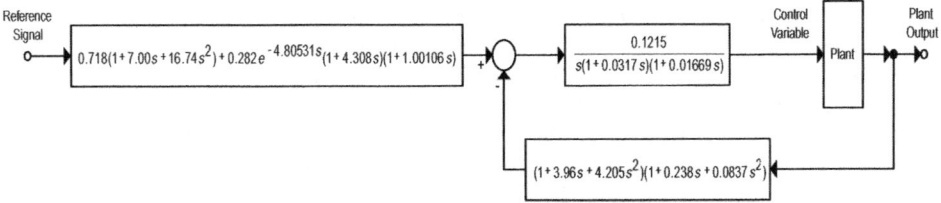

Figure 1 Best-of-run controller from generation 129 for three-lag plant with five-second delay.

Table 1 compares the fitness of the best-of-run controller from generation 129 and the Astrom and Hagglund 1995. Two of the entries are divided by the special weight $C = 5.0$. All 12 entries are better for the genetically evolved controller than for the Astrom and Hagglund 1995 controller.

Table 1 Fitness of two controllers for three-lag plant with five-second delay.

Element	Step size (volts)	Plant internal Gain, K	Time constant, τ	Best-of-run generation 129	Astrom and Hagglund controller
0	1	0.9	1.0	13.7	27.4
1	1	0.9	0.5	25.6	38.2
2	1	1.0	1.0	34.0 / 5 = 6.8	22.9
3	1	1.0	0.5	18.6	29.3
4	1	1.1	1.0	4.4	25.4
5	1	1.1	0.5	16.3	22.7
6	10^{-6}	0.9	1.0	13.2	27.4
7	10^{-6}	0.9	0.5	25.5	38.2
8	10^{-6}	1.0	1.0	30.7 / 5 = 6.1	22.9
9	10^{-6}	1.0	0.5	18.5	29.3
10	10^{-6}	1.1	1.0	4.3	25.4
11	10^{-6}	1.1	0.5	16.2	22.7
Disturbance	1	1	1	302	373

References

Angeline, Peter J. and Kinnear, Kenneth E. Jr. (editors). 1996. *Advances in Genetic Programming 2*. Cambridge, MA: The MIT Press.

Astrom, Karl J. and Hagglund, Tore. 1995. *PID Controllers: Theory, Design, and Tuning*. 2nd Edition. Research Triangle Park, NC: Instrument Society of America.

Banzhaf, Wolfgang, Daida, Jason, Eiben, A. E., Garzon, Max H., Honavar, Vasant, Jakiela, Mark, and Smith, Robert E. (editors). 1999. *GECCO-99: Proceedings of the Genetic and Evolutionary Computation Conference, July 13-17, 1999, Orlando, Florida USA*. San Francisco, CA: Morgan Kaufmann.

Banzhaf, Wolfgang, Nordin, Peter, Keller, Robert E., and Francone, Frank D. 1998. *Genetic Programming – An Introduction*. San Francisco, CA: Morgan Kaufmann and Heidelberg: dpunkt.

Banzhaf, Wolfgang, Poli, Riccardo, Schoenauer, Marc, and Fogarty, Terence C. 1998. *Genetic Programming: First European Workshop. EuroGP'98. Paris, France, April 1998 Proceedings. Paris, France. April l998.* Lecture Notes in Computer Science. Volume 1391. Berlin, Germany: Springer-Verlag.

Bennett, Forrest H III, Koza, John R., Shipman, James, and Stiffelman, Oscar. 1999. Building a parallel computer system for $18,000 that performs a half peta-flop per day. In Banzhaf, Wolfgang, Daida, Jason, Eiben, A. E., Garzon, Max H., Honavar, Vasant, Jakiela, Mark, and Smith, Robert E. (editors). 1999. *GECCO-99: Proceedings of the Genetic and Evolutionary Computation Conference, July 13-17, 1999, Orlando, Florida USA.* San Francisco, CA: Morgan Kaufmann. 1484 - 1490.

Boyd, S. P. and Barratt, C. H. 1991. *Linear Controller Design: Limits of Performance.* Englewood Cliffs, NJ: Prentice Hall.

Holland, John H. 1975. *Adaptation in Natural and Artificial Systems.* Ann Arbor, MI: University of Michigan Press.

Kinnear, Kenneth E. Jr. (editor). 1994. *Advances in Genetic Programming.* Cambridge, MA: The MIT Press.

Koza, John R. 1992. *Genetic Programming: On the Programming of Computers by Means of Natural Selection.* Cambridge, MA: MIT Press.

Koza, John R. 1994a. *Genetic Programming II: Automatic Discovery of Reusable Programs.* Cambridge, MA: MIT Press.

Koza, John R. 1994b. *Genetic Programming II Videotape: The Next Generation.* Cambridge, MA: MIT Press.

Koza, John R., Banzhaf, Wolfgang, Chellapilla, Kumar, Deb, Kalyanmoy, Dorigo, Marco, Fogel, David B., Garzon, Max H., Goldberg, David E., Iba, Hitoshi, and Riolo, Rick. (editors). 1998. *Genetic Programming 1998: Proceedings of the Third Annual Conference.* San Francisco, CA: Morgan Kaufmann.

Koza, John R., Bennett III, Forrest H, Andre, David, and Keane, Martin A. 1999. *Genetic Programming III: Darwinian Invention and Problem Solving.* San Francisco, CA: Morgan Kaufmann. Forthcoming.

Koza, John R., Bennett III, Forrest H, Andre, David, Keane, Martin A., and Brave Scott. 1999. *Genetic Programming III Videotape: Human-Competitive Machine Intelligence.* San Francisco, CA: Morgan Kaufmann.

Koza, John R., Deb, Kalyanmoy, Dorigo, Marco, Fogel, David B., Garzon, Max, Iba, Hitoshi, and Riolo, R. L. (editors). 1997. *Genetic Programming 1997: Proceedings of the Second Annual Conference* San Francisco, CA: Morgan Kaufmann.

Koza, John R., Goldberg, David E., Fogel, David B., and Riolo, Rick L. (editors). 1996. *Genetic Programming 1996: Proceedings of the First Annual Conference.* Cambridge, MA: MIT Press.

Koza, John R., Keane, Martin A., Yu, Jessen, Bennett, Forrest H III, and Mydlowec, William. 2000. Automatic creation of human-competitive programs and controllers by means of genetic programming. *Genetic Programming and Evolvable Machines.* (1) 121 - 164.

Koza, John R., and Rice, James P. 1992. *Genetic Programming: The Movie.* Cambridge, MA: MIT Press.

Langdon, William B. 1998. *Genetic Programming and Data Structures: Genetic Programming + Data Structures = Automatic Programming!* Amsterdam: Kluwer.

Poli, Riccardo, Nordin, Peter, Langdon, William B., and Fogarty, Terence C. 1999. *Genetic Programming: Second European Workshop. EuroGP '99. Proceedings.* Lecture Notes in Computer Science. Volume 1598. Berlin: Springer-Verlag.

Quarles, Thomas, Newton, A. R., Pederson, D. O., and Sangiovanni-Vincentelli, A. 1994. *SPICE 3 Version 3F5 User's Manual.* Department of Electrical Engineering and Computer Science, Univ. of California. Berkeley, CA. March 1994.

Ryan, Conor. 1999. *Automatic Re-engineering of Software Using Genetic Programming.* Amsterdam: Kluwer Academic Publishers.

Spector, Lee, Langdon, William B., O'Reilly, Una-May, and Angeline, Peter (editors). 1999. *Advances in Genetic Programming 3.* Cambridge, MA: MIT Press.

Sterling, Thomas L., Salmon, John, Becker, D. J., and Savarese, D. F. 1999. *How to Build a Beowulf: A Guide to Implementation and Application of PC Clusters.* Cambridge, MA: MIT Press.

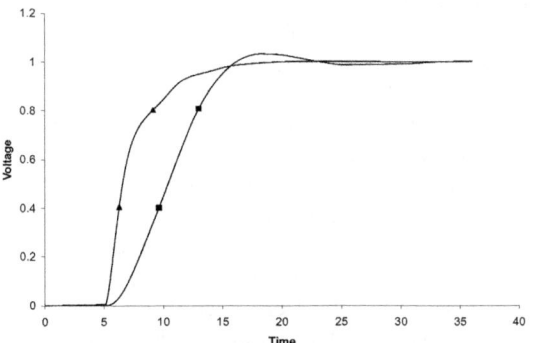

Figure 2 Comparison for step input.

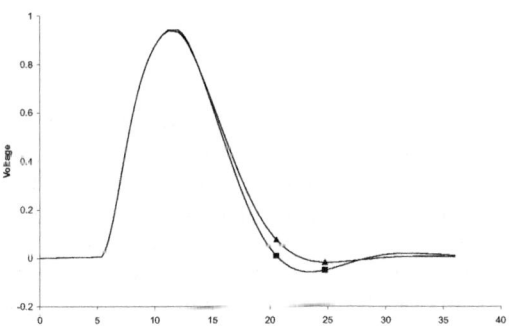

Figure 3 Comparison for disturbance rejection.

Automatic Design of Multivariable QFT Control System via Evolutionary Computation

K. C. Tan, T. H. Lee, and E. F. Khor

Department of Electrical Engineering
National University of Singapore
10 Kent Ridge Crescent Singapore 119260
{eletankc, eleleeth, engp8626}@nus.edu.sg

Abstract. This paper proposes a multi-objective evolutionary automated design methodology for multivariable QFT control systems. Unlike existing manual or convex optimisation based QFT design approaches, the 'intelligent' evolutionary technique is capable of automatically evolving both the nominal controller and pre-filter simultaneously to meet all performance requirements in QFT, without going through the conservative and sequential design stages for each of the multivariable sub-systems. In addition, it avoids the need of manual QFT bound computation and trial-and-error loop-shaping design procedures, which is particularly useful for unstable or non-minimum phase plants for which stabilising controllers maybe difficult to be synthesised. Effectiveness of the proposed QFT design methodology is validated upon a benchmark multivariable system, which offers a set of low-order Pareto optimal controllers that satisfy all the required closed-loop performances under practical constraints.

1   Introduction

Quantitative Feedback Theory (QFT) is well-known as an efficient frequency domain controller design methodology that utilises Nichols chart to achieve a desired robust design over specified ranges of structured plant parameter uncertainties with and without control effector failures [1-3]. The basic idea of QFT is to convert design specification on closed-loop response and plant uncertainty into robust stability and performance bounds on open-loop transmission of the nominal system as shown in Fig. 1. A fixed structure controller $G(s)$ and pre-filter $F(s)$ is then synthesized using gain-phase loop-shaping technique so that the two-degree-freedom output feedback system is controlled within specification for any member of the plant templates. For multi-input multi-output (MIMO) systems, conventional QFT method requires the design process to be turned into a sequence of multi-input single-output (MISO) problems before any QFT design procedure can be performed [1-3]. Giving a combined solution of controller $G(s) = diag[g_i(s)]$ and pre-filter $F(s) = [f_{ij}]$, $\forall\, i, j = 1$, 2, ..., m for an m input and m output control problem, the solution of the first set of MISO problem is the first transfer function of the diagonal controller $g_1(s)$ and $f_{11}(s)$.

S. Cagnoni et al. (Eds.): EvoWorkshops 2000, LNCS 1803, pp. 178-194, 2000.

Then a new set of MISO problem is defined based on the plant, $g_1(s)$, $f_{11}(s)$, $f_{12}(s)$,..., $f_{1m}(s)$ which results to the latter solution of the controller $g_2(s)$ and pre-filter $f_{21}(s)$, $f_{22}(s)$, ..., $f_{2m}(s)$ and etc., Hence, this design method leads to an overall design of m^2 MISO equivalent loops and its size grows exponentially to the m number of inputs/outputs, which can be very tedious to the designer [3].

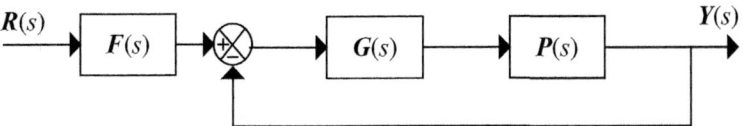

Fig. 1. A typical output feedback multivariable control system in QFT

As pointed out in [4], there are no other design methods than the manual trial-and-error process could be employed to determine these series of loops. Moreover, this sequential MISO design procedure may be conservative since the solution of a set of MISO is highly dependent and restricted to the former sets of MISO solution. The user may thus have to repeat or re-start the design procedure from the first MISO loop synthesis, if there exist any set of MISO solutions that is unfeasible due to improper or overly design of the previous sets of MISO loops. Besides, QFT bounds in Nichols chart for each specification of all frequency points must be acquired before the design, which is often an exhaustive trial-and-error process. The reason is that, for every frequency point with sufficiently small frequency interval, the template needs to be manually shifted up or down on the Nichols chart until the gain variation of the template is equal to the gain variation allowed for any particular robust specification at that frequency. In addition, only the controller can be synthesized via QFT bound computation using the conventional loop-shaping method. Another independent design task has to be accomplished in order to obtain the pre-filter within a two-stage design framework for each set of MISO solution.

The acquisition of an optimal QFT controller is in fact a multi-objective multi-modal design optimisation problem that involves simultaneously determining multiple controller and pre-filter parameters to satisfy different competing performance requirements, such as sensitivity bounds, cross-coupling bounds, robust margin and etc., To solve these problems, a few analytical/mathematics oriented optimisation or 'automatic design' techniques have recently been investigated and developed [5-7]. These convex based optimisation approaches, however, impose many unpractical or unrealistic assumptions that often lead to very conservative designs or are hard in finding the global Pareto optimal solutions in the multi-objective multi-dimensional design space. With the view of tackling these drawbacks and automating the QFT design procedure, computerised 'intelligent' trial-and-error based methodology based on evolutionary optimisation has been proposed and successfully applied to industrial or benchmark applications [8-10].

This paper further develops the multi-objective 'intelligent' automated QFT design methodology to MIMO control system using a high performance evolutionary algorithm toolbox [11]. Unlike existing methods, the evolutionary QFT design approach is capable of concurrently evolving the controller and pre-filter for the entire set of MISO sub-systems to meet all performance requirements in QFT, without going through the conservative and sequential design stages for each of the MISO sub-systems. Besides, the evolutionary toolbox is built with comprehensive user interface and powerful graphical displays for easy assessment of various simulation results or trade-offs among the different design specifications. The paper is organized as follows: The various QFT design specifications and the role of the MOEA toolbox in the multivariable QFT design are given in Section 2. Validation of the proposed methodology against a benchmark MIMO system is illustrated in Section 3. Conclusions are drawn in Section 4.

2.   Evolutionary Automated Multivariable QFT Design

2.1 Multi-Objective QFT Design Specifications

As mentioned in the Introduction, there are a number of usually conflicting design objectives need to be satisfied concurrently in multivariable QFT designs. In contrast to the conventional two-stage loop-shaping approach, these performance requirements are formulated as a multi-objective design optimisation problem here. The aim is thus to concurrently design the nominal controller $G(s)$ and pre-filter $F(s)$ in order to satisfy all the required specifications as described below:

(i) Stability (RHSP)

The cost of stability, *RHSP*, is included to ensure stability of the closed-loop system, which could be evaluated by solving the roots of the characteristic polynomial. Clearly, a stable closed-loop system for all the plant templates $\wp$ requires a zero value of *RHSP*,

$$RHSP = \sum_i Nr\left\{real\left[pole\left(\frac{P_iGF}{I+P_iG}\right)\right]\right\} > 0, \ \forall P_i \in \wp \tag{1}$$

In order to ensure internal stability and to guarantee no unstable pole and non-minimum phase zero cancellations, it is desired that a minimum phase and stable controller be designed. This implies that the search range for a polynomial coefficient set is limited to either the first or the third 'quadrant', i.e., all coefficients in the numerator or denominator must be of the same sign [10]. The poles and zeros of the controllers can be calculated explicitly to avoid RHP cancellations or alternatively, the Horowitz method for QFT design of unstable and non-minimum phase plants can be used, i.e., QFT bounds for an unstable/non-minimum phase nominal plant can be translated to those for a stable and minimum phase plant, if necessary.

(ii) Robust Upper and Lower Tracking Performance (ERRUT & ERRLT)

The cost of upper tracking performance of the ith diagonal element of MIMO closed-loop transfer function, given by $ERRUT_{(i,i)}$, is included to address the specification of upper tracking bound as shown in Fig. 2. It is computed as the sum of absolute error at each frequency point as given by,

$$ERRUT_{(i,i)} = \sum_{k=1}^{n}\left|e_{(i,i)ut}(\omega_k)\right| \tag{2}$$

where n is the total number of interested frequency points; $e_{(i,i)ut}(\omega_k)$ is the difference between the upper bound of the (i,i) element of the closed-loop transfer function $CL_{(i,i)U}$ and the pre-specified upper tracking bound $T_{(i,i)U}$ at frequency ω_k, if the upper bound of the closed-loop system is greater than the pre-specified upper tracking bound or less than the pre-specified lower tracking bound $T_{(i,i)L}$; otherwise, $e_{(i,i)ut}(\omega_k)$ is equal to zero as illustrated in Fig. 2, for which the length of the vertical dotted lines at each frequency ω_k represents the magnitude of $e_{(i,i)ut}(\omega_k)$.

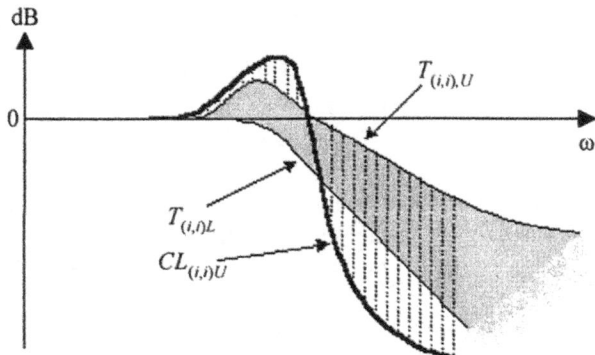

Fig. 2. Computation of upper tracking performance for the ith diagonal element

The cost for lower tracking performance of the ith diagonal element of the closed-loop transfer function, given by $ERRLT_{(i,i)}$, can be defined as the sum of absolute error at each frequency point,

$$ERRLT_{(i,i)} = \sum_{k=1}^{n}\left|e_{(i,i)lt}(\omega_k)\right| \qquad (3)$$

where n is the number of frequency points; $e_{(i,i)lt}$ is the difference between the lower bound of the closed-loop system $CL_{(i,i)L}$ and the pre-specified lower tracking bound $T_{(i,i)L}$, if the lower bound of the closed-loop system is greater than the pre-specified upper tracking bound $T_{(i,i)U}$ or less than the pre-specified lower tracking bound $T_{(i,i)L}$; Otherwise, $e_{(i,i)lt}$ is equal to zero as illustrated in Fig. 3.

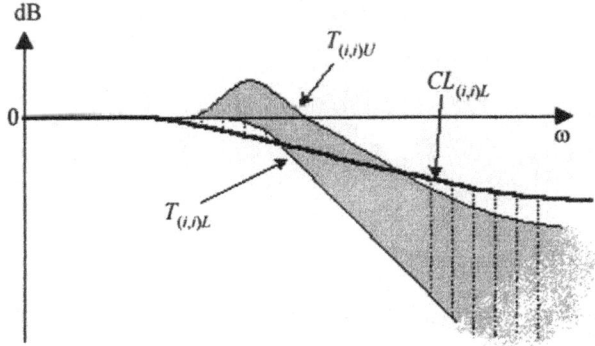

Fig. 3. Computation of lower tracking performance for the ith diagonal element

(iii) Cross-coupling Performance (ERRUC)

Apart from addressing the tracking performance of diagonal elements, it is also essential to reduce the coupling effect of the off-diagonal transfer functions for all the plant templates in MIMO QFT control system design. Since the objective is to reduce the gain and bandwidth of the off-diagonal transfer function, only the upper bounds in the frequency response need to be prescribed [4]. The upper bound of coupling effect represented by transfer function $T_{(i,j)U}$ for the off-diagonal transfer functions (i,j) where $i \neq j$ can be defined according to the allowable gain K and the bandwidth between w_1 and w_2, which is shown in Fig. 4 and takes the form of

$$T_{(i,j)U}(s) = \frac{K(\dfrac{1}{w_1}s)}{\left(\dfrac{1}{w_1}s+1\right)\left(\dfrac{1}{w_2}s+1\right)} \tag{4}$$

The cost of cross-coupling effect for the off-diagonal elements of MIMO closed-loop system, given by $ERRUC_{(i,j)}$, is include to address the specification of upper cross-coupling bound and is computed as the sum of absolute error at each frequency point,

$$ERRUC_{(i,j)} = \sum_{k=1}^{n}\left|e_{(i,j)uc}(\omega_k)\right|, \text{ for } i \neq j \tag{5}$$

where n is the number of frequency points; $e_{(i,j)uc}(\omega_k)$ at ω_k is the difference between the upper bound of the closed-loop system $CL_{(i,j)U}$ and the pre-specified upper cross-coupling bound $T_{(i,j)U}$, if the upper bound of the closed-loop system is greater than the pre-specified upper cross-coupling bound $T_{(i,j)U}$ as illustrated in Fig. 4; Otherwise, $e_{(i,j)uc}$ is equal to zero as given by,

$$e_{(i,j)uc}(\omega_k) = \begin{cases} CL_{(i,j)U}(\omega_k) - T_{(i,j)U}(\omega_k) & , CL_{(i,j)U}(\omega_k) > T_{(i,j)U}(\omega_k) \\ 0 & otherwise \end{cases}, \text{ for } i \neq j \tag{6}$$

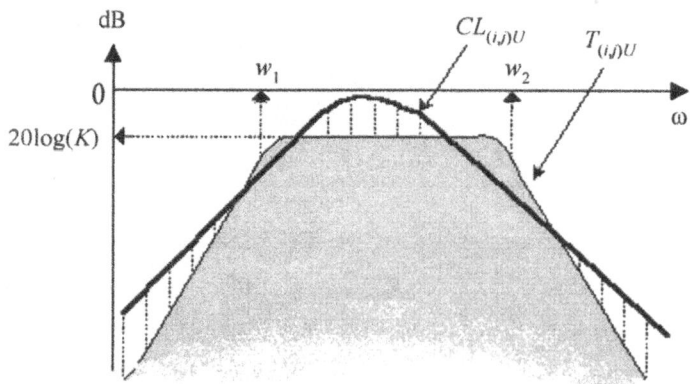

Fig. 4. Computation of upper cross-coupling performance for off-diagonal elements

(iv) Robust Margin (RM)
Practical control applications often involve neglected uncertainties or unmodelled dynamics at the high frequency. Because of these missing dynamics, the nominal model used in control system design is often inaccurate either through deliberate neglect or due to the lack of understanding of the physical process [12]. In order to address these unmodelled uncertainties, the uncertain feedback system with inverse multiplicative plant uncertainty, $P_{ip}(s) = P(s)\{I + W_{il}(s)\Delta_{il}\}^{-1}$ as shown in Fig. 5 is consdired in the QFT design. The robust margin specification that addresses the closed-loop stability due to the inverse multiplicative plant uncertainty for an uncertainty weighting function W_{il} can be defined as [12],

$$RM_{IM} = \left| \frac{1}{I + L_i(j\omega)} \right| < \frac{1}{W_{il}(j\omega)}, \quad \forall \omega \tag{7}$$

where $L_i(j\omega)$ is the ith open-loop transfer function with the jth loop being closed in an MIMO system, which is simply the loop transmission $P(j\omega)G(j\omega)$ in an SISO system.

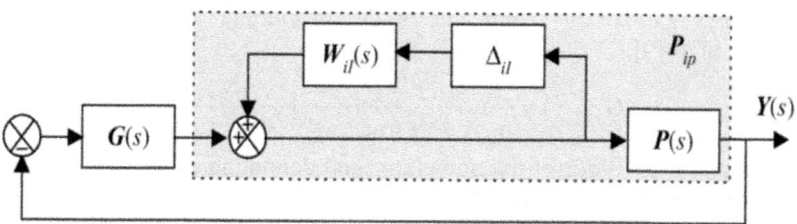

Fig. 5. Feedback system with inverse multiplicative uncertainty

(v) Sensitivity Rejection (RS)
The robust sensitivity rejection is to find a QFT controller that minimises the maximum amplitude of the regulated output over all possible disturbances of bounded magnitude. A general structure to represent the disturbance rejection is given in Fig. 6, which depicts the particular case where the disturbance enters the system at the plant output. The mathematical representation is given by,

$$S = \frac{Y}{D} = \{I + P(s)G(s)\}^{-1} \tag{8}$$

The matrix $S(s)$ is known as the *disturbance rejection*. The maximum singular values of S determines the disturbance attenuation since S is in fact the closed-loop transfer from disturbance D to the plant output Y.

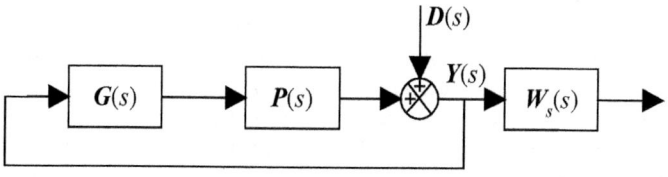

Fig. 6. Formulation of a sensitivity rejection problem

The disturbance attenuation specification for the closed-loop system may thus be written as,

$$\overline{\sigma}(S) \leq \left\| W_s^{-1} \right\|_\infty \quad \Rightarrow \left\| W_s S \right\|_\infty < 1 \qquad (9)$$

where $\overline{\sigma}$ defines the largest singular value and W_s the desired disturbance attenuation factor, which is a function of frequency to allow a different attenuation factor at each frequency.

(vi) High Frequency Gain Roll-off (HFG)
The high frequency gain performance, *HFG*, is included to reduce the gain of loop transmission $L(s)$ at the high frequency in order to avoid the high-frequency sensor noise and the unmodelled high-frequency dynamics/harmonics that may result in actuator saturation and instability. The high frequency gain of loop transmission $L(s)$ is given as,

$$\lim_{s \to \infty} s^r L(s) \qquad (10)$$

where r is the relative order of $L(s)$. Since only the controller in the loop transmission is to be optimised, this performance requirement is equivalent to the minimization of high frequency gain of the controller or the magnitude of b_n/a_m for a controller structure given as [9],

$$G_{(i,i)}(s) = \frac{b_{(i,i)n}s^n + b_{(i,i)n-1}s^{n-1} + \cdots b_{(i,i)0}}{a_{(i,i)m}s^m + a_{(i,i)m-1}s^{m-1} + \cdots a_{(i,i)0}} \qquad (11)$$

where n and m is the order of the numerator and denominator for the (i,i) element of diagonal controller $G(s)$, respectively.

2.2 Evolutionary Algorithm Toolbox and Its Role in QFT Design

Although the multi-objective optimisation based QFT design method has the merit of avoiding conventional independent two-stage controller synthesis or the tedious sequential designs to determine the series of loops for each of the MISO sub-system, the approach needs to search for multiple optimised controller and pre-filter coefficients to satisfy a set of non-commensurable and often competing design specifications. Such an optimisation problem is often semi-infinite and generally not everywhere differentiable [9]. It is thus hard to be solved via traditional numerical approaches that often rely on a differentiable performance index, which forms the major obstacle for the development of a generalised numerical optimisation package for QFT control applications.

This paper proposes an evolutionary automated design methodology for the multi-objective QFT control optimisation problem. Fig. 7 shows a general architecture for the computer aided control system design (CACSD) automation of MIMO QFT control system using a multi-objective evolutionary algorithm (MOEA) toolbox. The design cycle accommodates three different modules: the interactive human decision-making module (control engineer), the optimisation module (MOEA toolbox) and the QFT control module (system and specifications). According to the system performance requirements and *a-priori* knowledge on the problem on-hand if any, control engineers may specify or select the desired QFT specifications as discussed in previous sections to form a multi-objective function, which need not necessary be convex or differentiable. Based on these design specifications, responses of the control system consists of the set of input/output signals, the plant template as well as the candidate controller $G(s)$ and pre-filter $F(s)$ recommended from the optimisation module are simulated as to determine the different cost values for each design specification in the multi-objective function.

According to the evaluation results of the multi-objective function in the control module and the design guidance such as goal or priority information from the decision-making module, the optimisation module (MOEA toolbox) automates the QFT design process and intelligently searches for the 'optimal' controller and pre-filter parameters simultaneously that best satisfy the set of QFT performance specifications. On-line optimisation progress and simulation results, such as the design trade-offs or convergence are displayed graphically and feedback to the decision-making module. In this way, the overall QFT design environment is supervised and monitored effectively, which helps control engineers to make appropriate actions such as examining the competing design trade-offs, altering the design specifications, adjusting goal settings that are too stringent or generous, or even modifying the QFT control and system structure if necessary. This man-machine interactive design and optimisation process maybe proceeded until the control engineer is satisfied with the required performances or after the design specifications have been met. Such an evolutionary automated approach allows the QFT design problem as well as the interaction with optimisation process to be closely linked to the environment of that particular application. Control engineer, for most of the part, is not required to deal with any details that are related to the optimisation algorithm or to go through the manual trial-and-error two-stage and sequential design as adopted in conventional QFT design methods.

The MOEA toolbox [11] has been developed under the Matlab [13] programming environment, which is effective for global optimisation and assessment of multi-objective design trade-off scenarios, aiding at decision-making for an optimal solution that best meets all design specifications. It is also capable of handling problems with constraints and incorporating advanced goal and priority information with logical AND/OR operations for higher-decision support. Besides, it is fully functioned with graphical user interface (GUI) and is ready for immediate use with minimal knowledge on evolutionary computing or Matlab programming. The toolbox also allows the different representation of simulation results in various formats, such as text files or graphical displays for the purpose of on-line viewing and analysis. With the toolbox, designer merely needs to give a model file relating to his/her particular optimisation problem, and configures the problem based on a few simple GUI setups. Further descriptions of the toolbox and GUIs maybe referred to [11] or the tutorials in the toolbox, which is freely available for downloading at http://web.singnet.com.sg/~kaychen/moea.htm.

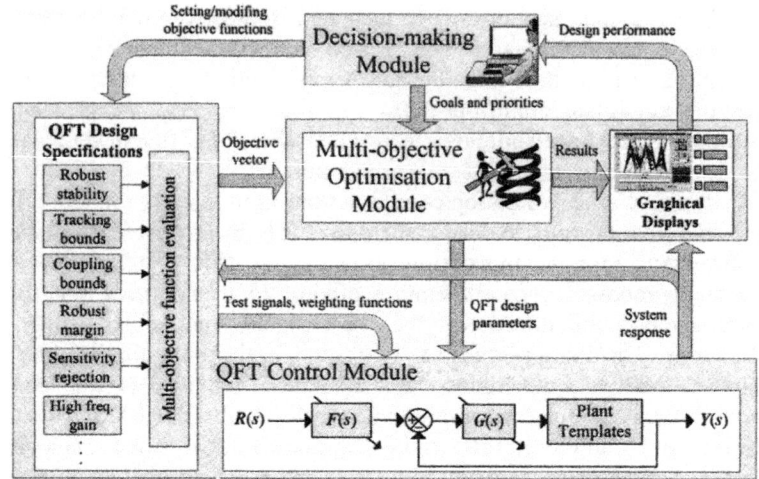

Fig. 7. A general evolutionary design automated QFT control framework

3.   A Benchmark MIMO QFT Design Problem

The benchmark MIMO QFT control problem given in [14] is studied in this section, which is shown in Fig. 1 with the MIMO uncertain plant sets given as,

$$P(s) = \begin{bmatrix} \dfrac{a}{\Lambda(s)} & \dfrac{3+0.5a}{\Lambda(s)} \\ \dfrac{1}{\Lambda(s)} & \dfrac{8}{\Lambda(s)} \end{bmatrix} \tag{12}$$

where $\Lambda(s) = s^2 + 0.03as + 10$ and $a \in [6, 8]$. Apart from the few design speciciations studied by [14], additional performance requirements such as robust tracking and cross-coupling specifications are included here for wider consideration of the QFT design objectives, which subsequently adds to the design difficulty and complexity. The specification of high frequency gain [9, 10] is also incorporated to avoid any high-frequency sensor noise and unmodelled high-frequency dynamics/harmonics. The various closed-loop performance requirements for this MIMO QFT design are formulated as follows:

(i)   Robust Tracking Bounds for diagonal transfer functions:

$$T_{(i,i)L}(\omega) \le \left| CL_{(i,i)}(j\omega) \right| \le T_{(i,i)U}(\omega) \text{, for } i = 1, 2 \tag{13}$$

Upper Tracking Model:

$$T_{(1,1)U}(\omega) = \left| \frac{1.9 \times 10^4 (j\omega) + 6.4 \times 10^5}{(j\omega)^3 + 2.3 \times 10^2 (j\omega)^2 + 1.9 \times 10^4 (j\omega) + 6.4 \times 10^5} \right| \tag{14a}$$

$$T_{(2,2)U}(\omega) = \left| \frac{6.4 \times 10^3 (j\omega) + 3.4 \times 10^5}{(j\omega)^3 + 1.5 \times 10^2 (j\omega)^2 + 8 \times 10^3 (j\omega) + 3.4 \times 10^5} \right| \tag{14b}$$

Lower Tracking Model:

$$T_{(1,1)L}(\omega) = \left| \frac{1 \times 10^6}{(j\omega)^3 + 3 \times 10^2 (j\omega)^2 + 3 \times 10^4 (j\omega) + 1 \times 10^6} \right| \tag{15a}$$

$$T_{(2,2)L}(\omega) = \left| \frac{2.5 \times 10^5}{(j\omega)^3 + 2.3 \times 10^2 (j\omega)^2 + 1.5 \times 10^4 (j\omega) + 2.5 \times 10^5} \right| \tag{15b}$$

(ii) Robust Cross-Coupling Bounds for off-diagonal transfer functions:

$$\left| CL_{(i,j)}(j\omega) \right| \le T_{(i,j)U}(\omega) \quad , \text{for } i \ne j, \text{ and } i, j = 1,2 \tag{16}$$

where,

$$T_{(1,2)U}(\omega) = \left| \frac{0.0032(j\omega)}{[0.016(j\omega)+1][0.016(j\omega)+1]} \right| \tag{17a}$$

$$T_{(2,1)U}(\omega) = \left| \frac{6.3 \times 10^{-3}(j\omega)}{[0.016(j\omega)+1][0.016(j\omega)+1]} \right| \tag{17b}$$

(iii) Robust Sensitivity Rejections for full matrix transfer functions:

$$\left| S_{i,j}(j\omega) \right| < a_{i,j}(j\omega) , \text{ for } \omega < 10 \tag{18}$$

where, $a_{i,j} = 0.01\,\omega$, for $i = j$; $a_{i,j} = 0.005\,\omega$, for $i \ne j$

(iv) Robust Stability Margin:

$$\left| \frac{1}{1 + L_{i,i}(j\omega)} \right| < 1.8 \quad , \text{for } \forall \, i = 1,2 , \text{ and } \omega > 0 \tag{19}$$

The performance bounds of QFT are computed within a wide frequency range of 10^{-2} rad/s to 10^3 rad/s. Without loss of generality, the structure of the diagonal controller $G(s)$ is chosen in the form of a general transfer function [9] as given by,

$$G_{i,i}(s) = \frac{\displaystyle\sum_{m=0}^{4} b_m s^m}{\displaystyle\sum_{n=0}^{4} a_n s^n} \quad , \forall \, b_m, a_n \in \Re^+, \text{ for } i = 1, 2 \tag{20}$$

Note that the controller can also be designed by refining position of poles and zeros directly or by using other structures such as the realisable (non-ideal) PID structure if desired. The filter is fixed to a full matrix first-order transfer function as it is relevant to the tracking and cross-coupling bound in the frequency response. Since the resultant pre-filter must satisfy $\lim_{s \to 0}[F(s)] = 1$ for a step forcing function [9], the structure of pre-filter $F(s)$ is chosen as a full matrix first-order transfer function as given by,

$$F_{i,j}(s) = \frac{1}{1 + \displaystyle\sum_{j=n}^{2} c_n s^n} \quad \forall \, c_n \in \Re^+, \text{ for } \forall \, i, j = 1, 2 \tag{21}$$

Apart from most default settings, the evolutionary toolbox has been configured with a population and generation size of 200 and 100, respectively. To guide the evolutionary QFT design optimisation process, goal and priority for each of the performance requirements maybe included optionally as shown in Fig. 8. Although determination of the goal and priority maybe a subjective matter and depends on the performance requirements, it maybe unnecessary and can be ignored for a 'minimum-commitment' design [9]. In principle, any number or combination of QFT performance specifications can be added to the design using the multi-objective evolutionary optimisation approach if necessary.

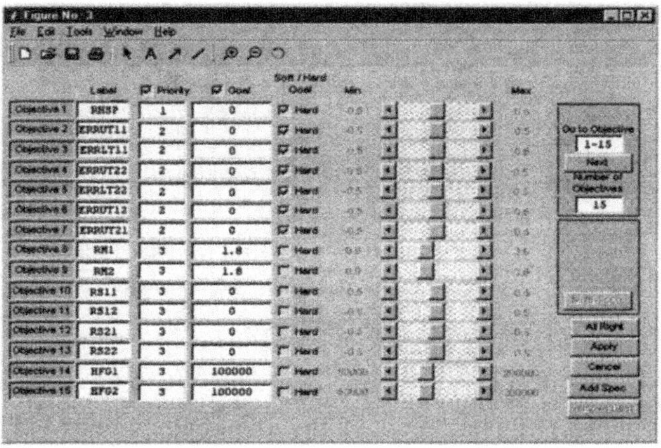

Fig. 8. Settings of the MOEA toolbox for the benchmark QFT design problem

A powerful feature of the evolutionary QFT design is that it allows on-line examination of different trade-offs among the multiple conflicting specifications, modification of existing objectives and constraints, and zoom into any region of interest before selecting one final set of controller and pre-filter for real time implementation. The trade-off graph of the resultant QFT control system is shown in Fig. 9, where each line representing a solution found by the evolutionary optimisation. The cost of objectives such as stability (*RHSP*), robust tracking and cross coupling performances (*ERRUT* and *ERRLT*) are labelled as objectives 1-7, which are all equal to zero as desired according to the goal settings in Fig. 8. The x-axis shows the design specifications, the y-axis shows the normalised cost for each objective and the cross-mark shows the desired goal setting for each performance requirement. Clearly, trade-offs between adjacent specifications results in the crossing of the lines between them, whereas concurrent lines that do not across each other indicating the specifications do not compete with one another. For example, the robust sensitivity objective of 12 (*RS*21) and 13 (*RS*22) are not competing with each other, whereas the robust margin objective 8 (*RM*1) and 9 (*RM*2) appear to compete heavily, as expected. The information contained in this trade-off graph also suggests that lower goal settings for robust sensitivity (objectives 10-13) are possible, which can be further optimised to arrive at an even better robust performance.

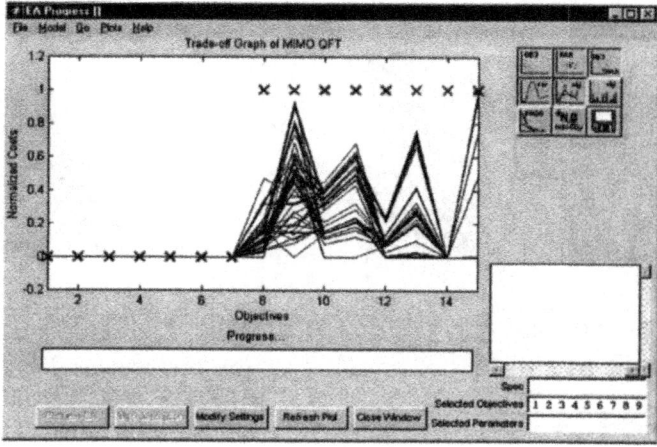

Fig. 9. Trade-off graph of the evolutionary designed QFT control system

Note that the evolutionary QFT design also allows engineers to divert the evolution to other focused trade-off region or to modify any preferences on the current specification settings after observation for a number of generations. For example, the designer can change his preference and decide to reduce the 9th goal setting for robust margin (*RM2*) from 1.8 to 1.3. Fig. 10 illustrates the behaviour of the evolution upon the modification of this goal setting after the evolutionary QFT design in Fig. 9. Due to the sudden change of a tighter goal setting, initially none of the individuals manage to meet all the required specifications as shown in Fig. 10(a). After continuing the evolution for 2 generations, the population moves towards satisfying the objective of *RM2* as shown in Fig. 10(b) at the performance expense of other objectives since they are highly correlated and competing to each other. The evolution continues and again leads to the satisfaction of all the required goal settings including the stricter setting of objective *RM2* as shown in Fig. 10(c). Clearly, this man-machine interactive design approach has enabled QFT designers to divert the evolution into any interested trade-off regions or to modify certain specifications and preferences on-line, without the need of restarting the entire design process as required by conventional QFT design methods.

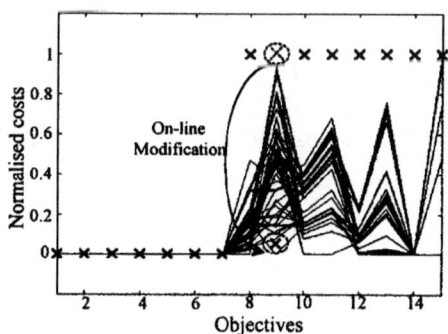

(a) On-line goal modification of robust margin objective (*RM2*)

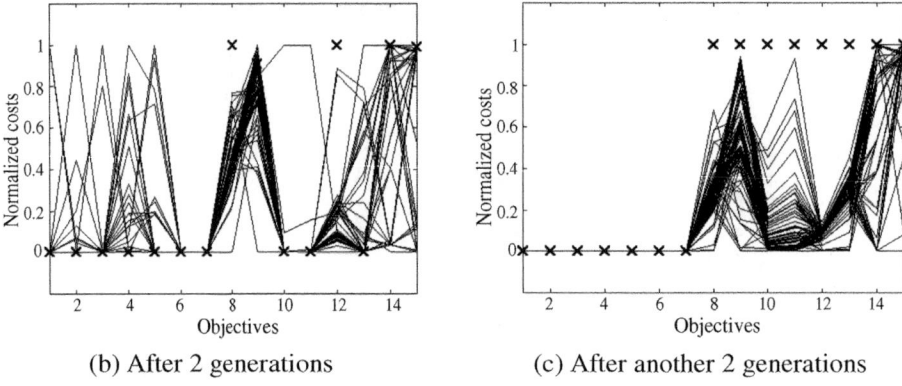

(b) After 2 generations   (c) After another 2 generations

Fig. 10. Effects of the evolution upon the on-line modification of goal setting

Fig. 11 shows the robust tracking performances in the frequency domain for the two diagonal elements of the closed-loop system. It can be seen that all the frequency responses of CL_U and CL_L for both the diagonal channels are located successfully within their respective pre-specified tracking bounds of T_U and T_L. Besides, the coupling effect from the off-diagonal elements of the closed-loop system for all the plant templates has also been reduced satisfactory and successfully bounded by the upper coupling bound with minimal gain and bandwidth of the off-diagonal transfer functions as shown in Fig. 12.

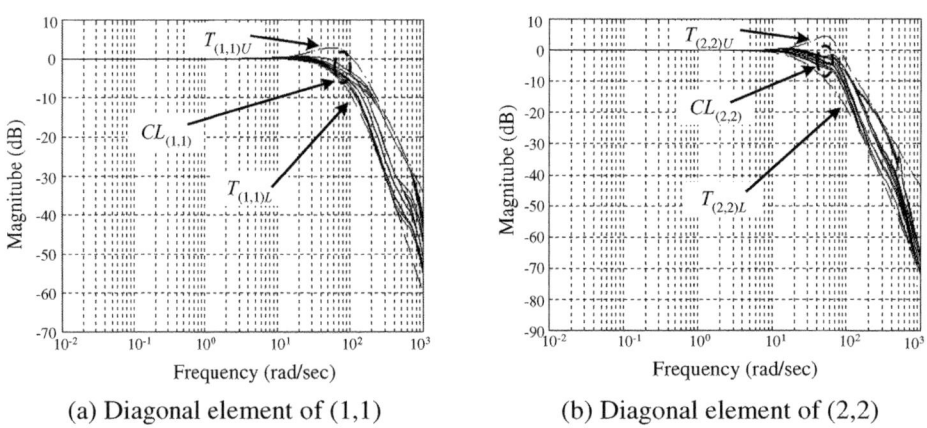

(a) Diagonal element of (1,1)   (b) Diagonal element of (2,2)

Fig. 11. The tracking performance in the frequency domain

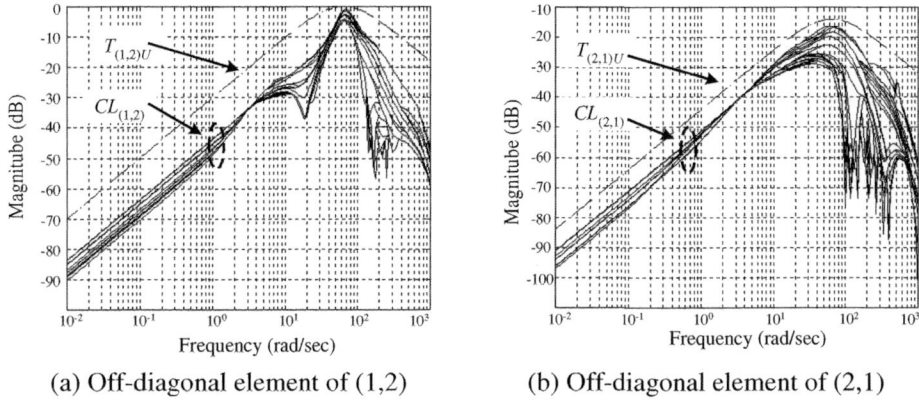

(a) Off-diagonal element of (1,2)   (b) Off-diagonal element of (2,1)

Fig. 12. The cross-coupling performance in the frequency domain

Figs. 13 and 14 show the unit step tracking and coupling performances for all the plant templates in the time domain for a random selected set of evolutionary designed controller and pre-filter. Clearly, all the time domain tracking and coupling performances have been satisfied successfully and within the required prescribed tracking bounds, as desired.

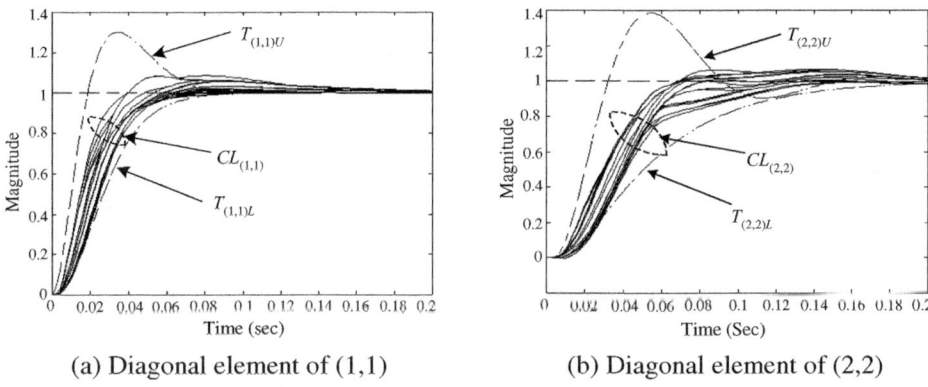

(a) Diagonal element of (1,1)   (b) Diagonal element of (2,2)

Fig. 13. The tracking responses of the diagonal elements in the closed-loop system

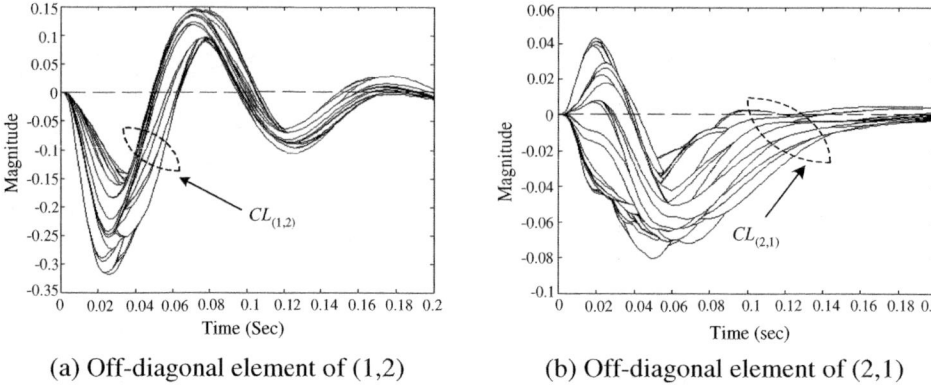

(a) Off-diagonal element of (1,2)   (b) Off-diagonal element of (2,1)

Fig. 14. The coupling responses of the off-diagonal elements in the closed-loop system

To illustrate robustness of the evolutionary designed QFT control system on disturbance rejection, a unit step disturbance signal was applied to the MIMO system. Output disturbance responses for all the final Pareto optimal controllers at each element of the closed-loop transfer matrix are illustrated in Fig. 15. Clearly, the unit step disturbance has been successfully attenuated to zero eventually for all the different values of parameter uncertainties, as quantified by the performance specification of robust sensitivity rejection.

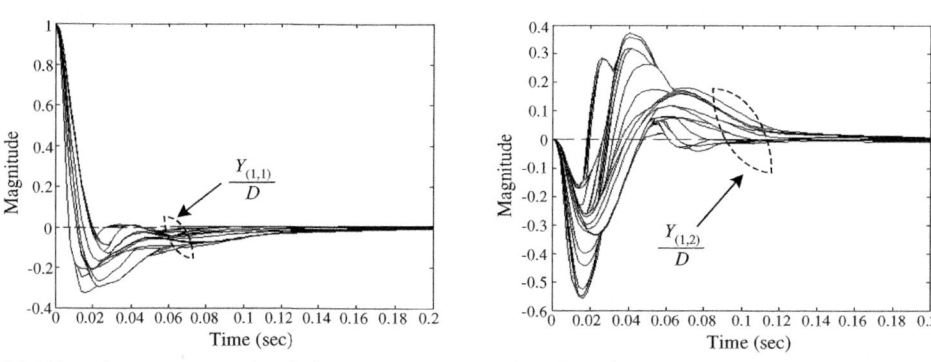

(a) Disturbance response of element (1, 1)   (b) Disturbance response of element (1, 2)

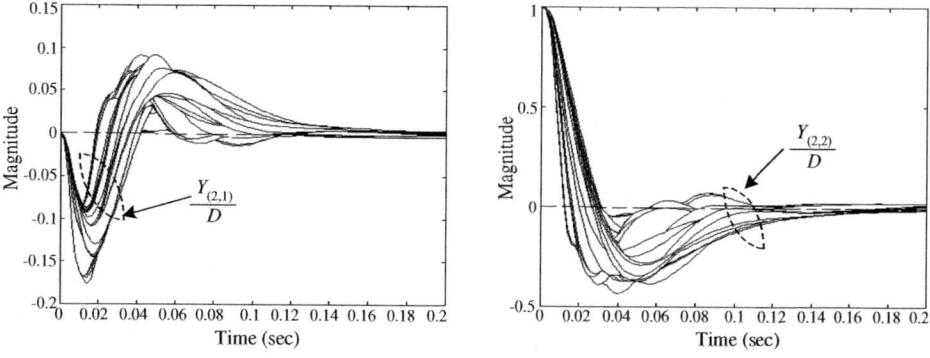

(c) Disturbance response of element (2, 1)   (d) Disturbance response of element (2, 2)

Fig. 15. Output responses for the unit step disturbance in the time domain

4   Conclusion

This paper has analysed difficulties in existing QFT design techniques for multivariable control systems. To address these design deficiencies, an automated multivariable QFT design methodology using a high performance MOEA toolbox has been proposed. Unlike existing design methods, the 'intelligent' design approach is capable of automatically evolving both nominal controller and pre-filter simultaneously to meet all performance requirements in QFT, without going through the sequential design stages for each of the multivariable sub-systems. Besides, the approach also avoids the need of manual QFT bound computation and trial-and-error loop-shaping procedures as required by conventional means. It is shown that control engineers' expertises as well as goal and priority information can be easily included and modified on-line according to the evolving trade-offs, instead of repeating or restarting the whole design process. It is obvious that the proposed evolutionary QFT design framework is fully expandable to on-line design optimisation and implementation. This can be realised either via the hard and software systems such as dSPACE [15] or MIRCOS [16] for graphical programming and real-time operation to provide necessary linkages between the toolbox and the physical environments. Apart from the developments for on-line adaptation, the multi-objective evolutionary QFT design paradigm is currently being extended to robust control of nonlinear systems and to incorporate other design specifications such as economical cost consideration. Progress and results will be reported in due course.

References

1. Yaniv, O., Horowitz, I.: A Quantitative design method for MIMO linear feedback system having uncertain plants. *Int. J. Control*, vol. 43, no. 2, pp. 401-421, 1986.
2. Yaniv, O., Schwartz, B.: A Criterion for loop stability in the Horowitz Synthesis of MIMO feedback systems. *Int. J. Control*, vol. 53, no. 3, pp. 527-539, 1990.
3. Houpis, C. H.: Quantitative feedback theory (QFT) technique. In Levine, W. S., ed. (1996). *The Control Handbook*, CRC Press & IEEE Press, pp. 701-717, 1993.

4. Snell, S. A., Hess, R. A.: Robust, decoupled, flight control design with rate saturating actuators. *Conf. and Exhibit. On AIAA Atmospheric Flight Mechanics*, pp. 733-745, 1997.
5. Thompson, D. F., and Nwokah, O. D. I.: Analytical loop-shaping methods in quantitative feedback theory, *J. Dynamic Systems, Measurement and Control*, vol. 116, pp. 169-177, 1994.
6. Bryant, G. F., and Halikias, G. D.: Optimal loop-shaping for systems with large parameter uncertainty via linear programming, *Int. J. Control*, vol. 62, no. 3, pp. 557-568, 1995.
7. Chait, Y.: QFT loop-shaping and minimisation of the high-frequency gain via convex optimisation, *Proc. Sym. Quantitative Feedback Theory and other Freq. Domain Method and Applications*, Glasgow, Scotland, pp. 13-28, 1997.
8. Chen, W. H., Ballance, D. J. Li, Y.: Automatic loop-shaping in QFT using genetic algorithms. *Proc. of 3rd Asia-Pacific Conf. on Cont. & Meas.*, pp. 63-67, 1998.
9. Tan, K. C., Lee, T. H. Khor, E. F.: Control system design automation with robust tracking thumbprint performance using a multi-objective evolutionary algorithm", *IEEE Int. Conf. Control Appl. and Sys. Design*, Hawaii, 22-26th August, pp. 498-503, 1999.
10. Chen, W. H., Ballance, D. J., Feng, W., and Li, Y.: Genetic algorithm enabled computer-automated design of QFT control systems, *IEEE Int. Conf. Control Appl. and Sys. Design*, Hawaii, 22-26th August, pp. 492-497, 1999.
11. Tan, K. C., Wang, Q. G., Lee, T. H., Khoo, T. T., and Khor, E. F.: *A Multi-objective Evolutionary Algorithm Toolbox for Matlab*, (http://vlab.ee.nus.edu.sg/~kctan/moea.htm), 1999.
12. Skogestad, S., Postlethwaite, I.: *Multivariable Feedback Control: Analysis and Design*. John Wiley & Sons Ltd, West Sussex. England, 1996.
13. The Math Works, Inc.: *Using MATLAB*, version 5, 1998.
14. Borghesani, C., Chait, Y. and Yaniv, O.: *Quantitative Feedback Theory Toolbox User Manual*, The Math Work Inc, 1995.
15. Hanselmann, H.: Automotive control: From concept to experiment to product, *IEEE Int. Conf. Contr. Appl. and Sys. Des.*, Dearborn, 1996.
16. Rebeschieß, S.: MIRCOS - Microcontroller-based real time control system toolbox for use with Matlab/Simulink, *IEEE Int. Conf. Contr. Appl. and Sys. Design*, Hawaii, USA, pp. 267-272, 1999.

Development of Power Transformer Thermal Models for Oil Temperature Prediction

W H Tang[1], H Zeng[2], K I Nuttall[1], Z Richardson[3], E Simonson[3], and Q H Wu[1]

[1] Department of Electrical Engineering and Electronics
The University of Liverpool, Liverpool, L69 3GJ, U.K.
q.h.wu@liv.ac.uk
[2] Electric Power Research Institute
Qinghe, Beijing 100085, P.R.China
[3] Engineering and Technology
The National Grid Company plc, U.K.

Abstract. This paper describes a new thermal model of oil-immersed, forced-air cooled power transformers and a methodology for model construction using intelligent learning applied to on-site measurements. The model delivers the value of bottom-oil and top-oil temperatures for thermal performance prediction and on-line monitoring of power transformers. The results obtained using the new thermal model are compared with the results of a traditional thermal model and the results derived from artificial neural networks.

1 Introduction

On-line monitoring of power transformers opens the possibility for extending the operating time of power transformers, reducing the risk of expensive failures and providing potential for changing the maintenance strategy [1] ~ [3]. The useful life of a transformer is determined in part by the ability of the transformer to dissipate the internally generated heat to its surroundings. Consequently, the comparison of actual and predicted operating temperatures can provide a sensitive measure of transformer condition and might indicate abnormal operation. Modeling transformer thermal dynamics is regarded as one of the most important issues and construction of an accurate thermal model is an important aspect of transformer condition monitoring. The generally accepted methods [5] [6], can be used to predict zones of excess temperature in a transformer. However, the conventional calculation of internal transformer temperature is not only a complicated and difficult task but also leads to a conservative estimate based on some assumptions of the operating conditions. Its ability to predict the transformer temperature under realistic loading conditions is therefore somewhat limited.

In this paper, two different intelligent learning methods, genetic algorithm (GA) and artificial neural network (ANN), are used to construct thermal models from the on-site measurements. The study shows that intelligent learning methods can predict online transformer temperatures in real time with greater

S. Cagnoni et al. (Eds.): EvoWorkshops 2000, LNCS 1803, pp. 195–204, 2000.

accuracy than that obtained using the traditional models. The developed thermal model could be used as the basis of an intelligent protection system, as well as an element in a remote supervising and control system.

2   Thermal Models

Considering the operating regimes of cooling systems and the actual measurements available, two important temperature measurements, bottom-oil temperature (BOT) and top-oil temperature (TOT), are chosen for the purpose of transformer condition monitoring. In the following, two thermal models for temperature prediction and condition monitoring of transformers are described.

2.1   Steady-state Models

For unpumped (or 'natural') oil cooling of power transformers (ON), the oil temperature at the top of windings is approximately equal to the TOT inside the tank. However, for forced oil circulation (OF), the TOT is the sum of the oil temperature at the bottom of the winding, BOT, and the difference between oil temperatures at the top and bottom of the winding [4] ∼ [7].

The BOT is described as follows:

$$\theta'_{BO} = \theta_a + \theta_{bo} \left(\frac{1 + dK^2}{1 + d} \right)^x . \tag{1}$$

and the TOT is represented by:

$$\theta'_{TO} = \theta_a + \theta_{bo} \left(\frac{1 + dK^2}{1 + d} \right)^x + (\theta_{to} - \theta_{bo})K^y . \tag{2}$$

where

θ_a = ambient temperature

θ'_{BO} = steady state oil temperature at the bottom of the winding with the operating load

θ_{bo} = oil temperature rise above ambient at the bottom of the winding under the rated load

d = the ratio of load loss (at rated load) to no-load loss

K = the ratio of the operating load current to rated load current, $K = \frac{I}{I_R}$

θ'_{TO} = steady state oil temperature at the top of the winding with the operating load

θ_{to} = oil temperature rise above ambient at the top of the winding under the rated load

x = exponent related to oil temperature rise due to total losses

y = exponent related to windings temperature rise due to the load current.

All the parameters involved in this model, such as d, x and y, are usually determined through experiment or by experience.

2.2   Transient-state Models Applicable to Fan Operation

In contrast to the winding temperature, the transient BOT and TOT (θ'_{bo} and θ'_{to} respectively) cannot immediately reach the corresponding steady-state values under changing loads, since their thermal time constants are in the order of hours [8]. We choose a recursive form of the model using the previous samples to represent the BOT and TOT respectively, which can reflect changes of the thermal time constants due to different operating conditions of the power transformer.

Fan assisted oil cooling of transformers is activated automatically according to the oil temperature. When the oil temperature increases and exceeds a certain value, θ'_0, fans will be switched on. To distinguish the different thermal dynamics appropriate to the periods when the fans are on and when they are off, a two-piece model is introduced. The following model is employed to predict the BOT and TOT that can be expressed by the combination of a recursive model and a fan function.

The BOT at time instant t is:

$$\theta'_{bo}(t) = C_b \theta_a(t-2) + D_b K(t-2)^{y_b} + \theta_{bo} \left(\frac{1 + d_b K(t-2)^2}{1 + d_b} \right)^{x_b}$$
$$+ A_b \theta'_{bo}(t-1) + B_b \theta'_{bo}(t-2) + E_b K(t-2) + H_b(t) . \tag{3}$$

where

$$H_b(t) = \begin{cases} F_b & \theta'_{bo}(t) > \theta'_0 \\ G_b & \text{otherwise} \end{cases}$$

The TOT is described as:

$$\theta'_{to}(t) = C_t K(t-2)(\theta_a(t-2) - D_t) + \theta_{to} \left(\frac{1 + d_t K(t-2)^2}{1 + d_t} \right)^{x_t}$$
$$+ A_t \theta'_{to}(t-1) + B_t \theta'_{to}(t-2) + H_t(t) . \tag{4}$$

where

$$H_t(t) = \begin{cases} E_t + F_t(t - t_{fan-on}) & \theta'_{to}(t) > \theta'_0 \\ G_t & \text{otherwise} \end{cases}$$

In the model (3) $\sim$ (4), the coefficients with subscript 'b' denote those associated with BOT and 't' with TOT. Each temperature is described by a two-piece model. The coefficients in the models are shown in Table 1, in which one set of coefficients are associated with the normal condition without fan operation and the other are related to the time period during which fans are switched on.

3   Modeling by Intelligent Learning

3.1   Genetic Algorithms

GA is a powerful numerical optimization technique, which is rooted in the mechanism of evolution and natural genetics. GAs derive their strengths by simulating the natural search and selection process associated with natural genetics. A set of genes in its 'chromosomes' determines every organisms identity, which is referred to as a 'string' in GA. Natural selection takes place in such a way that these characteristics are implicitly selected via the survival of the fittest criterion. This algorithm, the most popular format of which is the binary genetic algorithm, starts with setting objective functions based on the physical model of problems to calculate fitness values, and thereafter measure each binary coded string's strength with its fitness value. The stronger strings advance and mate with other stronger strings to produce offspring. Finally, the best survives. One of the important advantages is that GA could be able to find out the global minimum of fitness instead of a local solution. The GA has gained popularity in recent years as a robust optimization tool for a variety of problems in engineering, science, economics, finance, etc.[9] ~ [11]. The GA will typically employ three operators: reproduction, crossover and mutation [12]. Typically, each of these operators is applied to the population once per generation, and usually several generations are required to achieve satisfactory results.

Reproduction is a process in which an old string is carried through into a new population depending on its performance index (i.e. fitness) value. Those strings with higher fitness values tend to have a higher probability of contributing one or more offsprings to the next generation.

The crossover operator provides random information exchange. It is aimed towards evolving better building blocks (schemata with short defining lengths and high average fitness values). Crossover points are randomly chosen. The frequency of crossover is governed by a user selected crossover rate or probability of crossover. Increasing crossover rate increases recombination of building blocks, but with an increasing probability of loosing good strings.

The mutation operator is simply an occasional random alteration of a string position (based on a probability of mutation). In a binary code, this involves changing a 1 to a 0 and vice versa. The mutation operator helps in avoiding the possibility of mistaking a local minimum for a global minimum. Mutation is usually used sparingly. When mutation is used in conjunction with reproduction and crossover, it improves the global nature of the GA search.

The GA is employed to search for the optimal parameters of the thermal model. It is a binary coded implementation written in Visual C++ 5.0 and developed at the University of Liverpool. Its computation process is listed as follows:

(1) Initialize the population.
(2) Evaluate the fitness of every string in the population.
(3) Keep the best string in the population.
(4) Make a selection on the population at random.
(5) Make crossover on selected strings with probability p_c.
(6) Make mutation on selected strings with probability p_m.
(7) Evaluate the fitness of every string in the new population.
(8) Make elitism.
(9) Repeat (4) to (8) until the termination condition is met.

In the computation, the crossover probability, p_c, and the mutation probability, p_m, the size of the population and the maximum number of generations are selected 'a priori'.

3.2   Fitness Function

The fitness function should reflect the desired characteristics of the system being optimized. It is very important to use an appropriate fitness function since this drives the evolution process. To apply the GA to the power transformer thermal models formulated in Section 2, a fitness function and other relevant parameters of the binary genetic algorithm should be determined. The error between the measured variables and the model outputs is defined as fitness. Thus, the fitness function should at least contain two terms corresponding to the fitness of BOT and TOT respectively. For each individual (string) of a generation, its total fitness value is calculated as follows:

$$f = f_{bo} + f_{to} \tag{5}$$
$$= \sum_{k=1}^{n} \Delta\theta_{bo}(k)^{'2} + \sum_{k=1}^{n} \Delta\theta_{to}(k)^{'2}$$
$$= \sum_{k=1}^{n} (\theta_{bo}(k)^{'} - \theta_{mbo}(k))^2 + \sum_{k=1}^{n} (\theta_{to}(k)^{'} - \theta_{mto}(k))^2$$

where f_{bo} and f_{to} are the fitness for BOT and TOT components of the model respectively; $\Delta\theta_{bo}(k)^{'}$ and $\Delta\theta_{to}(k)^{'}$ are the errors between the real measurement and the model output under the appropriate service condition (ambient temperature, load ratio) respectively; $\theta_{mbo}(k)$ and $\theta_{mto}(k)$ represent a group of measurements and n is the total number of measurement groups. $\theta_{bo}(k)^{'}$ and $\theta_{to}(k)^{'}$ can be easily calculated by the above models.

4   Results and Analysis

4.1   Model Parameters

The calculation was based on the data provided by the National Grid Company (NGC). The data cover both natural oil, natural air (ONAN) and forced oil, forced air (OFAF) operating regimes which are switched over automatically. Two types of model have been investigated as described above: (1) Conventional steady-state model; (2) Transient-state model with recursion of calculated outputs and effect of fans, which actually fall into two groups. The first reflects the steady-state relationship between the TOT, BOT, ambient temperature and transformer load ratio; the second concerns the temperature variation over a period time as an inertial response.

As stated above, the TOT and BOT are selected as the model outputs. The difference among the models lies in the construction of the formulae for both the bottom-oil and top-oil temperatures. The measured parameters, provided by NGC, are ambient temperature θ_a, ratio of load K, TOT θ'_{to} and BOT θ'_{bo}, comprising 14000 groups of measurements with an sample interval of 1 minute. The model parameters have been optimized using the binary genetic algorithm based on 8400 groups of real measurements, and then verified on the remaining 5600 groups. The calculated results derived from model (1) and those from model (2) are listed in Table 1.

4.2   Model Evaluation

Steady-state Model The steady-state model has been tested first. It is assumed that the BOT and TOT can immediately reach their steady-state values and relate to only one service condition, expressed by equations (1) and (2) in section 2.1.

The model parameters to be determined involve d, x, y, rated BOT θ_{bo} and rated TOT θ_{to}. Normally, the parameters are obtained empirically. Therefore, together with load ratio K and ambient temperature θ_a, the responses of the empirical model can be calculated by equations (1) and (2). The comparisons between the model outputs and real outputs are shown in Figures 1 and 2 respectively. It may be noted that the steady-state model produces an unsatisfactory fitness (in Table 1) to the experimental data.

Transient-state Model Applicable to Fan Operation   As discussed in section 2.2, the BOT and TOT cannot immediately reach the corresponding steady-state values under changing loads, as their thermal time constants are of the order of hours. Therefore, their transient behaviour should be represented in a recursive form using previous samples.

The results obtained using the models relating to fan operation with recursion of calculated outputs are shown in Figures 3 and 4. As expected, these results present the best agreement with the measurements, compared with other models. If a more precise fan operation time could be provided, it is expected that an even more accurate result would be achieved.

A comparison with experimental measurements indicates the transient-state model with recursion of the outputs has a better performance than that of the steady-state model. Since it does not require the TOT and BOT measurements, this will certainly be of great practical value as a predictive tool under a wider variety transformer loading and ambient conditions.

5   Artificial Neural Network Models

ANN has been widely used to perform complex functions in various fields of application including pattern recognition, classification and control systems. The ANN technique is based on the theory of biological nervous systems and involves selection of input, output, network topology and weighted connections of the nodes [13]. In our case, the network inputs are chosen as load ratio and ambient environment temperature and the network output is the TOT or BOT of the transformer.

The training epochs are 1500. All the neuron functions are chosen as the sigmoid function except the output neuron which employs a pure linear function. 70% of total data are used to train the network and the remaining 30% are reserved for evaluation of the performance of the ANN mapping.

The direct mapping employs the ANN inputs:

$$[K(t-1),\ K(t-2),\ K(t-3),\ \theta_a(t-1),\ \theta_a(t-2),\ \theta_a(t-3)]$$

and the output $\theta'_{to}(t)$ or $\theta'_{bo}(t)$. A $6 \times 9 \times 1$ ANN is chosen for the direct mapping. The results are shown in Figures 5 and 6.

The ANN model provides a good mapping between the inputs and outputs. However it does not possess any physical meaning and it has been noted that, for direct mapping, the model only represents the input and output relationship accurately within the range covered by the training data. Outside this range, the model response and measured data do not agree with each other satisfactorily. It only demonstrates the possibility of one step prediction of the transformer temperatures. Also as the model does not have any physical meaning, it possesses little potential for condition monitoring.

6   Conclusions

The application of the models to the GA learning method, presented in this paper show that high precision results were attained when compared to the results derived from the models without recursion and the models of ANN. It can be seen that the transient-state model with recursion of calculated outputs, including the effect of fan operation, has a potential for representing accurately the real transformer thermal dynamics. Both the model parameters and responses in the model applicable to fan operation can be used for predicting thermal distribution and monitoring the service condition of transformers.

References

1. J. H. Provanzana, P. R. Gattens, Transformer condition monitoring realizing an intergrated adaptive analysis system, CIGRE 1992 session, 30 Agust-5 September.
2. T. Leibfried, W Knorr, K. Viereck, On-line montoring of power transformers-trends, new development and first experiences, CIGRE 1998 session.
3. I. J. Kemp, Partial discharge plant-monitoring technology: Present and future developments, *IEE Proceedings-Science, Measurement and Technology*, vol.142, no.1, January 1995, pp.4-10.
4. James, L.et al, Model-based monitoring of transformers, Massachusetts Institute of Technology, Laboratory for Electromagnetic and Electronic Systems, 1995.
5. Transformers Committee of the IEEE Power Engineering Society, *IEEE guide to loading mineral oil-immersed transformer, IEEE Std C57.115-1991*, The Institute of Electrical and Electronics Engineers, Inc., 1991, 345 East 47th Street, New York, NY 10017, USA.

6. International electrotechnical commission, *IEC354-loading guide for oil-immersed power transformers, equivalent to GB/T 15164-94 (Chinese Standard)*, 1991, Genève, Suisse.

7. Chinese standards board, *GB1094-85, power transformers-Part 2: Temperature rise*, 1985.

8. S. Q. Xu, Operation of power transformers, *Chinese Electric Power Publisher*, June 1993.

9. Q. H. Wu, J. T. Ma, Power system optimal reactive dispatch using evolutionary programming, *IEEE Trans on Power Systems*, vol.10, no.3, 1995, pp.1243-1249.

10. Q. H. Wu, Y. J. Cao, J. Y. Wen, Optimal reactive power dispatch using an adaptive genetic algorithm, *International Journal of Electrical Power and Energy Systems*, vol.20, no.8, 1998, pp.563-569.

11. J. T. Ma, Q. H. Wu, Generator parameter identification using evolutionary programming, *International Journal of Electrical Power and Energy Systems*, vol.17, no.6, 1995, pp.417-423.

12. P. Pillay, R. Nolan, Application of genetic algorithms to motor parameter determination for transient torque calculations, *IEEE Transactions on Industry Applications*, vol.33, no.5, September/October 1997, pp.1273-1282.

13. M. R. Zaman, Experimental testing of the artificial neural network based protection of power transformers, *IEEE Transactions on Power Delivery*, vol.13, no.2, Apirl 1998, pp.510-517.

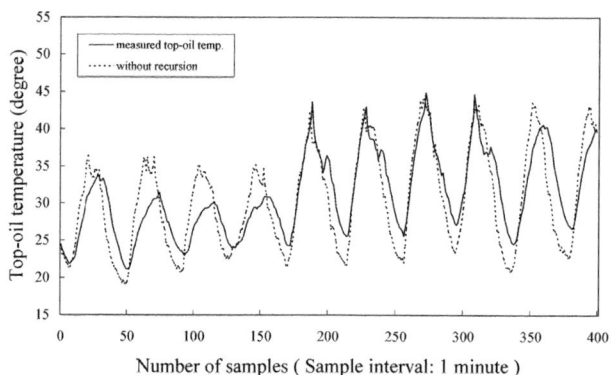

Fig. 1. Top-oil temperature of the model without recursion

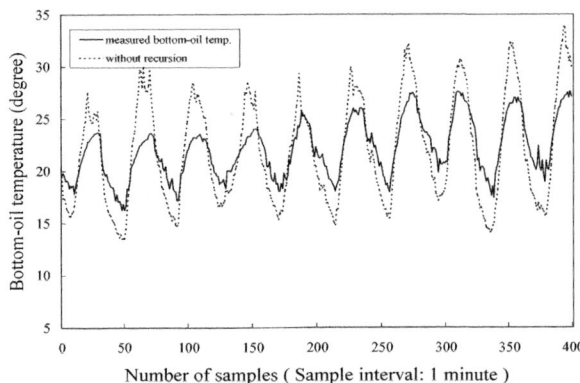

Fig. 2. Bottom-oil temperature of the model without recursion

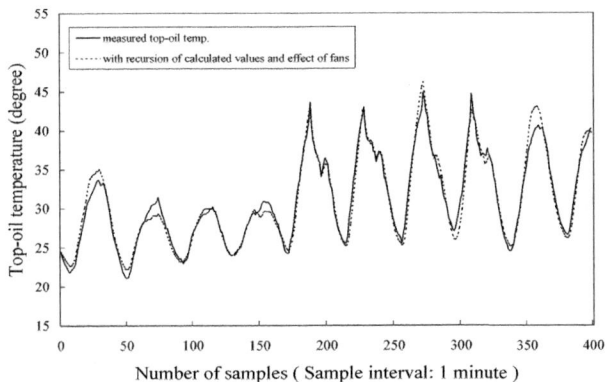

Fig. 3. Top-oil temperature of the model with recursion of calculated outputs and effect of fans

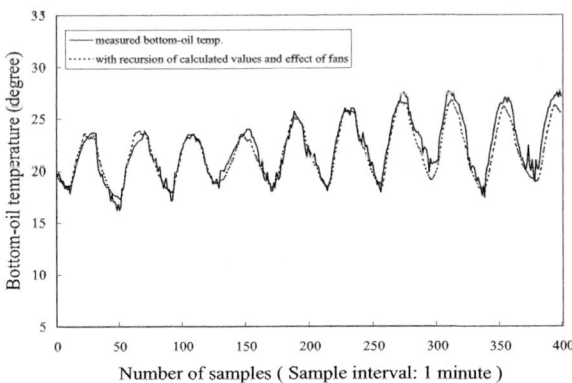

Fig. 4. Bottom-oil temperature of the model with recursion of calculated outputs and effect of fans

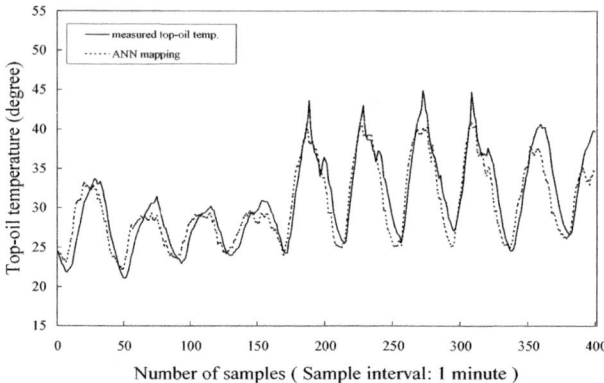

Fig. 5. Top-oil temperature of the ANN mapping

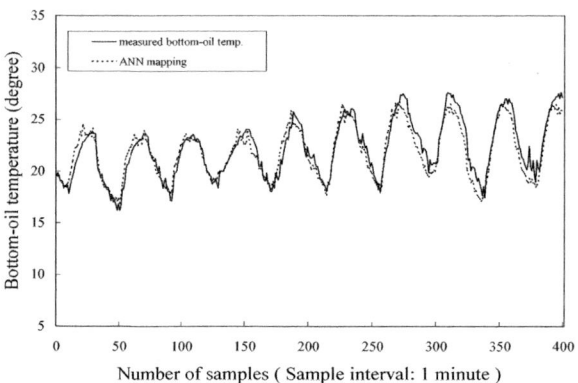

Fig. 6. Bottom-oil temperature of the ANN mapping

Table 1. Results of the models constructed by GA

Model(1) [a]	θ_{bo}	d	x	θ_{to}	y					Fitness
	4.01	0.03	0.83	15.74	0.68					158932
Model(2) [b]	θ_{bo}	d_b	x_b	A_b	B_b	C_b	D_b	E_b	F_b	
bottom	12.83	20.0	0.23	0.15	0.82	0.02	1.81	9.10	0.02	
	θ_{to}	d_t	x_t	A_t	B_t	C_t	D_t	E_t	F_t	
top	0.001	25.2	0.20	-0.003	0.99	0.02	-3.64	-0.14	0.02	8015
Empirical values			0.90		1.60					

[a] $p_c = 0.95$, $p_m = 0.08$. The size of population is 50 and the maximum number of generations is 2000. The total groups of data for calculation is 8400.

[b] The definition of parameters are shown in equations (3) ∼ (4).

Automatic Validation of Protocol Interfaces Described in VHDL

Fulvio Corno, Matteo Sonza Reorda, Giovanni Squillero

Politecnico di Torino
Dipartimento di Automatica e Informatica
Corso Duca degli Abruzzi 24 I-10129, Torino, Italy
{corno, sonza, squillero}@polito.it

Abstract. In present days, most of the design activity is performed at a high level of abstraction, thus designers need to be sure that their designs are syntactically and semantically correct before starting the automatic synthesis process. The goal of this paper is to propose an automatic input pattern generation tool able to assist designers in the generation of a test bench for difficult parts of small- or medium- sized digital protocol interfaces. The proposed approach exploit a Genetic Algorithm connected to a commercial simulator for cultivating a set of input sequence able to execute given statements in the interface description. The proposed approach has been evaluated on the new ITC'99 benchmark set, a collection of circuits offering a wide spectrum of complexity. Experimental results show that some portions of the circuits remained uncovered, and the subsequent manual analysis allowed identifying design redundancies.

1 Introduction

In the past years, the design flow of protocol interfaces, and Application Specific Integrated Circuits (ASICs) in general, experienced radical changes. Due to the maturity of automatic logic synthesis tools most of the design activity is now performed at high level of abstraction, such as register transfer level (RT), instead of low level such as gate. The new methodology dramatically increases designer productivity since high-level descriptions are more readable and considerably smaller.

One important step of the new design flow consists of *design validation*, i.e., the verification that the design is syntactically and semantically correct before starting automatic logic synthesis. Although many techniques have already been proposed in the CAD literature (e.g., static checks, formal verification [HuCh98], mutation testing [AHRo98]), none has gained enough popularity to compete with the current industrial practice of *validation by simulation*. Verification engineers resort to extensive simulation of each design, and of the complete system, in order to gain confidence over its correctness.

This situation is far from ideal, and designers need to face many difficulties. At the present days, simulation technology is effective enough for synthesized circuits. But when it comes to mixed-signal circuits, or to circuits containing embedded cores, or to

S. Cagnoni et al. (Eds.): EvoWorkshops 2000, LNCS 1803, pp. 205–214, 2000.

a complete system composed of a network of several interconnected interfaces, simulation is unable to provide the needed versatility.

Even restricting our attention to medium-sized, digital interfaces, the fundamental issue of measuring the test bench *quality* can be considered still unanswered. Many metrics have been proposed to evaluate the thoroughness of a given set of input stimuli, often adopted from the software testing domain [Beiz90], ranging from statement or branch coverage, state coverage (for finite state machine controllers), condition coverage (for complex conditionals), to the more complex path coverage. Many variants have been developed, mainly to cater for observability [DKGe96] and for the inherent parallelism of hardware descriptions [TAZa99], that are not taken into account by standard metrics. Since no well established metric is yet widely accepted for validation, some authors also propose to measure the quality of validation patterns with the stuck-at fault coverage.

Several products (normally integrated into existing simulation environments) are now available that provide the user with the possibility of evaluating the coverage of given input stimuli with respect to a selected metric. Designers can therefore pinpoint the parts of their design that are poorly tested, and develop new patterns specifically addressing them. Currently, this is a very time consuming and difficult task, since all the details of the design must be understood for generating suitable input sequences. The right trade-off between designer's time and validation accuracy is often difficult to find, and this often results in under-verified circuits. Moreover, in the generation of test vectors the designer may be "biased" by his knowledge of the desired system or module behavior, so that he often fails in identifying input sequences really able to activate possible critical points in the description.

When faced with this problem, the CAD research community traditionally invested in *formal verification* [GDNe91] [HuCh98], in the hope that circuits can be *proven* correct by mathematical means. Although formal verification tools give good results on some domains, they still have too many limitations or they require too much expertise to be used as a mainstream validation tool. Designers are left waiting for the perfect formal verification system, while few or no innovative tools help them with simulation-based validation.

The goal of this paper is to propose GIP-PI (Genetic Input Pattern generator for a Protocol Interface). GIP-PI is an automatic input pattern generation tool able to assist designers in the generation of a test bench for difficult parts of small- or medium-sized digital protocol interfaces. The proposed approach belongs to a brand new framework that can be called *approximate validation*, which explicitly relinquishes exactness in order to gain the ability of dealing with realistic designs. This philosophy has already been successfully applied in different areas: validation of the implementation of protocol interfaces [CSSq99c]; automatic test pattern generation [CSSq99a]; low-level [CSSq98] and mixed-level [CSSq99b] equivalence validation. Although the goal of this paper is completely different from the previous applications, all these methodologies share a common quality: being able to deal with *real* circuits exploiting an evolutionary algorithm.

GIP-PI employs a Genetic Algorithm, interacting with a VHDL simulator, for deriving an input sequence able to execute a given statement, or branch, in the high-level description. Whenever the test bench quality, as measured by one of the proposed metrics, is too low, our tool can be used to generate test patterns that are able to stimulate the parts of the design that are responsible for the low metric. The designer must manually analyze only those parts of the description that the tool failed to cover. Experimental results show that only a small fraction of "difficult" statements remain uncovered, and that many of them, upon closer inspection, indeed contain design errors or redundancies.

While no metric is yet widely accepted by validation teams, we aimed at evaluating the effectiveness of our approach using some pre-defined metric. The algorithm is quite easily adapted to different metrics, but for the sake of the experiments we adopted branch coverage as a reference. We developed a prototypical system for generating test patterns based on branch coverage, applicable to synthesizable VHDL descriptions. We aim at addressing moderately sized circuits, that usually can not be handled by formal approaches, and at working directly on the VHDL description, without requiring any transformation nor imposing syntax limitations.

The approach has been evaluated on the new ITC'99 benchmark set [ITC99], a collection of circuits described in high-level (RT) VHDL that offers a wide spectrum of complexity. Manually derived validation suites did not adequately cover all parts of the designs, and new sequences have been generated by the tool to increase the overall coverage. Experimental results show that some portions of the circuits remained uncovered, and the subsequent manual analysis allowed identifying design redundancies.

Section 2 gives an overview over the proposed approach for test bench generation, experimental are presented in Section 3 and Section 4 concludes the paper.

2   RT-level Test Bench Generation

The goal of test bench generation is to develop a set of input sequences that attain the maximum value of a predefined validation metric. Despite this implementation of GIP-PI is tuned for simulating high-level VHDL network interfaces only, the proposed method could be easily extended to deal with any simulable descriptions. For instance, given a protocol specification in ESTELLE, and with a reduced effort, GIP-PI could eventually generate a set of stimuli (events) to validate the protocol description itself.

2.1 Adopted Metric

Most available tools grade input patterns according to metrics derived from software testing [Beiz90]: statement coverage and branch coverage are the most widely known, but state/transition coverage (reaching all the states/transitions of a controller) and condition coverage (controlling all clauses of complex conditionals) are also used in hardware validation. Path coverage, although often advocated as the most precise one, is seldom used due to its complexity, and because it loses meaningfulness when multiple execution threads run concurrently in parallel processes. Some recent work extends those metrics to take also into account observability [DGKe96] and the structure of arithmetic units [TAZa99]. Those extensions are essential when the sequences have to be used as test patterns to cover stuck-at faults, but for validation they have lower importance since internal values are available.

The metric we adopt in this paper is branch coverage, although the tool can be easily adapted to more sophisticated measures. Also, since synthesizable VHDL is a structured language, complete statement coverage implies complete branch coverage, and the tool takes advantage of this simplification.

2.2   Overall Approach

The adopted approach is an evolution of the one presented in [CPSo97], where a Genetic Algorithm uses a simulator to measure the effectiveness of the sequences it generates. Instead of trying to justify values across behavioral statements, that would require solving Boolean and arithmetic constraints [FADe99], thanks to the nature of Genetic Algorithms we just need to simulate some sequences and analyze the propagation of values. Each sequence is therefore associated with the value returned by a *fitness function*, that measures how much it is able to enhance the value of the validation metric, and the Genetic Algorithm evolves and recombines sequences to increase their fitness.

The fitness function needs to be carefully defined, and accurately computed. In particular, the fitness function can *not* be just the value of the validation metric: it must also contain some terms that indicate how to *increase* the covered branches, not just to *count* the already covered ones. In a sense, the fitness function includes a dominant term, that measures the accomplished tasks (covered branches), and secondary terms, that describe sub-objectives to be met in order to cover new branches.

The computation of such function is accomplished by analyzing the simulation trace of the sequence, and by properly weighting the executed assignments, statements, and branches according to the target statements. In the implementation, to avoid arbitrary limitations in the VHDL syntax, simulation is delegated to a commercial simulator that runs an *instrumented* version of the VHDL code and records the simulation trace in the transcript file. Such trace is then interpreted according to control- and data-dependencies, that are extracted from a static analysis of the design description. Figure 1 shows a simplified view of the overall system architecture.

2.3   VHDL Analysis

The goal of the algorithm is to achieve complete coverage, but for efficiency reasons we do not consider each statement separately, and we group them into *basic blocks* [ASUl86]: a basic block is a set of VHDL statements that are guaranteed to be executed sequentially, i.e., they reside inside a process and do not contain any intermediate entry point nor any control statement (if, case, ...). All the operations required for code instrumentation, dependency analysis, branch coverage evaluation, and fitness function computation are performed at the level of basic blocks.

Since the Genetic Algorithm exploits the knowledge about data and control dependencies, we need to extract that information from the VHDL code: for this reason, we build a *database* containing a simplified structure and semantics of the design. The database is structured as follows:

- The hierarchy of component instantiations inside different entities is *flattened* (C1 and C2 in the figure 2). A dictionary of *signal equivalencies* is also built, that allow us to uniquely identify signals that span multiple hierarchical levels.

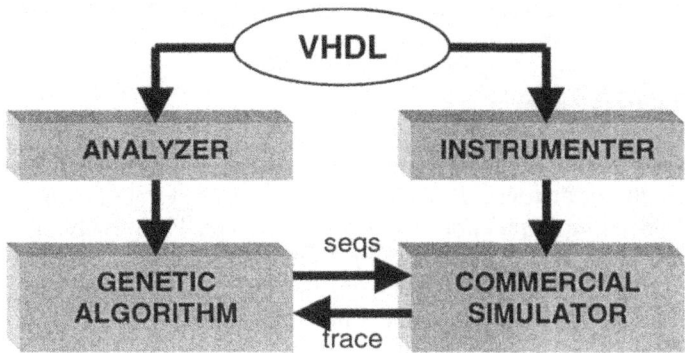

Figure 1: *System architecture*

- All VHDL *processes* occurring in the flattened circuit are given a unique identifier (P*i* in the figure 2). This operation also converts standalone concurrent statements into their equivalent process. The design is thus represented as a *network of processes interconnected by signals*.

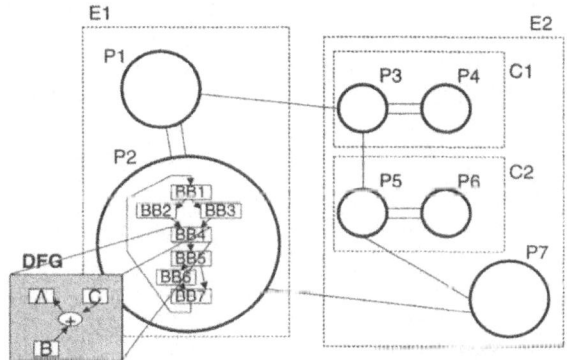

Figure 2: Abstract representation of RT-level designs

- Each process is analyzed to define its *interface*, in terms of signals that it reads and writes.
- The sequential part of each process is analyzed, its *control flow graph* (CFG) is extracted, and statements are grouped in *basic blocks* (BBs). The control structure of the process is described as a control flow of basic blocks (figure 2 reports the CFG for process P2).

- A *dependency matrix* between basic blocks is computed, by assigning a probability that a basic block will be executed, given that another block has just been executed. These correlation probabilities take into account the branching and looping nature of the control flow.
- Each basic block is entered, and the *data flow graph* (DFG) of the operations that occur inside each basic block is extracted. Since a basic block consists of multiple statements and/or conditions, multiple dependencies are associated to a single block. Fig. 2 shows the DFG for the basic block BB4 of process P2.

2.4   Genetic Algorithm

The Genetic Algorithm (GA) is based on encoding potential test sequences as variable length bit matrices. A number of such sequences are randomly generated and constitute an initial *population*: the goal of the GA is to evolve this population to increase its *fitness* value. The fitness function measures the closeness of a sequence to the goal. Currently, in GIP-PI the genetic algorithm is run several times, each time for a different target. Moreover, the fitness function assumes two different forms:

- In the initial phase, all basic blocks are considered simultaneously. The goal is to generate a set of sequences S that activate most of the blocks, for identifying easy-to-execute blocks. The fitness function is simply the number of activated blocks over the total number of blocks:

$$\text{fitness}(S) = \frac{activated\_blocks(S)}{tot\_blocks}$$

- Subsequently, GIP-PI is targeting a specific block T and the goal is to generate a sequence S able to cover it. In this phase sequences are targeted and the executed blocks are weighted by their correlation probability with respect to the target, measured as the weighted average of the execution counts of the basic blocks in the input cone (taking into account both control and data dependencies, thus potentially spanning several processes) of the target. The adopted weights take into account the probabilities of conditional execution that were statically computed in the database.

$$\text{fitness}(S,T) = \sum_{b \in covered\_bb(S)} \text{correlation}(b,T)$$

In both phases, a saturation mechanism prevents easy-to-execute but not-so-relevant statements from diverting the attention of the GA.

Moreover, during each run of the GA, an heuristic mechanism detects individuals that may be useful in a subsequent run. When such an individual is found, it is saved and later it is inserted in the initial population of the correct run of the GA. These predefined individuals may never exceed 5% of the initial population.

The GA in GIP-PI evolves a population of μ individuals and in each generation λ new sequences are first generated, then selection is performed on the whole set of $\mu + \lambda$ individuals. Individuals are selected for reproduction using a roulette wheel mechanism based on their linearized fitness. In p of the cases, the new individual is built mutating a single parent: the original sequence can be shortened, or enlarged, or some bits may be flipped. In $1-p$ of the cases, the new individual is built mating two different parents: the offspring sequence can inherit the beginning from one parent and the end from the other, or some entire bit column from each parent.

3   Experimental results

To test the effectiveness of the tool in generating test benches, we selected a set of VHDL benchmarks from the ITC'99 benchmark set [ITC99]. The first columns report some data about the RT-level descriptions, in terms of VHDL lines, VHDL processes (with hierarchy unflattened), and overall number of extracted basic blocks (BB). To have a better idea about circuit size, some characteristics of the synthesized netlists are reported in the last columns: number of Primary Inputs, Primary Outputs, Flip-Flops, and combinational gates.

These benchmarks have been publicly released in September 1999 at the IEEE International Test Conference, and there are no published results, yet, to compare with. We compared our results against a pure random approach, to evaluate the effectiveness of the GA, but it was so easily overcome that results are not reported.

The implementation consists of about 4,700 lines of C code for VHDL code analysis and instrumentation, linked to the LEDA LPI interface [LEDA95], and of 2,700 lines of C code for the Genetic Algorithm and the interface to the simulator. All experiments were run on a Sun Ultra 5 running at 333 MHz with 256MB of memory.

We adopt a population of $\mu = 30$ individuals, with $\lambda = 20$ new individuals in each generation. The mutation probability was set to $p = 0.3$.

Circ	VHDL			GATE			
	Lines	Proc	BB	PI	PO	FF	Gate
b01	111	1	28	2	2	5	46
b02	71	1	17	1	1	4	28
b03	142	1	27	4	4	30	149
b04	103	1	23	11	8	66	597
b05	333	3	94	1	36	34	963
b06	129	1	25	2	6	9	60
b07	93	1	21	1	8	49	420
b08	90	1	14	9	4	21	167
b09	104	1	16	1	1	28	159
b10	168	1	38	11	6	17	189
b11	119	1	37	7	6	31	481
b12	570	4	118	5	6	121	1,036
b13	297	5	74	10	10	53	339
b14	510	1	244	32	54	245	4,775
b15	672	3	171	36	70	449	8,893
b20	1,085	3	491	32	22	490	9,419
b21	1,089	3	491	32	22	490	9,803

Table 1: *Benchmark characteristics*

In Table 2, we report the experiments we obtained with our prototypical tool in terms of percent number of covered branches, number of generated vectors, and required CPU time. These data demonstrate that for most descriptions, our method is able to reach a complete or very high branch coverage. There are a few circuits (e.g., b05 and b12) where the obtained coverage is low: this is due to the specific characteristics of these circuits, which include highly nested conditional statements that the current version of our algorithm can hardly go through.

It is worth noting that a manual analysis of the branches left uncovered proved that many of them were effectively unreachable, in most cases due to `else` or `default` statements that are required by the synthesis tool not to infer sequential logic, but are redundant since all cases have already been considered in previous tests.

CIRCUIT	Cov %	#VECT	CPU [s]
b01	100,00	259	439.0
b02	100,00	114	41.3
b03	100,00	174	55.5
b04	100,00	83	425.1
b05	52,13	68	2014.9
b06	100,00	125	52.4
b07	95,24	351	920.6
b08	100,00	1,005	971.1
b09	100,00	958	511.5
b10	100,00	364	122.1
b11	94,59	1,222	1,410.2
b12	36,44	155	1,022.3
b13	100,00	3,303	4,203.9
b14	93,03	4,597	7,875.5
b15	91,81	2,838	9,369.2
b20	93,48	7,784	27,286.5
b21	93,69	6,376	28,878.9

Table 2: *Experimental results*

4   Conclusions

This paper presented an automatic input pattern generation tool able to assist designers in the generation of a test bench for difficult parts of small- or medium-sized digital protocol interfaces. The approach resorts to a Genetic Algorithm that interacts with a simulator to generate new sequences able to increase the coverage of the test bench with respect to a predefined validation coverage metric.

The methodology has been tested on the new ITC'99 benchmark set [ITC99]. Experimental results prove that the method is able to increase the quality of the validation process both over manual simulation and pseudo-random sequence generation. However, the proposed method could be easily extended to deal with any simulable descriptions, like the ESTELLE specification of a network protocol.

The tool results have also been useful as a feedback for better understanding the most difficult parts of the design from the validation point of view.

5   References

[AHRo98]   G. Al-Hayek, C. Robach: *From Design Validation to Hardware Testing: A Unified Approach*, JETTA: The Journal of Electronic Testing, Kluwer, No. 14, 1999, pp. 133-140

[ASUl86]   A.V. Aho, R. Sethi, J.D. Ullman, *Compilers, Principles, Techniques, and Tools*, Addison-Wesley Publishing Company, 1986

[Beiz90]   B. Beizer, *Software Testing Techniques (2nd ed.)*, Van Nostrand Rheinold, New York, 1990

[CPSo97]   F. Corno, P. Prinetto, M. Sonza Reorda: *Testability analysis and ATPG on behavioral RT-level VHDL*, Proc. IEEE International Test Conference, 1997, pp. 753-759

[CSSq98]   F. Corno, M. Sonza Reorda, G. Squillero, *VEGA: A Verification Tool Based on Genetic Algorithms*, Intl. Conf. on Circuit Design, 1998, pp. 321-326

[CSSq99a]   F. Corno, M. Sonza Reorda, G. Squillero, *Improved Test Pattern Generation on RT-level VHDL descriptions*, ITSW'99: International Test Synthesis Workshop, 1999

[CSSq99b]   F. Corno, M. Sonza Reorda, G. Squillero, *Simulation-Based Sequential Equivalence Checking of RTL VHDL*, ICECS'99: 6th IEEE Intl. Conf. on Electronics, Circuits and Systems, 1999

[CSSq99c]   F. Corno, M. Sonza Reorda, G. Squillero, *Approximate Equivalence Verification for Protocol Interface Implementation via Genetic Algorithms*, Evolutionary Image Analysis, Signal Processing and Telecommunications First European Workshops, EvoIASP'99 and EuroEcTel'99, 1999, pp. 182-192

[DGKe96]   S. Devadas, A. Ghosh, K. Keutzer: *An Observability-Based Code Coverage Metric for Functional Simulation*, Proc. ICCAD'96

[FADe99]   F. Fallah, P. Ashar, S. Devadas: *Simulation Vector Generation from HDL Descriptions for Observability-Enhanced Statement Coverage*, Proc. 36th DAC, New Orleans, 1999, pp. 666-671

[FDKe98]   F. Fallah, S. Devadas, K. Keutzer: *OCCOM: Efficient Computation of Observability-Based Code Coverage Metrics for Functional Verification*, Proc. 35th DAC, 1998

[GDNe91]   A. Ghosh, S. Devadas, A.R. Newton, *Sequential Logic Testing and Verification*, Kluwer, 1991

[HuCh98]   S.-Y. Huang, K.-T. Cheng, *Formal Equivalence Checking and Design Debugging*, Kluwer, 1998

[ITC99]   http://www.itctestweek.org/benchmarks.html

[LEDA95]   LVS System User's Manual, LEDA Languages for Design Automation, Meylan (F), April 1995

[TAZa99]   P.A. Thaker, V.D. Agrawal, M.E. Zaghloul: *Validation Vector Grade (VVG): A New Coverage Metric for Validation and Test*, VTS'99: IEEE VLSI Test Symposium, 1999, pp. 182-188

Evolutive Modeling of TCP/IP Network Traffic for Intrusion Detection

Filippo Neri

DSTA - University of Piemonte Orientale, Corso Borsalino 54,
15100 Alessandria (AL), Italy
neri@di.unito.it, neri@al.unipmn.it

Abstract. The detection of intrusions over computer networks can be cast to the task of detecting anomalous patterns of network traffic. In this case, patterns of normal traffic have to be determined and compared against the current network traffic. Data mining systems based on Genetic Algorithms can contribute powerful search techniques for the acquisition of patterns of the network traffic from the large amount of data made available by audit tools.

In this paper we compare models of data traffic acquired by a system based on a distributed genetic algorithm with the ones acquired by a system based on greedy heuristics. Also we discuss representation change of the network data and its impact over the performances of the traffic models.

1 Introduction

The rise in the number of computer break-ins, virtually occurring at any site, determines a strong request for exploiting computer security techniques to protect the site assets. A variety of approaches to intrusion detection do exist [Denning, 1987]. Some of them exploit signatures of known attacks for detecting when an intrusion occurs. Thus they are based on a model of possible misuses of a resource [Kumar and Spafford, 1994]. Other approaches to intrusion detection characterize the normal usage of the resources under monitoring. An intrusion is suspected when a significant shift from the resource's normal usage is detected. This approach seems to be more promising because of its potential ability to detect unknown intrusions. However, it also involves major challenges because of the need to acquire a model of the normal use general enough to allow authorized users to work without raising alarms, but specific enough to recognized unauthorized usages [Lane and Brodley, 1997,Ghosh et al., 1999,Lee et al., 1999]. For the scope of this work, we focus to the task of detecting intrusions by analyzing logs of networks traffic.

Many aspects of deploying in-practice approaches based on data mining, however, are still open [Lane and Brodley, 1998].

We concentrate here on selecting informative data representation and classification performances. As learning methods, we exploited two rule based systems: a heuristic one, RIPPER [Cohen, 1995], and a stochastic one (based on genetic

S. Cagnoni et al. (Eds.): EvoWorkshops 2000, LNCS 1803, pp. 214–223, 2000.

algorithms), REGAL [Giordana and Neri, 1995,Neri and Saitta, 1996]. The first system has been selected because of its previous use [Lee et al., 1999]; it will thus act as benchmark. The second system has been selected because its intrinsically stochastic behavior and should allow the acquisition of more robust models [Neri and Saitta, 1996].

In the following, a description of the system REGAL (Section 2) and of the experiments performed in the IES and DARPA contexts (Section 3 and Section 4) are reported. Finally, the conclusions are drawn.

2   The Systems Regal and Ripper

For space reason we will provide here an abstract description of both learning systems REGAL and RIPPER as their full descriptions have already been published. The system REGAL is available for free from the author.

REGAL [Giordana and Neri, 1995,Neri and Saitta, 1996] is a learning system, based on a distributed genetic algorithm (GA). It takes as input a set of data (training instances) and outputs a set of symbolic classification rules characterizing the input data. As usual, learning is achieved by searching a space of candidate classification rules.

The language L used to represent classification rules is a Horn clause language in which terms can be variables or disjunctions of constants, and negation occurs in a restricted form [Michalski, 1983]. An example of an atomic expression containing a disjunctive term is *color(x,[yellow, green])*, which is semantically equivalent to *color(x,yellow) or color(x,green)*. Such formulas are represented as bitstrings that are actually the population individuals processed by the GA. Classical genetic operators, operating on binary strings, with the addition of task oriented *specializing* and *generalizing* crossovers are exploited, in an adaptive way, inside the system (for details see [Giordana and Neri, 1995].

REGAL is a distributed genetic algorithm that effectively combines the Theory of Niches and Species of Biological Evolution together with parallel processing. The system architecture is made by a set of extended Simple Genetic Algorithms (SGA) [Goldberg, 1989], which cooperates to sieve a description space, and by a Supervisor process that coordinates the SGAs efforts by assigning to each of them a different region of the candidate rule space to be searched. In practice this is achieved by dynamically devising subsets of the dataset to be characterized by each SGA.

The system RIPPER [Cohen, 1995] is based on the iterated application of a greedy heuristic, similar to the Information Gain measure [Quinlan, 1993], to build conjunctive classification rules. At each iteration, those training instances correctly classified by the found rules are removed and the algorithm concentrate on learning a classification rule for the remaining one. The system outputs an ordered list of classification rules (possibly associated to many classes) to be applied in that same order to classify a new instance. An interesting features of the method is that it exploits on-line rule pruning while incrementally building a new classification rule to avoid overfitting.

3   Intrusion detection in the Information Exploration Shootout contest

An evaluation of REGAL over an intrusion detection task by exploiting data from the Information Exploration Shootout Project (IES) is reported in this section. The IES made available network logs produced by 'tcpdump' for evaluating data mining tool over large set of data. These logs were collected at the gateway between an enterprise LAN and the outside-network (Internet). In the IES context, detecting intrusions means to recognize the possible occurrence of unauthorized ('bad') data packets interleaved with the authorized ('good') ones over the network under monitoring. The IES's project makes available four network logs: one is guarantee not to contain any intrusion attempts, whereas the other ones do include both normal traffic and intrusions attempts. In the IES context, no classification for each data packet is requested, instead an overall classification of a bunch of the network traffic, as containing attacks or not, is desired. An approach to intrusion detection, based on anomaly detection, has been selected.

We proceed as follows. IES data can be partitioned, on the basis of their IP addresses, into packets exiting the reference installation (Outgoing), entering the installation (Incoming) and broadcasted from host to host inside the installation (Interlan). Three models of the packet traffic, one for each direction, have been built from the intrusion-free dataset. Then, these models have been applied to the three datasets containing intrusions. We expect to observe a significant variation in the classification rate between intrusion-free logs and logs containing intrusions because of the *abnormal* characteristics of the traffic produced by the intrusive behavior. If this would actually occur, we could assert that the learned traffic models correctly capture the essential characteristics of the intrusion-free traffic. Note that whether this approach should work, we could conclude that an intrusive attempt is happening because of a different *global traffic pattern*. Experiments have been performed both with RIPPER and REGAL.

When RIPPER is applied to the IES data, the classification rate appearing in Table 1 becomes evident [Lee et al., 1999]. Each table entry represents a classification error as measured over the four network logs, one for each line, and with respect to the three class labels: Outgoing, Interlan, and Incoming. The correct classification rates can be obtained by subtracting an entry's value from 1. Thus the first row show the misclassified normal traffic packets, whereas the other ones shows the misclassified packets during an intrusion attempt.

These results have been obtained by applying RIPPER to the data as available from the tcpdumped files (see Appendix A). No preprocessing over the data, such as feature construction, has been applied. The experimental findings shows that the acquired models do not exhibit very different classification rate when applied to logs containing intrusions with respect to intrusion-free logs. These findings may suggest that the exploited data representation is too detailed with respect to the capability of the learning system. In turn, this causes the learned models to miss the information characterizing intrusion-free traffic.

Table 1. Experimental results of applying RIPPER to IES datasets using the raw data representation. Each table entry states a classification error.

Dataset	interlan	incoming	outgoing
normal traffic	0.04	0.04	0.04
intrusion1	0.23	0.07	0.04
intrusion2	0.09	0.07	0.05
intrusion3	0.08	0.14	0.04

Table 2. Experimental results of applying RIPPER to IES datasets using a compressed data representation.

Dataset	interlan	incoming	outgoing
normal traffic	0.02	0.05	0.04
intrusion1	0.11	0.11	0.21
intrusion2	0.03	0.13	0.12
intrusion3	0.11	0.21	0.12

Following this observation, we develop a more compact representation for the packets that consists in mapping a subset of feature's values into a single value, thus reducing the cardinality of possible features values (see Appendix B). Exploiting this representation, RIPPER's performances become the ones reported in Table 2 and REGAL's performances exploiting the same compact data representation appear in Table 3. The observed figures show a more stable classification behavior of the models across different traffic conditions. Also a more distinct classification performance between the intrusion-free log and the logs including intrusions is evident. A compression-based representation is then a valuable way of increasing classification performances without introducing complex feature that may involves additional processing overhead. An evaluation of the effect caused by the addition of complex features to the raw network data representation has been performed in [Lee et al., 1999].

For the sake of clarity, an example of rule characterizing intrusion-free Incoming packets, learned by REGAL, appears in Figure 1. The Incoming packets are characterized in term of the values of the features from their TCP/IP header. This rule successfully covers 7349 Incoming packets without being fooled by any

Table 3. Experimental results of applying REGAL to IES datasets using a compressed data representation.

Dataset	interlan	incoming	outgoing
normal traffic	0.02	0.04	0.04
intrusion1	0.12	0.15	0.11
intrusion2	0.06	0.11	0.12
intrusion3	0.12	0.15	0.11

IF   srcprt(x,[[0,20],[40,100],[150,200],[>500]]) and
        dstprt(x,[>1024]) and flag(x,[FP,pt]) and
        seq1(x,[[100,150],[200,300],[500,5000],[>10000]]) and
        seq2(x,[[50,100],[200,300],[500,20000]]) and
        ack(x,[[0,3000],[5000,10000]]) and
        win(x,[[0,2000],[>3000]]) and
        buf(x,[<=512])
THEN   IncomingPacket(x)
Coverage: (Interlan, Incoming, Outgoing) = (0, 7349, 0)

Fig. 1. Example of a rule characterizing part of the incoming traffic. The rule describes 7349 incoming packets without confusing them with any outgoing or interlan packet.

Interlan or Outgoing ones. A description of the predicates appearing in the rule is provided in Appendix A.

4   Intrusion detection in the 1998 DARPA Intrusion Detection Evaluation Programme

We also performed an additional evaluation of our approach over network logs from 1998 DARPA Intrusion Detection Evaluation Programme [Lippmann et al., 1999] whose objective was to survey and evaluate research in intrusion detection. A standard set of data to be audited, which includes a wide variety of intrusions simulated in a military network environment, was provided. We exploited data available from the KDD'99 Intrusion Detection Contest[1].

The raw training data was about four gigabytes of compressed binary TCP dump data from seven weeks of network traffic. This was processed into about five million connection records. Similarly, the two weeks of test data yielded around two million connection records. A connection is a sequence of TCP packets starting and ending at some well defined times, between which data flows from a source IP address to a target IP address under some well defined protocol. Each connection is labeled as either normal, or as an attack, with exactly one specific attack type. Each connection record consists of about 100 bytes. Attacks fall into four main categories:

DOS: denial-of-service, e.g. syn flood;

R2L: unauthorized access from a remote machine, e.g. guessing password;

U2R: unauthorized access to local superuser (root) privileges, e.g., various "buffer overflow" attacks;

Probe: surveillance and other probing, e.g., port scanning.

In practice two datafiles containing classified connections are available: one has to be used for acquiring a model of the traffic and the other one for testing its

---

[1] Information about KDD'99 Intrusion Detection Contest is available on-line at http://www.epsilon.com/kdd98/task.html.

performances. The distinction is important because the test file contains attack types not occurring in the learning file. This is intended to make the task more realistic. In figure 2 and figure 3, performances of RIPPER plus Meta-Learning

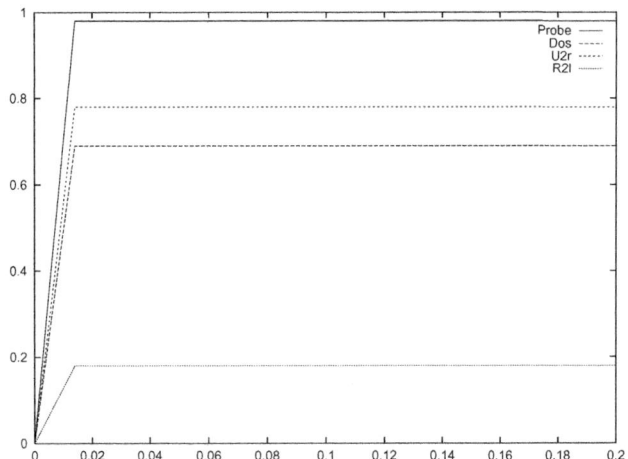

Fig. 2. Detection performances exhibited by RIPPER plus Meta-Learning on the DARPA test data. An extended representation of the data and a complex learning approach (meta-level learning) have been exploited.

(as used in [Lee et al., 1999]) and REGAL over DARPA's data are respectively shown. In this case performances are shown by exploiting Receiver Operating Curves as done when evaluating data mining tools. In the figures, the x axis represents the false alarm rate, i.e. the percentage of 'Normal' connections labeled as intrusions, whereas the y axis represents the detection rate, i.e. the percentage of intrusions that have been correctly recognized. Each line is associated to a different attack type. This kind of graph is used to show how a classifiers behavior degrades when "relaxing" its matching conditions: generally more intrusion are detected while also covering more false alarms. However in the case of a set of symbolic classification rules (it is actually better to say in the case of RIPPER and REGAL) there are non conditions to be relaxed that is the reason of the flat line at the top.

The reported performances have been obtained on the connections occurring in the test file. The reported graphs show similar detection performances, between the models acquired by the systems, for Probe and Remote-To-Local (R2l) attacks types. Instead, REGAL's model performs slightly better on DOS type attacks but worst on User-To-Root (U2r) attacks.

Let consider, now, the modeling approaches exploited by the two systems. Lee and Stolfo [Lee et al., 1999] run RIPPER over an extended data representation of the tcp connection including, in addition to the basic tcp features, derived information such as: the number of connections to the same host in the past

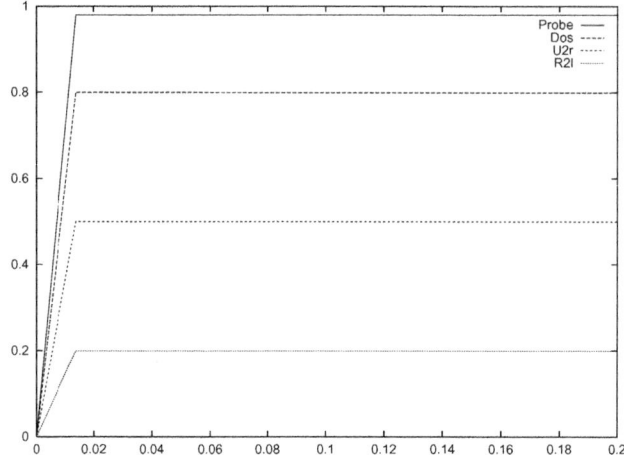

Fig. 3. Detection performances exhibit by REGAL on DARPA test data (no additional Meta-Learning has been used). A compressed data representation has been exploited.

two seconds ('count'), the number of connections to the same service, as the current connection, in the past two seconds ('srv-count'). These features have been chosen on the basis of the authors expertise. A preprocessing of the raw network logs is required in order to exploits this features. Several classifiers (rule sets) for each attack type have been obtained. Eventually meta-learning, i.e. learning at the classifier level, has been applied to produce the reported performances.

REGAL, on the contrary, has been run after applying a compression mapping to the feature values, as described in Appendix B. Only the basic features of a TCP connection have been considered such as: 'duration', stating the length (number of seconds) of the connection, 'protocol-type', stating the type of the protocol (e.g. tcp, udp, etc.), or 'src-bytes', stating the number of data bytes from source to destination. No additional meta-learning phase is necessary.

5   Conclusions

We report experiments that show the potentiality of a distributed genetic learner to the modeling of network data. Two different set-ups to deal with detecting intrusions have been explored.

We analyzed a data packet representation exploiting compression of the feature's values in the effort to reduce the complexity of acquiring model of the traffic. We believe this being an important requisite for the automatic modeling and the on-line deployment of intrusion detection system.

The experimental results support use of the compression of the feature values as a valuable method to increase detection performances while avoiding the use of derived and complex features that involve additional computational overhead.

Acknowledgements

Many thanks to the anonymous reviewers whose useful comments helped me in improving the paper.

A   Appendix. The Information Exploration Shootout raw data representation

The IES data (available on line at http://iris.cs.uml.edu) have been collected by means of the TCPDUMP utility. Taking into account privacy concerns, the data portion of each packet has been dropped. For each packet in the datasets the following attributes are available:

time - converted to floating pt seconds .. hr*3600+min*60+secs.

addr and port - (just get rid of x.y.256.256.port) The first two fields of the src and dest address make up the fake address, so the converted address was made as: x + y*256.

flag - added a "U" for udp data (only has ulen) X - means packet was a DNS name server request or response. The ID# and rest of data is in the "op" field. (see tcpdump description) XPE - means there were no ports... from "fragmented packets".

seq1 - the data sequence number of the packet.

seq2 - the data sequence number of the data expected in return.

buf - the number of bytes of the receiver buffer space available.

ack - the sequence number of the next data expected from the other direction on this connection.

win - the number of bytes of receive buffer space available from the other direction on this connection.

ulen - if a udp packet , the length.

op - optional info such as (df) ... do not fragment.

Particular attention has to be taken when dealing with fields like 'op' that contains a large amount of values.

B   Appendix. The compressed feature representation of IES data

Some features of the IES data may assume a large set of values either continuous or discrete. These large sets do impact over classification performances of the learned models because of the intrinsic difficulty of acquiring rule having a general scope. Then, a reduction of the range of potential values is desirable to increase both the generality of the learned model and to reduce the learning computational complexity.

An alternative approach to this problem consists in adding/building more complex features, combining the basic ones, to the original data representation. We do not follow this approach in this work, because we believe that the previous approach is simpler and should be the first to be analyzed.

Original Value	New Value
0≤srcport<50	srcport=0
50≤srcport<100	srcport=0
<... skipped test ... >	<... skipped text ... >
srcport>20000	srcport=10
<... skipped text ... >	<... skipped text ... >
op contains "DF"	op=1
op contains "NXDomain"	op=2
op contains ANY OTHER VALUE	op=3

Table 4. Compression mapping applied when dealing with IES network data.

As an instance of reducing the range of the feature values, considers that the feature 'srcport' (see Appendix A for a description) may virtually assume any integer number from 0 to 65536. Also, the feature 'op' may assume hundreds of discrete values. Taking into account basic knowledge about the domain, we manually developed the reduction mapping shown in Table 4. This mapping is not to be considered as the best one but as a proof that a simple reduction of the feature values may positively impact over the recognition capabilities.

References

[Cohen, 1995] Cohen, W. (1995). Fast effective rule induction. In *Proceedings of International Machine Learning Conference 1995*, Lake Tahoe, CA. Morgan Kaufmann.

[Denning, 1987] Denning, D. (1987). An intrusion detection model. *IEEE Transaction on Software Engineering*, SE-13(2):222–232.

[Ghosh et al., 1999] Ghosh, A., Schwartzbard, A., and Schatz, M. (1999). Learning program behavior profiles for intrusion detection. In *USENIX Workshop on Intrusion Detection and Network Monitoring*. USENIX Association.

[Giordana and Neri, 1995] Giordana, A. and Neri, F. (1995). Search-intensive concept induction. *Evolutionary Computation*, 3 (4):375–416.

[Goldberg, 1989] Goldberg, D. (1989). *Genetic Algorithms in Search, Optimization, and Machine Learning*. Addison-Wesley, Reading, Ma.

[Kumar and Spafford, 1994] Kumar, S. and Spafford, E. (1994). A pattern matching model for misuse detection. In *National Computer Security Conference*, pages 11–21, Baltimore.

[Lane and Brodley, 1997] Lane, T. and Brodley, C. (1997). An application of machine learning to anomaly detection. In *National Information Systems Security Conference*, Baltimore.

[Lane and Brodley, 1998] Lane, T. and Brodley, C. (1998). Approaches to online learning and conceptual drift for user identification in computer security. Technical report, ECE and the COAST Laboratory , Purdue University, Coast TR 98-12.

[Lee et al., 1999] Lee, W., Stolfo, S., and Mok, K. (1999). Mining in a data-flow environment: experience in network intrusion detection. In *Knowledge Discovery and Data Mining KDD'99*, pages 114–124. ACM Press.

[Lippmann et al., 1999] Lippmann, R., Cunningham, R., Fried, D., Graf, I., Kendall, K., Webster, S., and Zissmann, M. (1999). Results of the DARPA 1998 offline intrusion detection evaluation. In *Recent Advances in Intrusion Detection 99, RAID'99*, W. Lafayette, IN. Purdue University.

[Michalski, 1983] Michalski, R. (1983). A theory and methodology of inductive learning. In Michalski, R., Carbonell, J., and Mitchell, T., editors, *Machine Learning, an Artificial Intelligence Approach*, volume I, pages 83–134. Morgan Kaufmann, Los Altos, CA.

[Neri and Saitta, 1996] Neri, F. and Saitta, L. (1996). Exploring the power of genetic search in learning symbolic classifiers. *IEEE Trans. on Pattern Analysis and Machine Intelligence*, PAMI-18:1135–1142.

[Quinlan, 1993] Quinlan, J. R. (1993). *C4.5: Programs for Machine Learning*. Morgan Kaufmann, California.

Multimodal Performance Profiles on the Adaptive Distributed Database Management Problem

M. Oates[1], D. Corne[2], and R. Loader[2]

[1] British Telecom Adastral Park, Martlesham Heath, Suffolk, England, IP5 3RE
[2] Department of Computer Science, University of Reading, Reading, RG6 6AY

Abstract. Previous publications by the authors have demonstrated a bimodal performance profile for simple evolutionary search on variants of the Adaptive Distributed Database Management Problem (ADDMP) and other problems over a range of evaluation limits. This paper examines an anomaly seen in one of these profiles and together with results from a range of other problems, shows that with sufficiently high evaluation limits, a multimodal performance profile is apparent in search spaces with significant numbers of deceptive local optima. This is particularly apparent in the performance profile of the Hierarchial If and only If problem (H-IFF) where the regular structure of the search space produces several distinct peaks and troughs in the performance profile, possibly indicative of a range of specific 'fitness barriers' which are surmountable by specific rates of mutation. This observation could prove important in general EA parameter tuning over a range of problems with similar characteristics. Further, the existence of optimal mutation rates inducing a minimum in standard deviation of run-time, is of critical importance in the application of EAs to real-time, real-world problems.

1 Introduction

Many real world, real time applications of Evolutionary Algorithms (EAs) [1,3,4,6] require the search process to reliably produce quality solutions in a fixed number of evaluations. To increase the likelihood of this, it is often essential to tune the algorithm by selection of suitable parameter values such as population size and mutation rate. One such application which has been studied extensively by the authors [11,12,13,14,15,16,17] is the Adaptive Distributed Database Management Problem (ADDMP) which attempts to balance a number of user loads onto a range of available servers over a communications network to maximise a given quality of service metric. The problem is reduced to one of combinatorial optimisation, where a 'solution vector' generated by the EA defines for each client node (determined by locus), which server to use (determined by allele value) using a natural k-ary representation (where alleles are integers in the range 1 through k). This effectively determines a route through the communications network, and the combined delay imposed by the server processing time and the communications latency is calculated by a performance model utilising the principles of MM1 queuing and Little's Law. The problem has been likened to a form of complex 'bin-packing' and further details are available in [12,13] with example code and datasets available at http://www.dcs.napier/ac/uk/evonet part

S. Cagnoni et al. (Eds.): EvoWorkshops 2000, LNCS 1803, pp. 224–234, 2000.
© Springer-Verlag Berlin Heidelberg 2000

of the ECTELNET website (ECTELNET is the Telecommunications subgroup of the European Network of Excellence in Evolutionary Computation)

Client :	1	2	3	4	5	6	7	8	9	10
Which Server to use :	1	4	3	4	3	2	2	3	4	1

Fig. 1. Example solution vector.

Many instances of the ADDMP exist, with different user load patterns, different server profiles and different communications network topologies, and some of these are discussed in [13]. Some of these instances have been shown to present considerable difficulty to a range of algorithms [11,14] and in general, an Evolutionary algorithm, correctly tuned, has been shown to give the most consistent performance. Previously published studies of one of these instances [16,17] show a three dimensional performance profile for a simple generational Breeder EA [7] using 50% elitism, uniform crossover [19], and uniformly distributed allele replacement mutation at a fixed rate per gene. Experiments were run over a wide range of mutation rates, from 1E-7 to 0.83, the latter effectively degenerating the EA into random search, and a wide range of population sizes typically 10 to 500 members in steps of 10. Each run was allowed 20,000 evaluations and each experiment was repeated 50 times. During each run, a note was made of the fitness of the best solution found, and the evaluation number at which it was first found. For the 50 runs all with the same experimental parameters, the mean of the 'time to best solution' was plotted against population size and mutation rate, and the resulting graph is shown in Figure 2. As was commented on in [16], this produces, for a given, low population size, a bimodal performance profile, where the number of evaluations taken to first find the best solution to be found in the run first rises, then falls, then rises again before falling into random search. These features are interpreted as follows : the low values of evaluations used at low mutation rates represent premature convergence on poor solutions, rapidly exhausting the limited diversity available in the initial population which cannot be supplemented by mutation due to its low rate; as mutation rates increase, the EA is able to progress further with its search, as depleted allele values are re-injected into the population with increasing frequency by the increasing mutation rate; a point is reached where there is a sufficiently high rate of mutation to allow the search to typically utilise in excess of 80% of the available evaluations in terms of fitness improvement – this is the left hand peak on Figure 2; further increases in mutation rate allow good solutions to the problem to be found in fewer evaluations, leading to a point where the mutation rate allows good solutions to be found in a minimum of evaluations – the trough in Figure 2; as mutation rate is increased further, performance deteriorates as the mutation rate begins to destroy progress almost as fast as it is made – the right hand peak on Figure 2; finally at very high rates of mutation, in excess of 40%, performance deteriorates to random search with the mean number of evaluations used tending to 50% of those allowed. This interpretation is supported by examination of the mean fitnesses of solutions found as mutation rates are increased which show : poor fitness at low mutation rates; a steady improvement in fitness for mutation rates above the first peak in Figure 2; a plateau of good fitness coinciding with the trough of minimal evaluations in Figure 2; and a deterioration in fitness as mutation rates approach the right hand side of Figure 2. This co-incidence of trough in evaluations and peak in fitness occurs at mutation rates around 1 / L (where L = binary chromosome length, here equivalent to around 2.5%), which is in support of theoretical studies by, amongst others, Baeck [1], Mühlenbein [8] and many empirical studies including [2,8,9,18] by Deb et al, van Nimwegen and Crutchfield, Oates et al, and this and other features are discussed in more detail in [16,17].

Of particular interest however in this paper is the anomaly which can be seen on the left hand edge of the right hand peak in Figure 2 at low population sizes (towards the back of the figure). Here, the rise in mean evaluation number with increasing mutation can be seen to be not smooth, with a ridge feature apparent over a range of lower population sizes. Such an anomaly was not seen at lower evaluation limits, nor on extensive investigations of simple 'unimodal' search spaces such as presented by the 'Max-Ones' problem. The remainder of this paper explores this phenomenon in more detail, with section 2 describing three other test problems and their experimental set up; section 3 describing the results at both 20,000 and 1 million evaluations; section 4 drawing some preliminary conclusions from these results and sections 5 and 6 acknowledging support and referenced material.

2 Method

To examine this anomaly in more detail, a series of experiments was conducted with a range of 'standard' test problems over similar ranges of experimental parameter values, and a second series of experiments conducted with a much higher evaluation limit and more refined scale of mutation rates.

The first problem to be looked at was the 'Max-Ones' problem where in a binary string, fitness is calculated to be the number of '1's present in the chromosome. With standard representation, steady state, single 3 way tournament selection [3], uniform crossover and simple per gene mutation, this presents a simple unimodal search landscape which can be ascended by even the simplest 'hillclimbing' algorithm. 50 trials were run for each parameter setting, which ranged from population sizes of 2 to 100 and per gene mutation rates from 1E-7 to .83. Mean evaluations to first find the best solution found in 20,000 evaluations were noted. Chromosome length for results shown here was 50 bits, however similar experiments have also been done at 33, 300 and 1630 bits, giving corresponding results.

The second problem to be investigated was a 64 bit implementation of Watson's H-IFF (Hierarchical If and only If) problem [20,21], which increasingly rewards ever larger aligned blocks of contiguous '1's or '0's and can be represented as :

$$f(B) = \begin{array}{ll} 1, & \text{if } |B| = 1 \\ |B| + f(B_L) + f(B_R), & \text{if } (|B| > 1) \text{ and } (\forall i \{b_i = 0\} \text{ or } \forall i \{b_i = 1\}), \\ f(B_L) + f(B_R), & \text{otherwise} \end{array}$$

where B is a block of bits, $\{b_1, b_2, \dots b_n\}$, $|B|$ is the size of the block=n, b_i is the ith element of B, and B_L and B_R are the left and right halves of B (i.e. $B_L = \{b_1, \dots b_{n/2}\}$, $B_R = \{b_{n/2+1}, \dots b_n\}$. N must be an integer power of 2.

H-IFF therefore has two global optima, one at all '1's and one at all '0's. There are secondary optima at strings of 32 '1's followed by 32 '0's and vice versa, and a range of sub-optima at combinations of aligned blocks of '1's and '0's each of length 16, 8, 4 and 2. The 'search landscape' can be considered to be rugged but in a highly structured fashion. Simple hillclimbers perform inadequately on this surface as the basin of attraction of the global optima is very small, and there are many local optima when seen by single point mutation alone. Even standard one-point and two-point crossover operators have been seen to require very high population sizes to achieve reasonable and consistent performance on this problem unless an appropriate diversity maintenance technique is employed.

The third problem reported on is this paper is a 50-8 Kaufman NK landscape [5] generated by defining a table of random numbers of dimension 50 by 512. For a binary chromosome of length N=50, starting at each of the 50 gene positions, K+1 (here 9) weighted consecutive genes are used to generate an index (in the range 0-511), and the 50 values so indexed from the table are summed to give a fitness value. For NK landscapes with a K value of 0, each gene position contributes individually to fitness, hence producing a unimodal landscape to single point mutation hillclimbers. However as K is increased, any single point mutation will affect K+1 indexes, rapidly introducing positional linkage into the problem and producing a rugged and increasingly unstructured landscape.

The ADDMP, Max-Ones, H-IFF and NK50-8 problems were all trialled at 20,000 evaluations in the first series of experiments, with results shown in Figures 2,3,4 and 5 respectively. A second series of experiments was then performed with a limit of 1,000,000 evaluations, for ADDMP (Figure 6) over a range of population sizes (10-300 in steps of 10) and for Max-Ones and H-IFF at a fixed population size of 20 (Figures 7 and 8). For easier comparison of results, the ADDMP 1,000,000 evaluation runs at a population size of 20 are shown more clearly in Figure 9. Plots showing plus and minus one standard deviation seen over the 50 trials and mean fitness are also shown in each of the fixed population size graphs.

Unless otherwise stated, all experiments (with the exception of ADDMP at 20K evals) were conducted with a steady state, 3 way single tournament GA using uniform crossover. The H-IFF experiments were conducted using one-point crossover due to the obvious inappropriateness of the uniform operator on this problem.

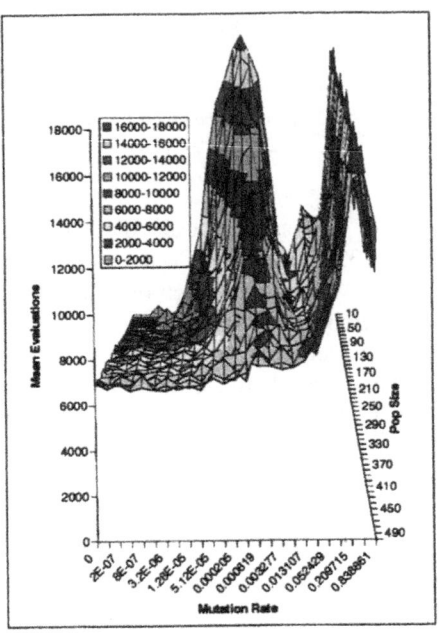

Fig. 2. ADDMP at 20K evaluations

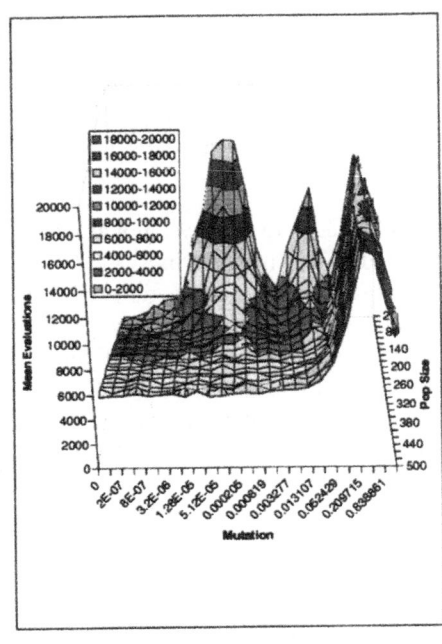

Fig. 4. H-IFF 64 at 20K evaluations

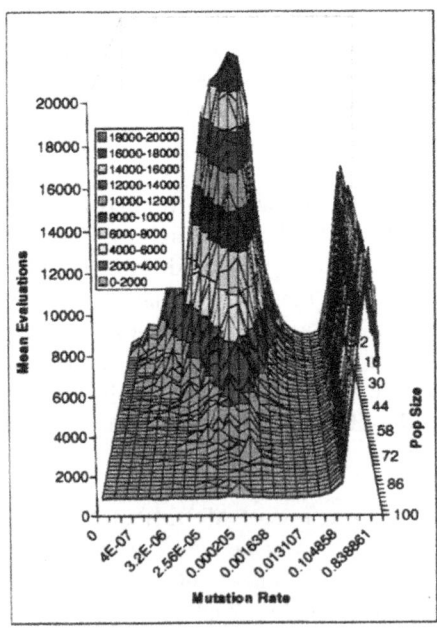

Fig. 3. One Max at 20K evaluations

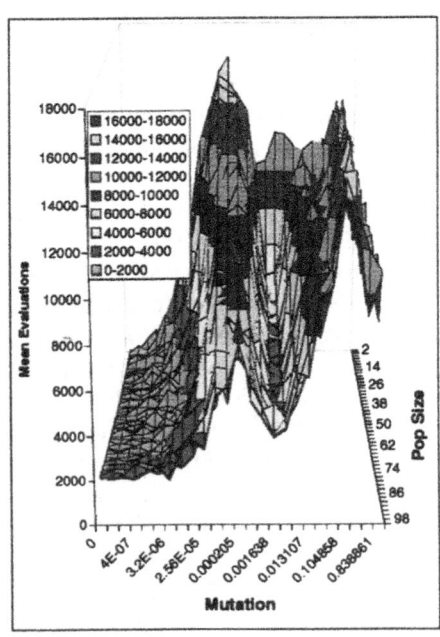

Fig. 5. NK 50-8 at 20K evaluation

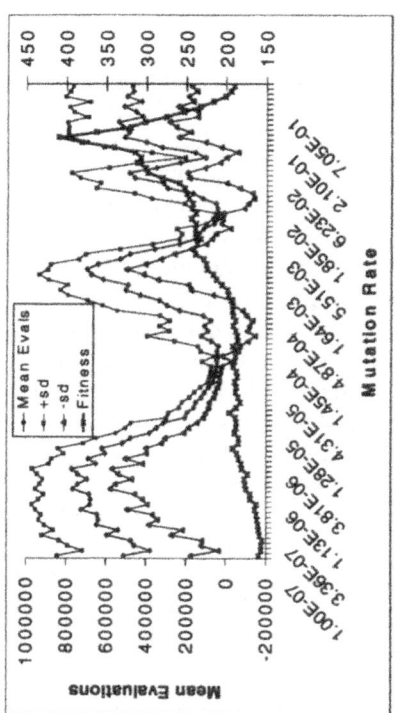

Fig. 8. H-IFF 64 at 1M evals, Pop size = 20

Fig. 9. ADDMP at 1M evals, pop size = 20

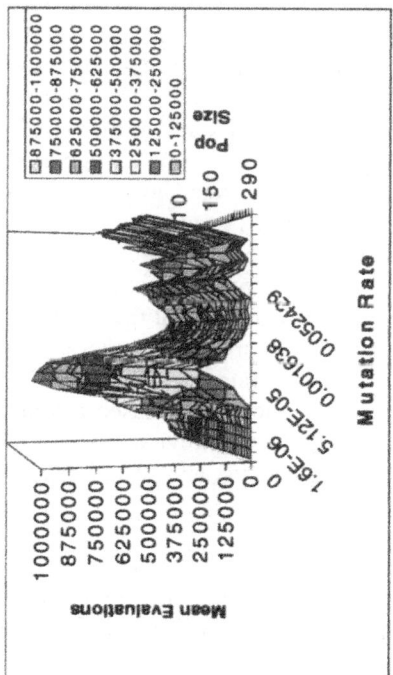

Fig. 6. ADDMP at 1M evaluations

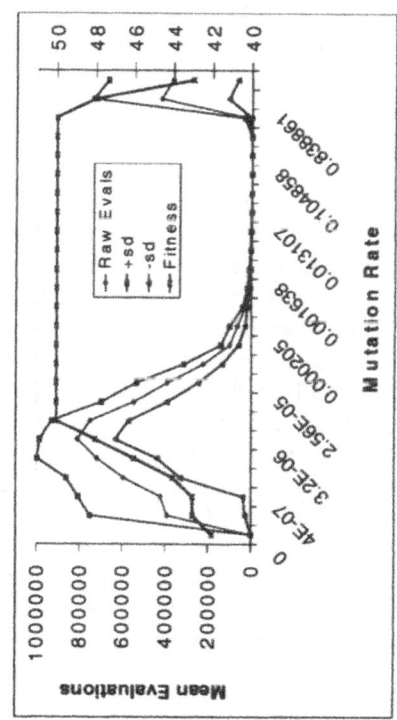

Fig. 7. One Max at 1M evals, pop size = 20

3 Results

An interpretation of Figure 2 has already been given in Section 1 of this paper, and a similar performance profile at 20,000 evaluations can also be seen for the Max-Ones problem in Figure 3. Again, at low population sizes and low mutation rates, the search is seen to stall after very few evaluations, as diversity in the initial population is rapidly depleted. Very low rates of mutation prevent re-occurrence of potentially useful alleles and thus the mean number of evaluations used is very low. As mutation rates increase, the search is able to progress to higher mean numbers of evaluations, until a point is reached where mutation rates allow, on average, most of the evaluations to be utilised in the search for good solutions. As mutation rates increase further, these good solutions are found in less and less evaluations, until the trough point in Figure 3 is reached where global optimum solutions are found in the minimum number of evaluations. Beyond this rate of mutation, performance deteriorates eventually degenerating into random search. However in contrast to ADDMP (Fig 2), this can be seen to deteriorate smoothly, with no anomaly on the left hand edge of the right hand peak. The mutation rate inducing the first peak can be seen to be effectively independent of population size (over its visible range) and occurs at a rate around .0001. The trough rate is centred around a mutation rate of .05 (= 5% chance of applying a New Random Allele (later referred to as NRA), this could be written as 2.5% chance of applying 'guaranteed flip' mutation ≈ 1 / 50).

Figure 4 shows the performance profile for the 64 bit H-IFF problem again at 20,000 evaluations. Here, in stark contrast to Figure 3, a third peak can be seen in the performance profile, particularly at low population sizes. It is believed that the left hand trough corresponds to a low mutation rate which represents a lower bound that is useful for finding a certain set of sub-optima in the problem space – for example blocks of size 4. However, as the mutation rate is increased, this rate becomes sub-optimal for this block size, and another mutation rate is found which proves useful in finding larger block sizes – possibly of size 8 or 16. This supposition is supported by the fact that a plot of mean fitnesses found against mutation rate shows marked step improvements between each of these troughs in evaluations used, implying that the EA was suddenly able to find better optima at each of these specific mutation rate boundaries. This observation can be seen in Figure 8 and is now the subject of further investigation which will be reported on in due course. It is likely that there exist other 'optimal' mutation rates for other block sizes in this problem, but these are not distinguishable at this scale of resolution on the mutation rate axis and this is again further explored in the discussion of Figure 8. As mutation is increased beyond an upper threshold, performance is seen to deteriorate, as in the previous two problem cases, into random search. The mutation rate inducing the first and second peaks are seen to be .0001 and .013 respectively.

Figure 5 shows the 20,000 evaluation performance profile for the NK50-8 problem. Here it can be seen, as was also visible in the ADDMP profile (Figure 2), that the deterioration of performance from the optimum trough is not smooth, with a 'ridge-like' anomaly again present on the left hand edge of the rightmost peak over a range of lower population sizes.

Figure 6 shows a performance profile for the same ADDMP instance as Figure 2, however this time the EA is allowed 1,000,000 evaluations over a population range of 10 through 300 in steps of 10. In fact results were obtained at each of fifty thousand, one hundred thousand, two hundred thousand and five hundred thousand evaluations, with the observed anomaly simply becoming more distinct as evaluation limit increased. Indeed, the mutation rate inducing the first peak, first trough and anomaly features were seen to decrease with increased evaluation limit, implying a relationship to total number of mutations utilised as well as actual mutation rate. This observation is also currently the subject of further investigation and will be reported on in due course.

As can clearly be seen in Figure 6, the mutation rate inducing the first peak has dropped from 4E-4 at 20,000 evaluations (Figure 2) to 1.3E-5 at 1 million evaluations. The secondary and tertiary peaks 'emerged' from the left hand edge of the right hand peak with higher evaluations limits until here at 1 million evaluations they can clearly be seen to be performance landscape features in their own right, persistent over a wide range of population sizes. Indeed, this plot shows similarities with results from the H-IFF problem (see Figures 4 and 8).

Figure 7 shows the 2 dimensional plot at a fixed population size of 20 on the 'Max-Ones' problem at 1,000,000 evaluations. Here, the position of the first peak can be seen to be at a considerably lower mutation rate than in Figure 3. The trough of low evaluations is seen to be much wider, indicating a wider range of mutation rates capable of finding the global optimum in a low number of evaluations. Figure 7 also shows plots of the mean number of evaluations plus and minus 1 standard deviation of results over the 50 runs. As can be seen, where mutation rates are too low, process variation is high, but as mutation rates approach optimum (the right hand edge of the left hand peak), this variation is reduced. Once mutation rates become counter productive (above 20% NRA), diversity is seen to increase again as the EA degenerates into random search. Also shown is the mean fitness of the best solution of each of the 50 runs at each mutation rate. The plateau representing runs finding the global optimum 50 times out of 50 is seen to coincide with the trough in evaluations indicating an ideally tuned EA.

Figure 8 shows a 2 dimensional performance profile at 1 million evaluations on the H-IFF 64 problem with a population size of 20. Here 3 distinct peaks and troughs can be seen before deterioration into random search, with specific peaks at a mutation rate of 1E-6, another at 1.6E-3 and a third at 4.4E-2. This is an increase in the number of features seen at 20,000 evaluations (Figure 4), and again it is noted that the features occur at lower mutation rates given the higher evaluation limit. It can also be clearly seen on the left most two troughs that the standard deviation in the troughs is at a distinct minimum on the left hand side of each trough in evaluations. Also, the average fitness plot clearly shows marked increases between the troughs showing that the EA is able to find step-like improvement as mutation rates are increased.

Finally, Figure 9 shows the 2 dimensional performance profile for our ADDMP scenario at 1 million evaluations with a population size of 20. Whilst the first peak and trough are clear (at mutation rates of 1.3E-5 and 8.2E-4 respectively, the second and third peak are less distinct, although clearly exist as features in Figure 6. There are clearly regions of low standard deviation corresponding to troughs at 8.2E-4, 2.5E-3 and 0.2, and a region of higher mean and much higher deviation between mutation rates of 2.5E-3 and 0.16. This relative disorder is to be expected as any structure in the search space is likely to be far less regular in the ADDMP scenario than in the case of H-IFF. Nonetheless, Figures 6 and 9 clearly show there to be some features for which key rates of mutation are either highly optimal or highly sub-optimal. The plot of mean fitness does not show such clearly defined 'step-like' improvement as in Figure 8, however once again it can be seen that there is general improvement towards higher mutation rates. Interestingly, and in contrast to other results shown here, fitness deterioration at exceedingly high rates of mutation is not as marked, again suggesting a lack of structure in the search space, favouring a 'random search' like process. Of course the marked increase in both mean and standard deviation of the number of evaluations taken to find these good solutions shows the relative inefficiency of random search with respect to a well tuned EA on this problem. The ideal mutation rate here being 0.31 (NRA) inducing high fitness solutions, in a minimum of evaluations with a minimum of process variation.

4 Conclusions

For the simple unimodal search space explored here, it can clearly be seen that a moderate rate of mutation (usually close to 1 / L) can produce optimal performance both in terms of consistent quality of solution found and minimum number of evaluations taken to find them. Where a higher number of evaluations is allowed, this range of suitable mutation rates extends towards ever lower values.

For the structured multi-modal search space, there appear to be a series of 'optimal' mutation rates, capable of exploiting certain features of the search space. In the case of H-IFF this is likely to be aligned blocks of constant length each separated by equal Hamming distance. Values around these 'optimal' rates are seen to be clearly sub-optimal in terms of process repeatability (increased mean and standard deviation). A 'globally optimum' mutation rate again seems to exist close to 1 / L which gives optimum fitness solutions, but in this case, not in an optimally minimum number of evaluations (however it is possible that further experimentation with a finer grained mutation rate range may show this not to be the case).

The performance profile of the ADDMP has been shown to exhibit some features in common with both structured and unstructured multi-modal search spaces. It is clearly multimodal over a range of population sizes where sufficient evaluations are permitted. The existence of key mutation rates inducing a low standard deviation in number of evaluations is highly significant when considering the application of EAs to this problem domain allowing a high degree of confidence in process repeatability. The fact that once again a mutation rate close to 1 / L is seen to give high fitness solutions in a low number of evaluations with low standard deviation, even when the allowed range of evaluations was considerably higher, is of critical importance to the real time application of these techniques to a real world problem domain.

5 Acknowledgements

The authors are grateful to British Telecommunications Plc for ongoing support for this research.

References

1. T Bäck, *Evolutionary Algorithms in Theory and Practice*, Oxford University Press, 1996

2. K Deb and S Agrawal : *Understanding Interactions among Genetic Algorithm Parameters.* in Foundations of Genetic Algorithms 1998, Morgan Kaufmann.

3. D Goldberg (1989), *Genetic Algorithms in Search Optimisation and Machine Learning*, Addison Wesley.

4. J Holland, *Adaptation in Natural and Artificial Systems*, MIT press, Cambridge, MA, 1993

5. Kauffman, S.A., The Origings of Order: Self-Organization and Selection in Evolution, Oxford University Press, 1993

6. Z Michalewicz, *Genetic Algorithms + Data Structures = Evolution Programs*, Springer, 1996.

7. H Mühlenbein and D Schlierkamp-Voosen (1994), *The Science of Breeding and its application to the Breeder Genetic Algorithm*, Evolutionary Computation 1, pp. 335-360.

8. H Mühlenbein, *How genetic algorithms really work: I. Mutation and hillclimbing*, in R.Manner, B. Manderick (eds), Proc. of 2nd Int'l Conterence on Parallel Problem Solving from Nature, Elsevier, pp 15-25.

9. E van Nimwegen and J Crutchfield : *Optmizing Epochal Evolutionary Search: Population-Size Independent Theory*, Santa Fe Institute Working Paper 98-06-046, also submitted to Computer Methods in Applied Mechanics and Engineering, special issue on Evolutionary and Genetic Algorithms in Computational Mechanics and Engineering, D Goldberg and K Deb, editors, 1998.

10. E van Nimwegen and J Crutchfield : *Optmizing Epochal Evolutionary Search: Population-Size Dependent Theory*, Santa Fe Institute Working Paper 98-10-090, also submitted to Machine Learning, 1998.

11. M Oates, D Corne and R Loader, *Investigating Evolutionary Approaches for Self-Adaption in Large Distributed Databases*, in Proceedings of the 1998 IEEE ICEC, pp. 452-457.

12. M Oates and D Corne, *QoS based GA Parameter Selection for Autonomously Managed Distributed Information Systems*, in Procs of ECAI 98, the 1998 European Conference on Artificial Intelligence, pp. 670-674.

13. M Oates and D Corne, *Investigating Evolutionary Approaches to Adaptive Database Management against various Quality of Service Metrics*, LNCS, Procs of 5th Intl Conf on Parallel Problem Solving from Nature, PPSN-V (1998), pp. 775-784.

14. M Oates, *Autonomous Management of Distributed Information Systems using Evolutionary Computing Techniques*, Computing Anticipatory Systems, AIP Conf Procs 465, 1998, pp. 269-281.

15. M Oates, D Corne and R Loader, *Skewed Crossover and the Dynamic Distributed Database Problem*, Artificial Neural Networks and Genetic Algorithms 1999, Dobnikar et al (eds), Springer pp 280-287.

16. M Oates, D Corne and R Loader , *Investigation of a Characteristic Bimodal Convergence-time/Mutation-rate Feature in Evolutionary Search*, in Procs of Congress on Evolutionary Computation 99 Vol 3, IEEE, pp. 2175-2182

17. Oates M, Corne D and Loader R, *Variation in Evolutionary Algorithm Performance Characteristics on the Adaptive Distributed Database Management Problem*, in Procs of Genetic and Evolutionary Computation Conference 99, Morgan Kaufmann, pp.480-487

18. M. Oates, J. Smedley, D. Corne, R. Loader, *Bimodal Performance Profile of Evolutionary Search and the Effects of Crossover*, in Procs of 1999 Evonet Summer School on Theoretical aspects of Evolutionary Computation.

19. G Syswerda (1989), *Uniform Crossover in Genetic Algorithms*, in Schaffer J. (ed), Procs of the Third Int. Conf. on Genetic Algorithms. Morgan Kaufmann, pp. 2 – 9

20. Watson RA, Hornby GS, and Pollack JB, *Modelling Building-Block Interdependency*, LNCS, Procs of 5th Intl Conf on Parallel Problem Solving from Nature, PPSN-V (1998), pp. 97-106.

21. Watson RA, Pollack JB, *Hierarchically Consistent Test Problems for Genetic Algorithms*, in Procs of Congress on Evolutionary Computation 99 Vol 2, IEEE, pp. 1406-1413

Protocol Construction Using Genetic Search Techniques

Nicholas Sharples & Ian Wakeman

University of Sussex
Falmer, Brighton, BN1 9QH
nichs@cogs.susx.ac.uk & ianw@cogs.susx.ac.uk

Abstract. The construction of transport protocols which offer reliable communication is a complicated task. The communicating agent must quickly adapt to changes in the network in order to maintain optimal performance. This adaptive element is an extremely difficult component to construct as the highly dynamic environment in which the protocol operates is difficult to predict. Our work attempts to automate the design process by converting it from a design problem to one of optimisation, in which genetic algorithms are used to search the space of possible protocol designs in an attempt to find the optimal solution. We present results from experiments in which we have evolved alternating bit protocols and also windowed flow control protocols, which have high channel utilisation.

1 Introduction

This work explores the possibility of applying evolutionary search strategies to the synthesis of network communications protocols. The principles we have adopted were originally pioneered in the field of Evolutionary Robotics [3] and [2], which uses genetic algorithms to search the design space of robot control architectures for systems which exhibit a desired behaviour. The ultimate goal of our research is to develop a reusable methodology, or conceptual framework, for the construction of the adaptive algorithms and protocols employed by transport layer protocols for packet switched networks. Specifically, we are focusing on the development of congestion control and avoidance algorithms such as those employed by reliable protocols such as the Transport Control Protocol(TCP). Since TCP is so extensively used, it is clear that any small improvement in its ability to utilise resources and adapt to change will greatly improve global network performance. However, hand-design of these concurrent processes, which must co-operate over an unreliable and continually changing communication medium, is a perplexing and formidable task. Design flaws are difficult to predict, and for the most part detect, as problems in the design will only become apparent when the algorithm is implemented at each node of the network.

At a purely software level, the Internet is composed of many small, interacting processes, each of which has a goal that is clearly defined at the level of that process: that of communicating data to a receiving peer process. At a higher level, the collective interaction of these processes is a direct result of the rules

S. Cagnoni et al. (Eds.): EvoWorkshops 2000, LNCS 1803, pp. 235–246, 2000.

that govern a single communication. However, unless we are careful at the lower level, the emergent dynamic at that higher level will be unpredictable, leading to network congestion and finally collapse. So, by careful development at this lower level we can control the higher-level emergent dynamic.

Developing adaptive transport control protocols for communications networks as large and highly dynamic as the Internet has inherent and conflicting problems. The primary function of reliably communicating data involves the coordination of remote processes. This is a difficult problem which is complicated further when the communication medium used exhibits unpredictable loss characteristics. Adding to this difficulty, communication must be performed in an efficient manner. A protocol must use the network resources in a friendly manner with respect to peer protocols, which is in conflict with the aim of efficiency. It must be able to detect changes in the network environment and consume either more or less of its resources depending on the circumstances around it. The protocol must satisfy all of these demands without global information about the state of the network, and further still, without global control over the path a packet may take.

The fundamental goal of this research is to develop a methodology which will allow the synthesis of adaptive protocols and provide insights into the design-space of such algorithms. Rather than a protocol designer hand-crafting the finite state machine at the heart of the protocol, he/she will specify the solution space and define a fitness function which will assess solutions against the protocol requirements. The designer will then use Genetic Search techniques to locate solutions which *optimise* the fitness function and so meet the requirements. In Section 2, we show how the various components of our methodology work together to evolve working protocols.

One of the most important functions of a communicating system is the ability to communicate data reliably using an unreliable medium. If a system can be developed in which reliable protocols can be synthesised, then it will go some way to proving that higher-level behaviours are also attainable using the same technique. In our first experiment, described in section 3.1, we show that reliable protocols using different techniques, such as acknowledgements and redundant transmissions, are evolved under different simulation conditions. When we factor channel capacity into our problem space, channel utilisation becomes an important measure of the fitness of the solution. In the second experiment, described in Section 3.2, we show that windowing protocols are evolved which have the ability to run the channel at 100% utilisation. We conclude with a discussion and give pointers as to how protocols which adapt to congestion can be evolved.

2   Methodology

A network can be viewed as a virtual environment. A protocol is a communicating entity of that environment whose actions are dictated by its underlying finite state machine. Since in this respect a protocol is just a control architecture, it is clear that in order to evolve adaptive communication protocols we can use the same evolutionary techniques as those used in the production of robot control architectures.

In order to evolve network protocols, we require the implementation of the following components:

- Representation of protocol
- Search strategy
- Encoding scheme
- Operators to manipulate encoding
- Network simulation
- Fitness function

2.1 Representation of protocol

The representation we have chosen to use is an extension of the finite state machine(FSM), called the communicating finite state machine(CFSM). We have chosen to use the CFSM as our representation for two important reasons. Firstly, protocols are easily represented using the CFSM and this has been adopted by the protocol research community as a standard representation [6], [4]. Second, the FSM has been used in computer science and artificial intelligence for many years, and its characteristics are well documented and understood. Researchers have used the FSM as a representation for optimisation problems for some time, using both traditional and evolutionary search strategies [5].

The CFSM differs from the FSM in that it can affect as well as react to the environment in which it operates. The CFSM passes messages or signals via bounded FIFO queues. Each machine has an input and an output queue and executes a two-step algorithm. During the first step, input signals are evaluated and a transition rule selected. In the second step, the output signal is updated and the machine changes state. If none of the rules for a state can be executed the machine remains in its current state.

2.2 Search Strategy

Fundamental to this work is the ability to view the protocol as a control architecture which can be represented in some symbolic form. This representation must be one that can be mutated over successive iterations, through the search space of possible solutions. This transforms the problem from one of controller development into one of controller *optimisation*. The task of generating complex control architectures using search has been accomplished, in other fields using Genetic algorithms(GAs). We will attempt to construct communication protocols using similar principles.

2.3 Encoding Scheme

The encoding scheme used to represent the solution in the problem space is closely tied to the search strategy employed. The obvious choice of representation when using genetic algorithms is a binary string in which each transition is represented by a sequence of bit values. In addition to allowing easy manipulation by the standard genetic operators, this representation will also lend itself easily to variable length encoding. The binary encoding of the CFSM is know as the genome and represents a solution in the search space. Prior to evaluation, the genome is transformed into the phenome, an instantiation of the solution in design space.

2.4   Operators to Manipulate Encoding

The genetic algorithm employed during this research uses the standard genetic operators i.e. mutation, crossover, addition and subtraction. For the sake of clarity, the following sections detail the actual implementation used.

Mutation. A figure is specified for each run, relating to the number of mutations per genome. During mating, each bit in the string is given an individual chance of mutation, each chance being an equal proportion of the mutation rate for this genome. Each bit in the string is then check against the recalculated mutation rate, and any mutation flips the bit value.

Crossover. Many forms of crossover exist and can operate on a per gene or per bit basis. For all the experiments carried out during this research, single point crossover has proved adequate. For the most part this has been implemented on a per gene basis, where the crossover point is restricted to gene boundaries.

Addition and Subtraction. The addition and subtraction operators allow for a thorough exploration of the search space, adding or removing genes from the genome randomly. Addition adds a new randomly constructed gene to the genome. Although no selection pressure for size is introduced by the fitness function, the redundancy in longer genomes creates robust solutions with regard mutation. This is because the working genes of the solution (the parts of the finite state machine that contribute to fitness) can hide themselves amongst the redundant genes of the genome.

2.5   Network Simulation

For linear optimisation problems, each candidate solution is evaluated against a static fitness function which provides a measure of its ability to solve the problem. For non-linear problems, such as those faced here, we must evaluate a protocol's functionality in its working environment. For many applications this is impractical due to the amount of time required for an evaluation, and a computer model must be used to simulate the agent's environment and its interaction with it. This is not the case for the application we propose; evaluation of a protocols performance could take place in a real network without unreasonable time overhead, although at this stage, real world evaluation would add unnecessary complexity. What is required then is a test bed in which to evaluate the performance of a protocol within its working environment, or as close to that environment as is possible.

The following experiments use a simple sender-receiver network simulation, an example of which can be seen in figure 1. To initiate a simulation, two phenomes are constructed: a sender and a receiver. The CFSM component of each phenome is generated from the genome to be evaluated. Both sender and receiver have an associated communication channel, the characteristics of which are dependent on the experiment being performed. Typically, each phenome also has a transmission buffer and a receive buffer. On the sending machine, the transmission buffer contains the message to transmit. In addition, each machine also has some form of current memory in which data items can be held temporarily before being placed into buffers or messages.

The simulation executes for a specific number of time steps. During each step, the sender and then the receiver phenomes get a chance to execute. During execution, each machine iterates through the transitions of its current state, from first to last. The action associated with the transition is evaluated, if the action can be performed the transition *fires*. In effect, each transition forms a boolean expression. If the expression holds then the machine makes the transition to the next state. At each time step in the simulation, every possible transition of the current state is attempted until an execution is capable of firing. At this point the machine changes state to the state indicated by the next-state field of the transition.

Future work will involve larger simulations in which many simulated network nodes will communicate across a single, shared channel, with each node employing the same protocol as in the real world. The evaluation will be a measure of the global traffic through-put across the channel. Continued work may use the current real-world Internet in place of the simulation.

2.6   Fitness Function

The purpose of the fitness function is to evaluate the performance of the protocol during its execution in the network simulation and provide some quantifiable measure of its functionality. The measure itself should include as few components as possible, since the greater the number of components the less predictable the population's convergence on the desired point in solution space. This measure can be calculated in two ways: the most difficult is to gather information during the simulation and calculate the fitness from the data collected; an easier option is to measure the fitness of the state of the network nodes after the simulation has finished.

We must bear in mind that assessing the value of a solution using only a single simulation does not produce an accurate measure of its ability, since it is possible that the protocol is exploiting some aspect of the test data used. In order to gain a fair measure of the fitness of a solution, each must be tested a number of times, using different test data. We must be careful when selecting the number of re-tests to perform, if we use too many re-tests the performance will degrade and if we use too few the solution will not be adequately tested. This leaves us some choice as to which value should be chosen: the highest, lowest or some form of mean. Through experimentation, we found that using the mean value generated valid machines in the least number of generations. Therefore, this is the value we have used in the following experiments.

3   Experiments

The following sections detail two of the experiments undertaken to date. In each case, the experiment was performed using a distributed population of four hundred individuals evolved for one thousand generations. The task was to communicate ten data items from the sender to the receiver.

3.1   Reliability

In our initial experiment we set out to synthesise a fully reliable communication protocol, and in doing so prove that the control architecture underlying a communication protocol can be generated using evolutionary search. To provide a measure of the quality of the synthesised protocol, each candidate solution was evaluated using a network simulation which emulates the environment for which the alternating bit protocol is an optimal solution. It was hoped that the system would produce a machine with similar state transitions and external behaviour. The alternating bit protocol is a simple protocol that provides reliable communication using an alternating validation bit, it indicates whether or not a message was lost during transmission. It is well documented in the literature and was first proposed in [1].

Network Simulation. To generate a sender similar to that of our benchmark machine, the simulation must closely model the operating environment of the original protocol. For this, the channel must have the capacity to hold only a single message. In accordance with the reference protocol; only messages can be lost and if a message is lost, the channel will flip the bit in the message header. For a schematic overview of the simulation see figure 1.

Actions. The following actions have been extrapolated from the machine specified in the original alternating bit protocol.

- Send : Send a message on the communications channel, taking data from memory if required.
- Receive : Receive a message from the communications channel and place the data into memory.
- Enqueue : Place the item in memory, into the receive buffer.
- Dequeue : Place the first element of the transmit buffer into memory.

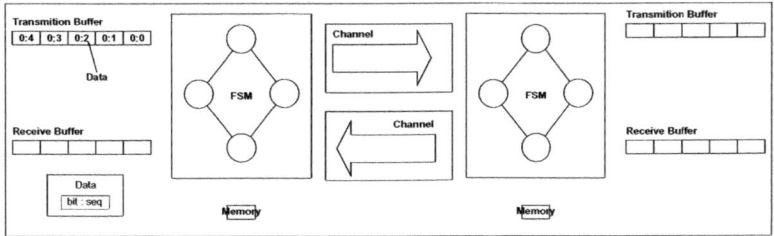

Fig. 1. Network simulation for evolution of reliable communicating finite state machine.

– Null : Null action always works

Each of these actions returns a true or false boolean value depending on its ability to complete the specified task. Only for the send action is this not the case, it will return true under all conditions to prevent unintentional feedback from the receiving machine. In the case where the channel already contains a message, the new message is discarded.

Fitness Function. The fitness for a single protocol is based on the proportion of a message received after a fixed number of time steps. To ensure a reliable protocol is produced, only the in-order portion of the received message is considered. A genome will score zero even if all but the first message was successfully received. If this was not the case, and the entire proportion of the message was considered, then the population would prematurely converge on an unreliable solution and become incapable of back-tracking through the search space to reliable solutions.

Results. The CFSM shown in figure 3 has been generated during a typical run of the system. As with a typical state transition diagram, each node represents a state. Each transition is represented as an arch from the state in which it is executed to the the next state. A transition is marked with the action to be taken and the parameters for that action. Some actions require no parameters, in which case this redundant encoding in the transition adds neutrality to the genome: increasing its robustness with regard to mutation. The transitions are marked with the order in which they are to be attempted, shown in square brackets.

As can be seen in the diagram, the sender continually sends a message with identifier "1". Only after receiving an acknowledgement from the receiver does it dequeue the next data item. This is a simple stop-go strategy and is the closest we have come to generating the alternating bit protocol. Note: each transition has the same number of parameters, although for Enqueue, Dequeue and Null these are superfluous and discarded during execution.

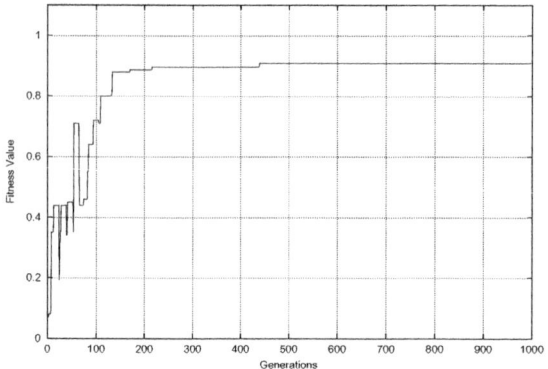

Fig. 2. Fitness value obtained by best individual during evolution of a reliable protocol.

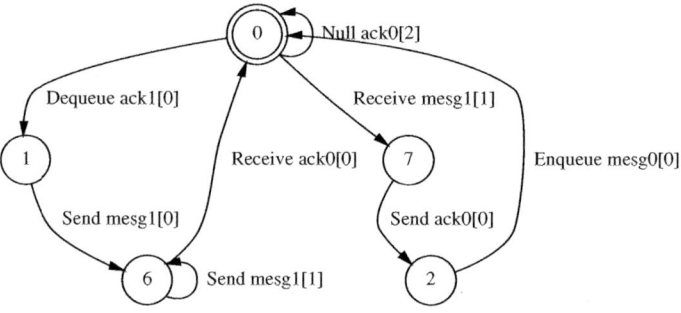

Fig. 3. The typical result of the reliable communication experiment, a stop go strategy.

3.2   Optimal Communication

In the second set of experiments we set out to develop protocols which attempt to make optimal use of the bandwidth available. Therefore, a good solution would be one which transmitted a new message during each time step. Ideally, we would like to generate a machine which exhibits a windowing behaviour.

Network Simulation. Of critical importance for this experiment is the latency encountered during communication. To simulate this, the communication channels of the phenome have been converted from the standard queue structure, used in the standard CFSM, into an *array* of size equal to the latency required. During each simulation time step, messages on the channel are moved one place along the array. As with the reliability experiments, loss is calculated during the receive phase of communication. If a message is lost during communication it is not passed to the receiving phenome. In the previous experiment, each machine was permitted one action per time step. However, initial experiments here revealed that solutions generated using this timing scheme overcome reliability by sending multiple packets. The time spent storing a packet in the retransmission buffer was used to send a second packet, successive packets having far less chance of loss than the previous. To force the development of a retransmission scheme, each transition has been assigned a parameter which specifies the length of time it may take to execute the action.

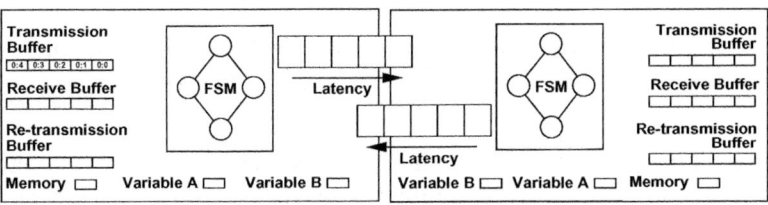

Fig. 4. Network simulation for evolution of reliable communicating finite state machine, which also communicates data in optimal time.

Actions.

- Send : Send a message on the communications channel, taking data from memory if required.
- Receive : Receive a message from the communications channel and place the data into memory.
- Enqueue : Place the item in memory into the receive buffer.
- Dequeue : Place the first element of the transmit buffer into memory.
- EnqueueRetransmit : Place the item in memory into the retransmit buffer.
- DequeueRetransmit : Place the first element of the retransmit buffer into memory.
- RemoveFromRetransmit : Remove the current item of data from the retransmission buffer.
- SetVariableA : Set variable A with the current item in memory.
- GetVariableA : Set the memory to the value of variable A.
- SetVariableB : Set variable B with the current item in memory.
- GetVariableB : Set the memory to the value of variable B.
- Null : Null action always works

Phenome. In addition to the standard transmit, receive and memory buffers of the phenomes used previously, the phenome used for this set of experiments is also equipped with a finite re-transmission buffer and two variables with which to manipulate data. The re-transmission buffer has been added to encourage solutions to operate on *segments* of the overall message. The extra memory variables enable temporary storage of data when the re-transmission buffer becomes full.

Fitness Function For this experiment, the fitness function must assess the solution's ability to reliably communicate the data. In addition, it must also provide some measure of the time taken for the communication to take place; solutions which communicate most quickly score a higher fitness value. To determine this value, we have combined the original function used to develop the reliable protocol with an additional value which indicates the proportion of the simulation-time taken to complete the data exchange.

$$\frac{1}{2} \left(\frac{MaxTimeSteps - TimeTaken}{MaxTimeSteps} + \frac{ItemsCommunicated}{MessageLength} \right)$$

Results. The graphs presented in figure 5 show the average amount of simulation time taken to send the complete message. If the entire message was not received by thirty time steps the simulation was terminated. Also shown is the average number of data items successfully received during an evaluation. The graph in figure 7 shows the corresponding fitness value of the highest scoring individual at each generation.

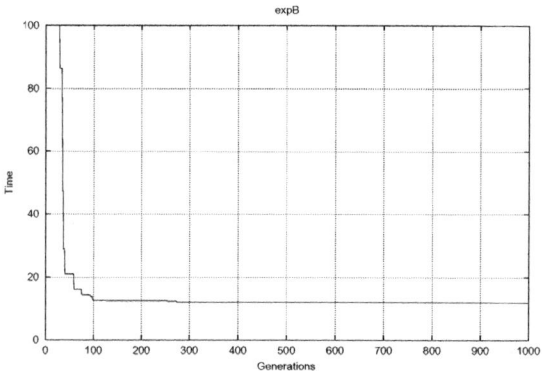

Fig. 5. Time Taken to Communicate Complete Message

4   Discussion and Future Work

This paper has shown that the search space of finite state machines for communication protocols can be explored using genetic search techniques.

Initially, we attempted to construct well known protocols such as the alternating bit protocol. Attempting to force the search process to generate specific protocols proved difficult, with the desired protocol never being evolved. However, valid protocols which performed the equivalent task *were* produced, thus

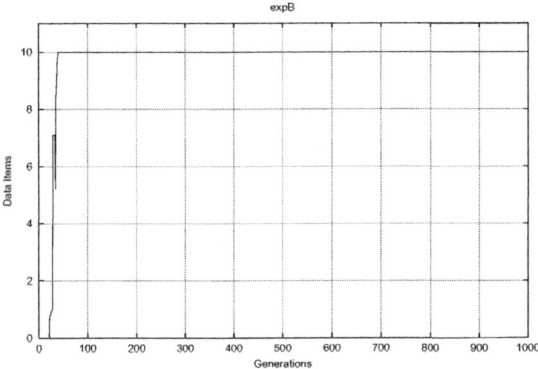

Fig. 6. Number of Data Items Communicated

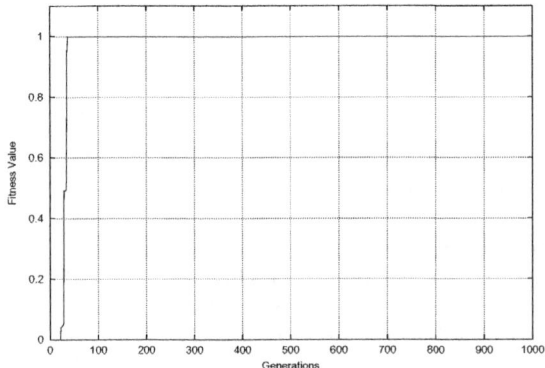

Fig. 7. Fitness value obtained by best individual during evolution of a optimal protocol.

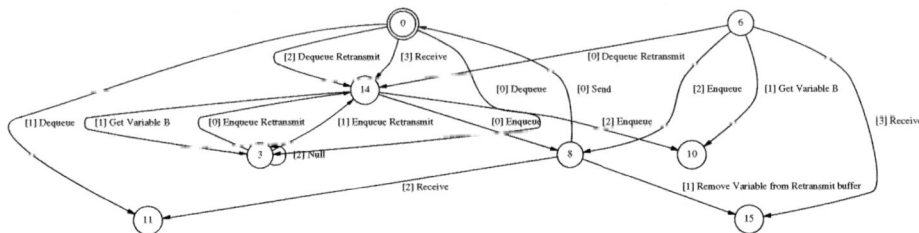

Fig. 8. The typical result of the optimal communication experiment.

proving that protocols exhibiting a desired behaviour can be generated using this technique.

The work undertaken here has shown that the the process of constructing valid network simulations is a difficult problem. However, for complex protocols which operate in highly dynamic environments, the difficulty in constructing the simulation will be negated by the difficulties of trying to predict the emergent dynamic. The advantage of using this approach is its ability to give us direct feedback on the protocol's performance in the simulated environment in order to drive the design process, since the fitness function uses a simulation of the network environment to determine the performance of the solutions. By extending the simulation space from a single pair of protocol entities to multiple communicating entities, we can assess the impact of a single communication strategy on the global network dynamic. By feeding-back simulation performance into the design process, we can ensure that the protocol produced is both robust and optimal with respect to the global network dynamic.

Our next experiments will have multiple entities communicating over a single channel using the same protocol. The fitness function will use some notion of fairness, as well as reliability and utilisation, to ensure that the evolved protocol reacts to congestion.

References

1. Scantlebury R.A. Bartlett K.A. and P.T. Wilkinson. A note on reliable full-duplex transmission over half-duplex lines. *Communication of the ACM*, 12(5):260–265, 1969.
2. D. Cliff, I. Harvey, and P.Husbands. Incremental evolution of neural network architectures for adaptive behaviour. Technical Report Cognitive Science Research Paper CSRP256, School of Cognitive and Computing Sciences, University of Sussex, Brighton BN1 9QH, England, UK, 1992.
3. D. Cliff, P. Husbands, and I. Harvey. Evolving visually guided robots. Technical Report Cognitive Science Research Paper CSRP220, School of Cognitive and Computing Sciences, University of Sussex, Brighton BN1 9QH, England, UK, 1992.
4. Brand D. and Zafiropulo P. On communicating finite state machines. *Journal of the ACM*, 30(2):323–342, 1983.
5. Lawrence J. Fogel. *Artificial intelligence through simulated evolution*. John Wiley and Sons Inc, 1966.
6. G.V.Bochmann. Finite state descriptions of communication protocols. *Computer Networks*, 2(4/5):361–372, 1978.

Prediction of Power Requirements
for High-Speed Circuits

F. CORNO, M. REBAUDENGO, M. SONZA REORDA, M. VIOLANTE

Dip. Automatica e Informatica
Politecnico di Torino
http://www.cad.polito.it

Abstract. Modern VLSI design methodologies and manufacturing technologies are making circuits increasingly fast. The quest for higher circuit performance and integration density stems from fields such as the telecommunication one where high speed and capability of dealing with large data sets is mandatory. The design of high-speed circuits is a challenging task, and can be carried out only if designers can exploit suitable CAD tools. Among the several aspects of high-speed circuit design, controlling power consumption is today a major issue for ensuring that circuits can operate at full speed without damages. In particular, tools for fast and accurate estimation of power consumption of high-speed circuits are required. In this paper we focus on the problem of predicting the maximum power consumption of sequential circuits. We formulate the problem as a constrained optimization problem, and solve it resorting to an evolutionary algorithm. Moreover, we empirically assess the effectiveness of our problem formulation with respect to the classical unconstrained formulation. Finally, we report experimental results assessing the effectiveness of the prototypical tool we implemented.

1. Introduction

Modern telecommunication systems must ensure high bandwidth and high performance to cope with the increasing amount of delivered data. Different solutions at different levels of abstraction can be exploited: improved communication channels, better communication protocol and high-speed communication equipment. Modern VLSI design methodologies and manufacturing technologies are making circuits every day faster. To cope with the complexity of designing high-speed circuits designers must exploit suitable CAD tools. Among the several aspects of high-speed circuit design, power consumption is today a major issue. In the last years, design for low-power consumption has became a widespread design paradigm. Design techniques to control power consumption are mandatory because excessive power dissipation can cause performance degradation, run-time errors, or device destruction due to overheating. Large instantaneous power dissipation may also cause local hot spots that have negative impact on circuit reliability. With increasing demands for reliable high-speed circuits accurate estimation of peak-power dissipation during design process is becoming essential.

S. Cagnoni et al. (Eds.): EvoWorkshops 2000, LNCS 1803, pp. 247–254, 2000.

In this paper we address the peak-power estimation problem for high-speed sequential circuits. Power consumption in sequential circuits is a non-linear function, P_f, of three variables: the circuit initial state, S, and two vectors (V_{t-1}, V_t) that are applied to the circuit inputs at time t-1 and t. We represent them as a 3-tuple (S, V_{t-1}, V_t). The problem of computing the peak-power consumption of sequential circuits can be formulated as the problem of finding the 3-tuple (S, V_{t-1}, V_t) that maximizes the function P_f.

Several techniques have been proposed to solve this problem resorting to either exact approaches [1] or techniques based on Automatic Test Pattern Generators [2]. But only in [3] and [4] an algorithm is proposed able to cope with large sequential circuits. In particular the peak-power problem is formulated as an unconstrained optimization problem and a technique based on Genetic Algorithms is proposed. This technique is strongly approximated since the reachability property of the state S is neglected during the optimization process. As a consequence, this approach could lead to solutions where the 3-tuple (S, V_{t-1}, V_t) embeds a state S that the sequential circuit never reaches. To make the 3-tuple a feasible solution, the authors of [3] and [4] propose to replace at the end of the optimization algorithm the state S with a reachable one as close as possible to S.

In the following, we will analyze the behavior of this algorithm, and experimentally show that it can underestimate peak-power when the number of reachable states is a small fraction of all the possible states. We will then propose an approach that overcomes this drawback by exploiting the knowledge about the reachable state set during the optimization process. In this paper we propose and evaluate an implementation of our peak-power estimation algorithm that is based on the Selfish Gene algorithm [5], an evolutionary optimization algorithm we developed. We support our approach by reporting experimental results assessing the effectiveness of our approach.

The algorithm presented here is based on a previous work [8] where Genetic Algorithms were used to solve the peak-power problem. In [8] a Genetic Algorithm evolves a population where each individual is a 3-tuple composed of a reachable state and two vectors. To guarantee that the population contains only valid individuals, each time crossover and mutation operators are applied the state in the new individual is analyzed, and replaced with a reachable one when required. This approach can limit the search capability of Genetic Algorithms and thus the effectiveness of our peak-power estimation algorithm. By resorting to the peculiarities of Selfish Gene, state modification is no longer required, and thus the effectiveness of our algorithm can be improved.

The remainder of the paper is organized as follows. Section 2 presents a formal description of the problem. Section 3 describes the evolutionary algorithm we exploit in our tool, and Section 4 presents the peak-power estimation algorithm we developed. Section 5 reports experimental results on some benchmark circuits. Finally, Section 6 draws some conclusions.

2. Problem formulation

We assume that the sequential circuits we are analyzing are manufactured with CMOS technology. Power consumption in CMOS circuits is mainly due to the number of times the output of circuit gates switches from 0 to 1 and vice-versa. Peak-power consumption is defined as the maximum power consumption that takes place when a couple of input vectors are applied to the circuit inputs, starting from a given initial state. The procedure to measure peak-power can be outlined as follows:
1. the circuit memory elements are initialized to a given state S;
2. an input vector V_{t-1} is applied to the circuit, and a clock pulse is applied;
3. an input vector V_t is applied to the circuit, a clock pulse is applied and the power consumption is measured during the transient period.

The first two steps lead the circuit to a configuration where the circuit nodes hold known values (either 0 or 1). We refer to the power measured during the last step as P_f, and use to notation $P_f(S, V_{t-1}, V_t)$ to indicate that power is function of the 3-tuple (S, V_{t-1}, V_t). The classical peak-power problem formulation that can be found in [3] and [4] is the following:

$$PP = MAX(P_f(S, V_{t-1}, V_t)), S \in \mathrm{B}^n, V \in \mathrm{B}^m \tag{1}$$

where PP is the peak-power consumption, $\mathrm{B} = \{0,1\}$, n is the number of memory elements in the circuit and m is the number of circuit inputs. This formulation neglects the fact that in sequential circuits the number of reachable states is often less than 2^n. Therefore, algorithms that solve equation (1) could compute 3-tuples that correspond to PP values that cannot be obtained, since the memory elements can never reach the configuration S during circuit operations. Power figures attained by solving equation (1) are overestimation of actual peak-power consumption, and if adopted during design could lead to unnecessary expensive design solutions. The authors of [3] and [4] suggest replacing an unreachable state in the solution of (1) with the closest reachable one. This approach weakens the correlation between S and (S, V_{t-1}, V_t) and usually lead to peak-power figures that underestimate the actual power consumption. As a consequence, the obtained peak-power prediction could lead to wrong design solutions.

The problem formulation we propose is the following:

$$PP = MAX(P_f(S, V_{t-1}, V_t)), S \in \Sigma, V \in \mathrm{B}^m \tag{2}$$

where $\Sigma \subseteq \mathrm{B}^n$ is the set of states that the sequential circuit under analysis can reach starting from the reset state, i.e., the state where all the memory elements are set to 0. By solving equation (2) and therefore by considering reachability of the state S as a dimension of the search space, we compute peak-power figures more accurate than what equation (1) provides.

3. The Selfish Gene algorithm

The *Selfish Gene algorithm* (SG) is an evolutionary optimization algorithm based on a recent interpretation of the Darwinian theory. It evolves a population of individuals seeking for the fittest one. In the *selfish gene* biological theory, population itself can be simply seen as a *pool of genes* where the number of individuals, and their specific identity, are not of interest. Therefore, differently from other evolutionary algorithms, the SG resorts to a statistical characterization of the population, by representing and evolving some statistical parameters only. Evolution proceeds in discrete steps: individuals are extracted from the population, collated in tournaments and winner offspring is allowed to spread back into the population.

An individual is identified by the list of its genes. The whole list of genes is called *genome* and a position in the genome is termed *locus*. Each locus can be occupied by different genes. All these candidates are called the gene *alleles*. In the context of an optimization problem, looking for the fittest individual corresponds to determine the best set of genes according to the function to be optimized.

Since the SG "virtual" population is unlimited, individuals can be considered to be unique, but some genes would certainly be more frequent than others might. At the end of the evolution process, the *frequency* of a gene measures its *success* against its alleles. However, at the beginning of the evolution process, the frequency can be regarded as the gene *desirability*. When the majority of a population is characterized by the presence of a certain characteristic, new traits must harmonize with it in order to spread.

```
genome SG(VirtualPopulation P)
{
    genome BEST, G₁, G₂, winner, loser;

    iter = 0;
    BEST = select_individual(P);        // best so far
    do {
        ++iter;
        G₁ = select_individual(P);
        G₂ = select_individual(P);
        tournament(G₁, G₂); // identify winner and loser
        increase_allele_frequencies(P, winner);
        decrease_allele_frequencies(P, loser);
        if(winner is preferable to BEST)
            BEST = winner;
    } while(steady_state()==FALSE && iter<max_iter) ;
    return BEST;
}
```

Fig. 1: Selfish Gene algorithm pseudo-code.

The pseudo-code of the SG algorithm core is reported in Fig. 1. Further implementation details about the SG algorithm are available in [9], while biological motivations are better analyzed in [10]. An extension to the basic SG algorithm to deal with more complex fitness landscapes is described in [11].

4. Power estimation algorithm

The peak-power estimation algorithm we developed is composed of the following steps:

1. the set of the reachable states $\Sigma \subseteq B^n$ is computed, which cardinality is $|\Sigma|$. It can be computed either resorting to exact symbolic calculation techniques [7] or through logic simulation;
2. peak-power estimation is performed. The SG is run to solve equation (2) where the genome represents the 3-tuple (S, V_{t-1}, V_t) and is composed of $1 + 2 \cdot m$ loci. The first locus is an index ranging from 0 to $|\Sigma| - 1$ representing a reachable state $S \in \Sigma$. The remaining loci are binary values, coding the couple of vectors (V_{t-1}, V_t).

During step 2, the power consumption $P_f(S, V_{t-1}, V_t)$ is computed resorting to a unit-delay logic simulator. The adoption of a logic simulator is a well-known effective approach to measure power consumption in CMOS circuits [6], since it conjugates simulation speed with accuracy.

5. Experimental results

A prototypical version of our algorithm named SG-ALPS, Selfish Gene-based AnaLyzer of Power in Sequential circuits, has been written, which implements the above-introduced procedures. The tool consists of about 500 lines of ANSI C code and exploits the SG and logic simulation packages developed at our institution. Reachability analysis has been performed resorting to exact calculation techniques exploiting the BDD [7] package developed at our institution.

The subset of ISCAS'89 sequential circuits tractable with symbolic calculation techniques have been used to evaluate the performance of SG-ALPS: all the experiments have been performed on a Sun UltraSparc 5/333 with 256 MB RAM. To compare SG-ALPS with a state-of-the-art tool, we have re-implemented the algorithm proposed in [4]. Experimental results of our re-implementation of [4] are the same of [4]; we can therefore perform a fair comparison between the two tools.

Two sets of experiments have been performed. The first one aims at empirically showing that solving equation (1) as done by [4] could greatly underestimate peak-power consumption. Conversely, the second set of experiments aims at assessing the effectiveness of the approach we propose.

Table 1 reports results we gathered with our implementation of [4]. The first column reports the benchmark name, *PPU* and *PPR* report respectively the peak-power consumption obtained by solving equation (1) (which takes into account also unreachable states) and the power obtained by replacing the state in the solution with the closest reachable one. Column Δ reports the difference between these power figures, while column CPU reports the time requirements.

By observing the column Δ, one can observe that several circuits exist where peak-power strongly depends on the reachability of the initial state S and on the correlation between S and the vectors (V_{t-1}, V_t). Where a significant loss in peak-power is found, a large portion of the state S has been modified to make it reachable, thus a high difference exists between the ideal initial state and the selected reachable one.

Circ	PPU	PPR	Δ [%]	CPU [s]
s208	0.900	0.900	0.00	3.2
s298	1.007	0.833	-17.28	3.5
s344	1.573	1.552	-1.34	4.7
s349	1.020	0.869	-14.80	4.6
s382	0.961	0.790	-17.79	4.9
s386	0.855	0.855	0.00	4.6
s400	1.000	0.435	-56.50	5.2
s420	0.893	0.893	0.00	5.3
s444	1.077	0.551	-48.84	6.3
s499	0.595	0.230	-61.34	4.9
s510	0.860	0.860	0.00	6.7
s526	0.915	0.631	-31.04	7.0
s641	2.753	2.739	-0.50	17.0
s713	2.793	0.837	-70.02	16.5
s820	0.951	0.951	0.00	9.9
s832	0.927	0.927	0.00	9.9
s1196	1.035	1.020	-1.45	14.7
s1238	1.015	1.000	-1.47	15.3
s1488	1.235	1.156	-6.40	22.7
s1494	1.234	1.155	-6.39	22.6
Avg.			-15.96	10.4

Table 1. Analysis of algorithm [4]

Table 2 reports results obtained by running SG-ALPS when max_iter (Fig 1) is set to 1000. Column SG-ALPS reports the peak-power figures predicted by our algorithm, while the third column reports the best results attained by the algorithm proposed in [4]. We compare the two algorithms in column Δ. Finally, the CPU time requirements are reported.

As far as peak-power estimation accuracy is concerned, we can conclude that our approach is superior to [4]. Even if on the average SG-ALPS attains PP figures 21% higher than [4], several circuits exist where SG-ALPS computes power figures 50% higher than [4], thus showing the importance of considering reachability during the optimization phase.

As far as CPU time is concerned, the algorithm proposed in [4] is far more effective than SG-ALPS. This is mainly due to two factors: SG requires more time to converge than Genetic Algorithms, and guaranteeing state reachability during the optimization process is a time consuming operation.

We do point out that the CPU requirements of SG-ALPS, although higher than those of previously proposed approaches, still remain in the order of minutes, which is negligible with respect to the time required by most steps in the current circuit design flow. The increase in the CPU time is worth paid by the improvement in the attained estimation quality.

When compared with the method proposed in [8], implementing a similar approach using Genetic Algorithm, we observed that the algorithm proposed here provides comparable results. However, the adoption of the SG algorithm allows to more easily deal with optimization problems for which a heterogeneous solution encoding is more suitable.

Circ	SG-ALPS PP	[4] PP	Δ [%]	CPU [s]
s208	0.900	0.900	0.00	28.1
s298	0.833	0.833	0.00	56.6
s344	1.553	1.552	0.03	218.0
s349	1.535	0.869	76.65	231.9
s382	0.859	0.790	8.72	198.5
s386	0.872	0.855	2.01	37.5
s400	0.850	0.435	95.43	110.6
s420	0.893	0.893	0.00	27.1
s444	0.942	0.551	70.96	103.7
s499	0.230	0.230	0.00	15.5
s510	0.869	0.860	1.05	45.0
s526	0.737	0.631	16.80	399.7
s641	2.739	2.739	0.00	58.3
s713	2.155	0.837	157.36	98.2
s820	0.962	0.951	1.16	50.5
s832	0.967	0.927	4.35	71.4
s1196	1.044	1.020	2.37	168.8
s1238	1.050	1.000	5.00	595.6
s1488	1.157	1.156	0.04	117.8
s1494	1.156	1.155	0.08	116.4
Avg.			21.0	130.9

Table 2. SG-ALPS results

6. Conclusions

High-speed telecommunication equipment is required to cope with the market request for high-speed data delivering. When designing high-speed circuits for communication equipment, particular care must be posed to the control of power consumption. In particular, peak-power estimation is mandatory for designing today high-speed circuits. In this paper the peak-power prediction problem has been formulated as a constrained optimization problem, and an algorithm based on a new evolutionary paradigm has been proposed. The algorithm is designed for addressing sequential circuits and thanks to its ability of guaranteeing feasibility of solutions during the optimization process it is more effective than other approaches.

Moreover, in this work we experimented that SG is better suited to address optimization problems having heterogeneous solutions, i.e., solutions mixing components defined over different domains. As an example, in peak-power estimation the solution embeds a state that is naturally expressed as an index in a state table, and a couple of vectors, represented as a collection of binary values.

We are currently working toward an implementation of SG-ALPS where approximated reachability analysis is performed through logic simulation. Thanks to this improvement, we will be able to address large sequential circuit currently not tractable by symbolic calculation techniques.

7. References

[1]   S. Manne, A. Pardo, R. I. Bahar, G. D. Hachtel, F. Somenzi, E. Macii, M. Poncino, "Computing the maximum power cycles of a sequential circuit", Proc. of IEEE/ACM DAC, 1995, pp. 23-287

[2]   C.-Y. Wang, K. Roy, "Maximum Current Estimation in CMOS Circuits Using Deterministic and Statistical Techniques", IEEE Trans. on VLSI Systems, March 1998, pp. 134-140

[3]   M.S. Hsiao, E.M. Rudnick, J. Patel, "K2: An Estimator for Peak Sustainable Power of VLSI Circuits", Proc. of Int. Symp. on Low Power Electronics and Design, 1997, pp. 178-183

[4]   M.S. Hsiao, E.M. Rudnick, J. Patel, "Effects of Delay Models on Peak Power Estimation of VLSI Sequential Circuits", Proc. of IEEE/ACM ICCAD, 1997, pp. 45-51

[5]   F. Corno, M. Sonza Reorda, G. Squillero, "Optimizing Deceptive Functions with the SG-Clans Algorithm", CEC'99: 1999 Congress on Evolutionary Computation, Washington DC (USA). July 1999, pp. 2190-2195

[6]   A. Ghosh, S. Devadas, K. Kuetzer, J. White, "Estimation of average switching activity in combinational and sequential circuits", Proc. of IEEE/ACM DAC, 1992, pp. 253-259

[7]   R. E. Bryant, "Symbolic Boolean Manipulation with Ordered Binary Decision Diagrams," ACM Computing Surveys, Vol. 24, No. 3, 1992, pp. 293-318

[8]   F. Corno, M. Rebaudengo, M. Sonza Reorda, M. Violante, "ALPS: A Peak-Power Estimation Algorithm for Sequential Circuits", GLS-VLSI'99: 8th Great Lakes Symposium on VLSI, 1999, pp. 350-353

[9]   F. Corno, M. Sonza Reorda, G. Squillero, "The Selfish Gene Algorithm: a New Evolutionary Optimization Strategy", SAC'98: 13th Annual ACM Symposium on Applied Computing, 1998, pp. 349-355

[10]   F. Corno, M. Sonza Reorda, G. Squillero, "A New Evolutionary Algorithm Inspired by the Selfish Gene Theory", ICEC'98: IEEE International Conference on Evolutionary Computation, 1998, pp. 575-580

[11]   F. Corno, M. Sonza Reorda, G. Squillero, "Optimizing Deceptive Functions with the SG-Clans Algorithm", CEC'99: 1999 Congress on Evolutionary Computation, 1999, pp. 2190-2195

A Communication Architecture for Multi-Agent Learning Systems

N. Ireson, Y. J. Cao, L. Bull and R. Miles

Intelligent Computer Systems Centre
Faculty of Computer Studies and Mathematics
University of the West of England, Bristol, BS16 1QY, UK
Bristol, BS16 1QY, UK

Abstract. This paper presents a simple communication architecture for Multi-Agent Learning Systems. The service provided by the communication architecture allows each agent to connect to the user interface, the application and the other agents. The communication architecture is implemented using TCP/IP. An application example in a simplified traffic environment shows that the communication architecture can provide reliable and efficient communication services for Multi-Agent Learning Systems.

1 Introduction

Many researchers in the field of Distributed Artificial Intelligence are beginning to build agents that can work in a complex, dynamic multi-agent domains [1]. Such domains include virtual theater [2], realistic virtual training environments [1], RoboCup robotic and virtual soccer [3] and robotic collaboration by observation [4]. This is because that there is a realisation of the benefits of using problem-solving models based upon an interacting group of agents rather than a single agent and multi-agent systems can benefit from the inherent properties of distributed systems, i.e. parallelism, robustness, scalability.

Learning in multi-agent systems has been seen as important both in removing the need to "hard code" the agent behaviour, as for certain problems the appropriate behaviour is unknown [5, 6]. It is motivated by the insight that it is impossible to determine a-**priori** the complete knowledge that must exist within each component of a distributed, heterogeneous system in order to allow satisfactory performance of that system. Especially if we want to exploit the potential of modularity, such that it is possible for individual agents to join and leave the multi-agent system, there is a constant need for the acquisition of new and the adaptation of already existing knowledge, i.e., for learning.

Within this setting, different kinds of learning tasks must be investigated, such as 'traditional' single agent learning tasks, learning in teams, learning to act within a team, and learning to cooperate with other agents. To solve any of these tasks, **communication**, i.e., the existence of appropriate **information** that can be **communicated** to the learning agents is of primary importance.

S. Cagnoni et al. (Eds.): EvoWorkshops 2000, LNCS 1803, pp. 255–266, 2000.
© Springer-Verlag Berlin Heidelberg 2000

In this work, we consider the communication issues in the cooperative multi-agent learning systems. In a cooperative multi-agent system, the agents can cooperate either implicitly or explicitly. With implicit cooperation agents act selfishly to satisfy their individual goals but their actions can have beneficial effects upon other agents. With explicit cooperation, agents share information and rewards, thus perform actions which provide mutual benefit, involving some form of direct communication.

Sen *et al* showed that agents attempting to optimise the use of limited resources converge to optimal states faster with a limited view [7]. They also investigated the formation of agent coalitions, where agents within a coalition share information about their resource utilisation intentions. The use of such coalitions further improves the system's convergence to the optima. Prasad and Lesser examined an explicitly cooperative directly communicating system in which agents learning to communicate only relevant information dependant on situation specifics [8]. The learning algorithm uses the Instance-Based learning paradigm. At the end of a particular problem solving run, agents assess their coordination strategy according to four performance measures. Agents then derive new strategies and broadcast these to the other agents. Seredynski *et al* developed a coevolutionary multi-agent system showing that global behaviour evolves via only local cooperation between agents acting without global information about the system. Seredynski introduces an exchange process into the game, which redistributes the payoff [9, 10]. The players are placed in a ring and play a set number of games with a selected number of neighbours, i.e. local interactions. Three exchange schemes were tested:

1. no payoff exchange, no cooperation;

2. payoff is exchanged amongst interacting players, local cooperation sharing;

3. payoff is exchanged amongst all players, global cooperative sharing.

They considered two simple coevolutionary schemes: loosely coupled GA and loosely coupled classifier systems. The term "loosely coupled" refers to the individuals in a population being evaluated only on their local rather than a global fitness function. The experiments showed that with no payoff exchange the players evolve to the Nash equilibrium (both defecting), whilst with both the local and global payoff exchange players evolved to cooperate. In fact the payoff exchange alters the game's payoff matrix and encourage cooperation. Seredynski applied the technique of payoff exchange to a dynamic mapping and scheduling problem [9]. Bull *et al* applied the learning classifier system design to a multi-agent environment in which each of a number of classifier systems represents a co-operating communicating agent [11]. It is demonstrated that the evolution of multiple co-operating agents can give improved performance over an equivalent single agent model. The performances of the varying components, such as reinforcements, discovery system, in a multi-agent environment, were examined in detail in [12].

In this work, we extend the framework of the multi-agent systems in [10, 12] to a general version and aim at developing a communication architecture for these multi-agent learning systems. In these environments, each agent is defined by a classifier system which evolves a set of control rules and is concerned with agents as software rather than theoretical constructs. In this communication architecture, each agent is allowed to connect to user interface, the application

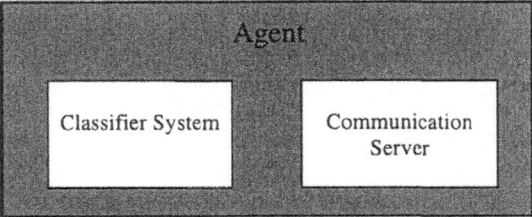

Figure 1: Structure of the agent

and the other agents. An application example is given to show that the communication architecture can provide reliable and efficient communication services.

2 The Framework of the Multi-Agent Learning System

Multi-agent learning systems usually consist of a number of collaborative agents and consider how these agents can interact to effectively cooperate in problem solving tasks. They have many applications in process control, network management, scheduling, etc. [9, 13, 17].

As with a number of contemporary fields on computer science, such as AI and Artificial Life (ALife) the definition of an agent ranges from a strong to a weak notion. On the stronger end of the scale an agent is deemed to possess properties akin to those found in humans, such as knowledge, belief, intention, etc. The weaker notion of an agent tends to be more pragmatic and associated with agent software engineering. Such agents can be defined as possessing less anthropomorphic properties, for example [18]:

- autonomy: operate without the direct intervention of others, and have some control over actions and internal states

- communication: interactions with other agents

- reactivity: perception of their environment and timely response to changes that occur in it.

- Pro-activeness: not merely reacting to events but exhibit pre-active goal-directed behaviours.

The multi-agent learning system considered in this paper contains a number of distributed, communicating agents, where each agent, as shown in Figure 1, has a learning classifier system (providing the rule base and control actions) and a communication server which is used to connect the agent to the user interface, the application and to other agents. These two elements of the agent are separate since; as messages are passed around the agent network, the communication server acts independently of the classifier system to route the message to its neighbours. Another reason for keeping the communication server distinct

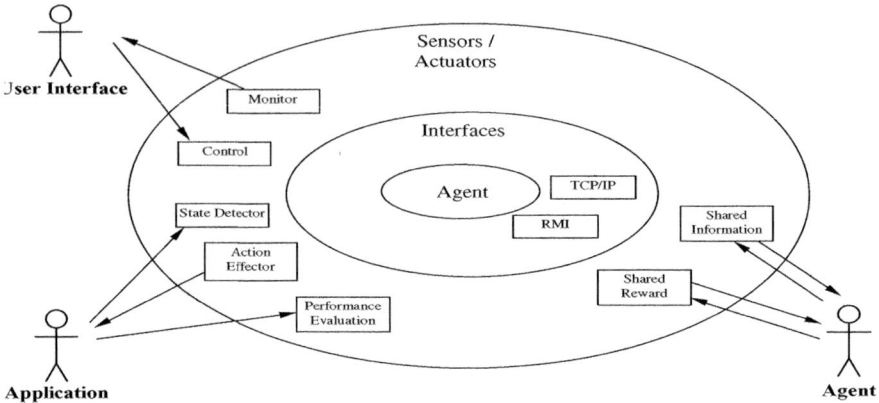

Figure 2: Interaction structure of the agent

from the classifier system is that communication is likely to be implementation specific (even in the test applications). Thus it is necessary when specifying the communication server to consider the general requirements of setting up and maintaining the communication in a multi-agent learning system rather than those in a specific software and hardware implementation.

The communication server must provide the following services:

- Interaction between the application and the agents: The interface between the agents and the application is defined by the agents' detectors and effectors. The application must package the system state into a message which is in the representation expected by the agents. The application receives an action from the detectors which is interpreted to effect the system.

- Interaction between the User Interface and the agents: The amount of performance information reported by the agents is parameterised. During the development of an application it is likely that the user will wish to monitor performance more closely to ensure the system has been correctly configured. Another consideration is whether information (performance statistics, warnings, error, etc.) is logged to a file or sent to the user interface directly.

- Inter-agent communication: The communication strategy between the agents is effected by the application and the learning strategy to be employed. Each agent can communicate with a specific other agent or group of agents, known as an agents neighbourhood, the user specifies the agents contained in each agents neighbourhood.

The Figure 2 shows a representation of the interactions between an agent and

Agent Network

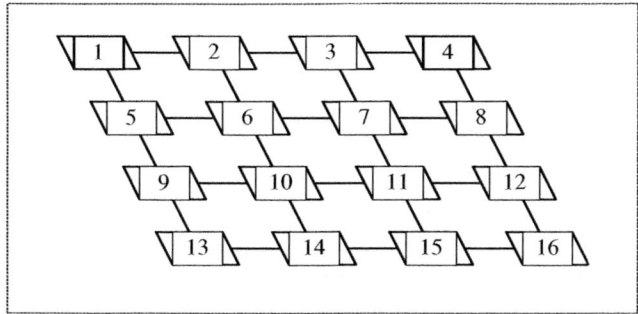

Figure 3: Structure of the agent network

its environment which contains three types of actors: the user, the application, and other agents.

While it is important to examine the research into agent languages (and implementations) such as KQML, such work provides functionality beyond the needs of these systems. The communication in such learning systems will have a limited syntax and have a well defined content, thus the language required will be fairly simplistic. We have specified the nature of the communication, which includes: utility measure (enabling "fitness" sharing); environmental state; actions performed or to perform (intentions); shared classifiers.

3 The Proposed Communication Architecture

3.1 The Whole Agent Network

An agent network is defined by the characteristics of the application, which determines the appropriate distribution and connections between agents. Each agent is allocated a specific address, and a port for each channel of communication, this includes communication with the user interface, application and other agents.

There are two basic protocols of communication. The simplest is to broadcast the message which is received by all the agents, the message can include a tag to identify the sender and nature of the message. It is the responsibility of the receiving agent to determine whether to utilise or ignore the information in the message. The second means of communication is to send messages to the agents in the local neighbourhood. In this approach each agent is the centre of a neighbourhood, those agents contained in the neighbourhood (from none, i.e. no communication, to all other agents) receive any messages sent by the central agent. As neighbourhoods might overlap an agent can be a member of a number of neighbourhoods, for example, for the agent network, shown in Figure 3, neighbourhoods are constructed from an agents nearest neighbours, thus sixteen groups, each centred on one agent, are formed:

Group 1: 1, 2, 5
Group 2: 2, 1, 3, 6
Group 3: 3, 2, 4, 7
...
Group 6: 6, 2, 5, 7, 10
Group 7: 7, 3, 6, 8, 11
...
Group 16: 16, 12, 15.

This approach is general known as multicast or one-to-many communication. There are a number of possible approaches to this form of communication in distributed systems from flooding where each node sends a copy of the incoming message to all the connected nodes (except for the message source node) to routing where messages follow a pre-specified path to their destination node.

The principal problem with flooding is that it causes a great deal of redundant communication, also as each node must check if it has already received each incoming message, with frequent communication this can lead to a communication overload. Routing requires a more complex initialisation process but minimises the communication traffic. Each node when it receives a message refers to a lookup table giving the nodes to which copies of the message are sent. Unlike flood broadcast there is no redundant communication, thus the severing of a communication link will cause at least one node to fail to receive messages. In practice the choice of method should reflect the constraints of the system, i.e. the trade-off between fault tolerance and communication load.

3.2   Initialisation of the Agent Network

The initialisation process firstly involves the setup and opening of communication channels from; agent to user interface, agent to application and agent to other agents. Once this has been successfully completed, the classifier system can be initialised, the agent then waits for the first message from the application to begin its control process.

The initialisation of communication involve each agent connecting to the user interface, application and other agents. All these channels might involve two-way communication. During the initialisation the agents open a communication channel and await a connection message. The channel is tested to ensure the communication is setup correctly as although the configuration parameters have been previously checked for consistency, the parameters may be inconsistent with the physical communication process, also this process might be faulty.

Although the term "socket" is used in the specification as the medium to connect communication channels in implementation other methods can be used, such as calls to remote objects, when using RMI or DCOM. The basis of the communication initialisation and run-time processes are not affected. The creation of the communication object and binding in a remote registry (on a given hostname and port) replaces the creation of a server socket and calls to the remote object replace read and write calls to the sockets.

Note that it is possible for the communication server to create separate processes to listen on the communication channel for messages, this allows the

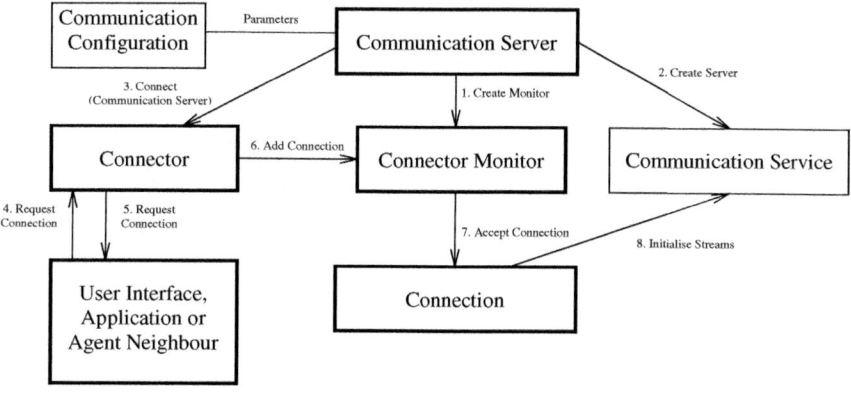

Figure 4: Structure of communication server

agent to be reactive to external messages. The communication with neighbours requires a single channel for incoming messages, and separate channels from sending to each neighbour (except if the messages are broadcast on sent via a proxy). The initialisation of the communication server, as shown in Figure 4, involves the following steps:

1. The Communication Server object creates the specific Communication Services (Application, User Interface or Neighbourhood) as specified by the configuration.

2. The Communication Server object create a monitor which maintains the list of current connections.

3. The Communication Server object passes the Communication Service object and connection configuration information to the Connector object which, for connection with the User Interface and Application and incoming channel from the neighbouring agents, opens a Server Socket on the specified port and waits for a request to connect. For the outgoing channel to the neighbouring agents the Connector object intermittently requests a connection to the neighbours specified port.

4. The User Interface, Application or Neighbouring Agent sends a request to connect.

5. The request to connect is accepted by the neighbour's server socket.

6. The Connector sends the Communication Service object and open socket to the Communication Monitor.

7. The Communication Monitor object tests the communication channel, if the test succeeds the Communication Service is passed to the Connection object, otherwise the socket is closed and the failure reported.

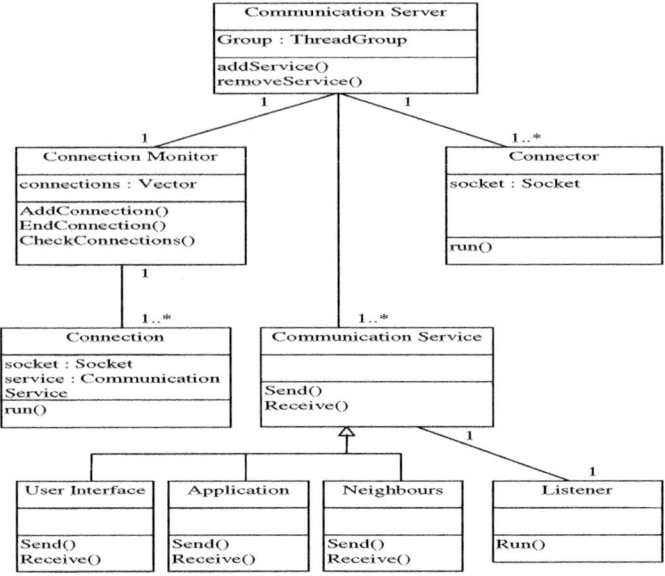

Figure 5: Class diagram of the communication

8. The Connection object starts the thread to handle the connection and passes the input and output streams to the Communication Service object.

3.3   Logical View of the Communication

A class diagram of the communication is given in Figure 5. The function of each class (object) in the communication is described briefly as follows:

- Communication Server: The Communication server supports a number of multi-threaded channels. It creates (or opens) each channel on a specified port allowing the Agent to communicate with the User Interface, Application and other Agents. It provides the ability to send messages whilst listening on the port for incoming messages which are passed to the appropriate objects.

- Connection Monitor: The Connection Monitor object maintains a list of the current connections. The thread waits to be notified if a connection terminates and updates the list.

- Connector: The Connector class either listens for a connection on a specified port (using a server socket) or connects to another agent's server socket. Once accepted the socket is sent to the Connection Monitor.

- Connection: Connection objects are created by the Connector thread using the Communication Monitor method *addConnection()*. It simply creates a thread to handle the connection.

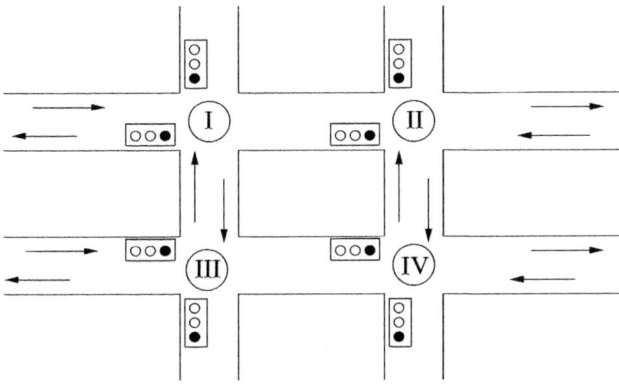

Figure 6: The simulated traffic environment

- Communication Service: A general class for each of the types of communication required.

- Listener: An object that listens on a communication channel for an incoming message.

- User Interface: The communication service require by the User Interface. This object handles messages received from or sent to the User Interface. Incoming messages are interpreted and, if necessary, call the appropriate function. Outgoing messages are packaged and sent to the User Interface.

- Application: The communication service require by the Application. This object handles messages received from or sent to the Application. Incoming messages (generally system state or rewards) are interpreted and, if necessary, sent to the appropriate objects (such as classifiers) in the classifier system. Outgoing messages (general actions) are packaged and sent to the Application.

- Neighbourhood: The communication service require by the Agent Neighbourhood. This object handles messages received from or sent to the Agent Neighbourhood. Incoming messages are interpreted and, if necessary, sent to the appropriate functions in the classifier system. Outgoing messages are packaged and sent to the neighbouring agents.

4   Application Example

Optimization of a group of traffic signals over an area is typical multi-agent type real-time planning problem without precise reference model given. To do this planning, each signal should learn not only to acquire its control plans individually through reinforcement learning but also to cooperate with each other. This requires communication between the agents. In this example, we developed a multi-agent learning system, which is aimed at learning the efficient control rules for the dynamic traffic environment and with the communication provided by

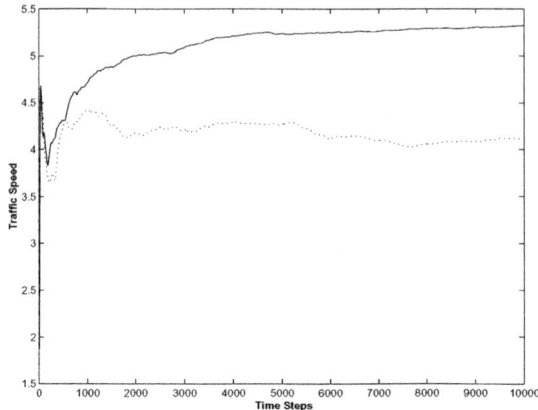

Figure 7: Performance comparison of different control strategies

the developed communication architecture. Each agent has a classifier system (providing the control strategy) and a communication server which is used to connect the agent to the user interface, the application and to other agents.

To control a traffic network, we associate an agent to each junction of the traffic network. The agents are initialised according to the traffic network configuration and user-specified parameters. For the simulated 2×2 traffic network, as shown in Figure 6, four agents, i.e., agents I, II, III, and IV, associating with junctions I, II, III, and IV, are need to provide comprehensive control of the network. Agent I has the neighbouring agents II and III, and agent II has the neighbouring agents I and IV, etc. The communication server in each agent provides the control actions of its neighbouring agents, and these information is used to construct control rules for its junction.

The classifier system employed is a version of Wilson's "zeroth-level" system (ZCS) [19], with some changes on the classifier representation [20]. The condition part of each classifier consists of six bits, which reflects the scalar level of queue length from each direction and the previous actions of the neighbouring agents. In this application, the scalar level of the queue length is set to 4, which ranges from 0 to 3, corresponding to the four linguistic variables, {zero, small, medium, large }. The action part indicates the required state of the signal. For instance, for junction I, the rule 130201:1 says that if the queue from directions east and west are small (1) and zero (0), but the queue from directions south and north are large (3) and medium (2), and the previous neighbourhood junction controllers' actions are vertically red (0) (junction II) and green (1) (junction III), then the traffic light stays green vertically (1) for a fixed period of time. The performance evaluation, reinforcement learning strategy, genetic algorithm and the simulated traffic environment are all similar to those used in [20].

For comparison purpose, two types of control strategies are employed: random control strategy and the developed multi-agent learning system (MALS) strategy. The random control strategy determines the traffic light's state (0 or 1) randomly at 50% of probability; whilst MALS strategy determines the traffic light's state according to the action of the winning classifier of the agent.

Experiments were carried out for three different types of traffic conditions. In these simulations, the mean arrival rates for the cars are the same but the number of cars in the area is limited to 30, 60, and 90, corresponding to a sparse, medium, and crowded traffic condition. In all cases, the MALS strategy is found to learn how to reduce the average queue length and improve the traffic speed in the network. For example, Figure 7 shows the average performances of the random control strategy and MALS strategy respectively over 10 runs in the crowded case, where the solid line represents MALS strategy and the dotted line represents random control strategy. It can be seen that the MALS strategy consistently learns and improves the traffic speed over 10,000 iterations.

5   Conclusion and Future Work

We have extended the framework of the multi-agent learning systems in [10, 12] to a general case and developed a simple communication architecture for these systems. The service provided by the communication architecture allows each agent to connect to the user interface, the application and the other agents. An application example shows that the communication architecture is reliable and efficient. Although the communication architecture is implemented using TCP/IP, it can also be implemented using RMI or DCOM, via binding the objects in the remote registry and making calls to the remote objects.

6   Acknowledgment

This work was carried out as part of the ESPRIT Framework V Vintage project (ESPRIT 25.569).

References

[1] Tamble, M., Rosenbloom, P. S.: RESC: An approach for real-time, dynamic agent tracking. In Proc. of the International Joint Conference on Artificial Intelligence, Montreal, Canada, (1995)

[2] Hayes-Roth, B., Brownston, L., Gen, R. V.: Multiagent collaboration in directed improvisation. In Proc. of International Conference on Multi-Agent Systems. USA (1995)

[3] Kitano, H., Asada, M. Kuniyoshi, Y., Noda, I., Osawa, E.: The robot world cup initiative. In Proc. IJCAI-95 Workshop on Entertainment and AI/Alife, Montreal, Canada (1995)

[4] Kuniyoshi, Y., Rougeaux, S., Ishii, M., Kita, N., Sakane, S., Kakikura, M.: Cooperation by observation: the framework and the basic task pattern. In Proc. IEEE International Conference on Robotics and Automation. (1994)

[5] Weiss, G. and Sen, S. (eds): Adaptation and Learning in Multi-Agent Systems. Springer-Verlag, Berlin, Heidelberg, New York, (1995)

[6] Sen, S. (ed): AAAI Spring Symposium on Adaptation, Coevolution and Learning in Multiagent Systems. AAAI Press, (1996)

[7] Sen, S. Sekaran, M. and Hale: Learning To Coordinate without Sharing Information. In Proceedings of the Twelfth National Conference on Artificial Intelligence, (1994) 426-431.

[8] Prasad, M.V.N. and Lesser, V.R.: Learning Problem Solving Control in Cooperative Multi-Agent Systems. Workshop on Multi-Agent Learning AAAI-97, (1997)

[9] Seredynski, F.: Coevolutionary Game-Theoretic Multi-Agent Systems: the Application to Mapping and Scheduling Problems Technical Report TR-96-045 Institute of Computer Science, Polish Academy of Sciences, Warsaw, Poland. (1996)

[10] Seredynski, F., Cichosz, P. and Klebus, G. P: Learning classifier systems in Multi-Agent Environments, In Proc. First IEE/IEEE International Conference on Genetic Algorithms in Engineering: Innovations and Applications, (1995) 287–292

[11] Bull, L., Fogarty, T. C., and Snaith, M.: Evolution in Multi-Agent Systems: Evolving Communicating Classifier Systems for Gait in a Quadrupedal Robot. In Eshelman, L. J. (ed): Proceedings of the Sixth International Conference on Genetic Algorithms, Morgan Kaufmann, (1995) 382–388

[12] Bull, L: On ZCS in Multi-Agent Environments. Parallel Problem Solving From Nature - PPSN V, Springer Verlag (1998) 471–480

[13] Fleury, G., Goujon, J., Gourgand, M. and Lacomme, P., Multi-agent approach and stochastic optimization: random events in manufacturing systems. Journal of Intelligent Manufacturing, **10**, (1), (1999) 81–102

[14] Cao, Y. J. and Wu, Q. H.: A mixed-variable evolutionary programming for optimisation of mechanical design. International Journal of Engineering Intelligent Systems, **7**, (2), (1999) 77–82

[15] Cao, Y. J. and Wu, Q. H.: An improved evolutionary programming approach to economic dispatch. International Journal of Engineering Intelligent Systems, **6**, (2), (1998) 187–194

[16] Cao, Y. J. and Wu, Q. H.: Optimisation of control parameters in genetic algorithms: a stochastic approach. International Journal of Systems Science, **20**, (2), (1999) 551–559

[17] Kouiss, K., Pierreval, H. and Mebarki, N., Using multi-agent architecture in FMS for dynamic scheduling. Journal of Intelligent Manufacturing, **8**, (1), (1997) 41–48

[18] Wooldridge, M. and Jennings, N.R.: Intelligent agents: theory and practice. In The Knowledge Engineering Review, **10** (2), (1995) 115-152.

[19] Wilson, S. W.: ZCS: A zeroth level classifier system. Evolutionary Computation, **2**, (1994) 1–18

[20] Cao, Y. J., Ireson, N. I., Bull, L. and Miles, R.: Design of Traffic Junction Controller Using a Classifier System and Fuzzy Logic. In Computational Intelligence: Theory and Applications, Reusch, B. (ed), Lecture Notes in Computer Sciences, **1625**, Springer Verlag, (1999) 342–353

An Ambulance Crew Rostering System

P. V. G. Bradbeer[†], C. Findlay[‡] and T.C Fogarty.[¶]

[†] Fife College of Further and Higher Education., pvgb@cit.fife.ac.uk
[‡] Fife Ambulance Service.
[¶] Napier University., t.fogarty@dcs.napier.ac.uk

Abstract

The production of a roster for the duties of ambulance crew is subject to a variety of practical, managerial and social constraints. This document describes the first steps of an investigation into the characteristics of the search space for such a problem and reports on the success to date of an evolutionary algorithm approach in finding an acceptable solution to the problem. The visualisation method described is suggested as a quicker way of testing the appropriateness of representations than having to perform multiple time consuming experiments.

Section 1 - Introduction

Recently, the decision was made to re-establish the ambulance service sub-station attached to the Victoria Hospital in Kirkcaldy. The previous arrangement was to provide all ambulance services from a central base in Glenrothes. 'Front Line' personnel for the new service comprises a group of eight staff, four of whom are designated paramedics and have received specialised training for their duties, and four staff who are designated medical technicians, also after appropriate training. This group of staff are required to provide twenty four hour a day, seven day a week coverage. Each ambulance is crewed at all times by a team consisting of one paramedic and one technician, working a twelve hour shift. Clearly such a coverage scheme requires careful arrangement to ensure that sensible rosters are produced for all of the individuals concerned. There are a number of constraints on when individuals are available to crew a vehicle.

Given the nature of the duties that crew members undertake it is not a practicable option to allow double shift working, for both the safety of users of the service and the crews themselves. This complement of staff implies that a crew member is required to perform an average of three and a half shifts per week. In practice this translates to either three or four shifts in one seven day period. Experience has shown that regularly working excess shifts in a week to be both physically and emotionally draining, and this is to be avoided as far as possible.

In addition to the practical constraints described, management have decided against forming permanent pairings. This means that each possible combination of paramedic and technician must work together at some stage in the schedule. This is seen to bring with it the benefits of promoting 'skill pull-through' and ensuring that the relationships between paramedics and technicians remain productive. In the interests of 'fairness' it is desirable that each of the sixteen (four paramedics and four technicians) possible teams are active an equal number of times.

A further complication arises from the ambulance staff themselves. Many of them have families, and are desirous of spending 'quality time' with them. This not only promotes development of the family unit, but also allows the staff to unwind after some of the more harrowing incidents that they are required to attend. Possibly the most important opportunity is to spend time with a family is during the weekend, when children are not at school.

It is clearly not possible to avoid weekend work, but the staff have a stated preference to either work a complete weekend, or to have no shifts in a weekend at all. The intermediate situation where only one shift is worked in a weekend is referred to in house as a 'ruined weekend'. There is thus a request that single shift weekends are avoided by any system used to generate ambulance staffing rosters.

There is also the notion of perceived 'fairness' within a roster. This is best viewed as a desire to have all members of staff working an equal number of day/night shifts and the same number of weekend/weekday shifts. This prevents any accusations of bias when slight imbalances are noticed in rosters.

Staff leave requires no special treatment, as staff are seconded from the central ambulance base in Glenrothes to fill in for scheduled holidays. This may cause slight imbalances in the frequency with which individuals are partnered, but this is not regarded as important.

Section 2 - Approach 1

Attempts to produce a roster by hand proved that the problem was quite difficult, and a computerised solution was sought. At first glance the development of a software system to automatically generate useable ambulance rosters within the constraints described in section 1 looks as though a simple depth first recursive tree search would be a feasible approach, as the choice of the 'next shift' is so highly constrained that considerable pruning of the search tree would be possible. Unfortunately the effect of some of the constraints can only be assessed when the tree has been grown to some depth. This has the effect of making exhaustive search an non-viable option even if very fast computational machinery were available.

This leads us into the area where a search based on evolutionary techniques suggests itself as a possibility. Even from the outset, the constraints placed on the system suggest that finding a solution will present a considerable challenge, but it was felt that given the past successes in solving schedule based problems with genetic algorithms (GAs), including those described in Langdon [Langdon1995] Wren and Wren [Wren1995] and Fang et. al. [Fang1993], there was a reasonable chance of success.

The alphabet initially chosen to represent the genetic material in the candidate solutions has sixteen characters, one for each of the different technician paramedic pairings. This was partly to allow easy checking to see if the partnership constraints were being met. The selection of the length of the genetic material (chromosome length) is also an issue, as the length of the roster is not specified. Under such circumstances an application of messy GAs [Goldberg1990] may prove particularly appropriate, but for an initial investigation a fixed length representation was chosen.

As there are sixteen teams to accommodate the fixed length was set at sixteen weeks, giving a pattern that repeats three times a year. Each of these sixteen week cycles would ideally have each partnership active fourteen times (seven on day shift and seven on nights). A year could thus comprise three of these cycles, leaving a gap of about four weeks. This is convenient for the disruption in pattern required for the Christmas/New Year period when the working requirements are subject to change. The festive period is a busy time for the emergency services.

Note that if working to these sixteen week rosters that the 'wraparound' effect from the end of one cycle to the beginning of the next must be taken into account. Failure to do so may result in doubleshift working between the end of a cycle, and the beginning of the next.

The first incarnation of genetic search was largely to gain insight into the characteristics of the space, and used a fairly traditional approach, in that each locus on the chromosome was allowed to take any of the allowable alleles, a low mutation rate, uniform crossover [Syswerda1989] and binary tournament selection [Brindle1981] (with 100% chance of the better of the two candidates progressing).

Following the success reported by Miller et. al. [Miller1995], Jones and Bradbeer [Jones1994] and Chisholm and Bradbeer [Chisholm1997] using small breeding pool sizes, a relatively small breeding pool of thirty was initially adopted. A random sample of five thousand individuals generated the distribution shown as figure 1. This suggests that a large portion of the population have poor fitness, and leaves us hoping that there are very long (if thin) tails to the distribution. This type of distribution, allied with the fact that there are many constraints on the search space reinforces the suspicion that this will be a difficult problem for this brand of GA. It is worth commenting that there was little code produced for this initial system that was not to be of use in subsequent implementations.

The fitness function used for evaluation purposes is based on the accrual of penalties, with different weights being assigned to breaches of different constraints. Each time a constraint is violated the integer representing the fitness of the candidate solution is increased. This means that in the reporting of results a low fitness number indicates a a good solution.

Initially the following penalties were allocated

Constraint breach	Penalty level
Day/night imbalance	1
48 hour+ per seven days	1
Double Shift	1
Ruined Weekends	1

Table 1: Initial Penalty Scheme

This apportionment of penalties implies that all breaches of constraints are equal, which is clearly an oversimplification of the problem. In reality, a sixteen week roster that produced one or two breaches of the 48 hour rule would probably be acceptable, whereas a scheme with the same number of double shifts would not be acceptable. Those constraints that must not be breached are usually referred to as *hard constraints*, while those for which minor or limited breaches could be tolerated are referred to as *soft constraints*. Usually a hard constraint violation would attract a higher penalty value than a soft penalty.

Michchalewicz [Michchalewicz1994] provides a discussion of some of the soft constraints that can appear in timetabling problems, and the contrasting hard constraints.

The apportionment of relative penalty levels is an area of interest in its own right, and can either be static, or dynamic as can be seen for example in Eiben et. al. [Eiben1998] where different approaches are compared.

As this is an initial study however we are more interested in the gross performance of the system, and merely note that an ideal solution to this problem would produce a zero penalty result.

For more detailed study, the exact weight given to each class of breach would need more attention, possibly using an approach such as that detailed by Paechter et. al. [Paechter1998] where a 'front panel' is attached to the system allowing dynamic alteration of the weights. It is certainly possible that different penalty levels could change the trajectory of the search.

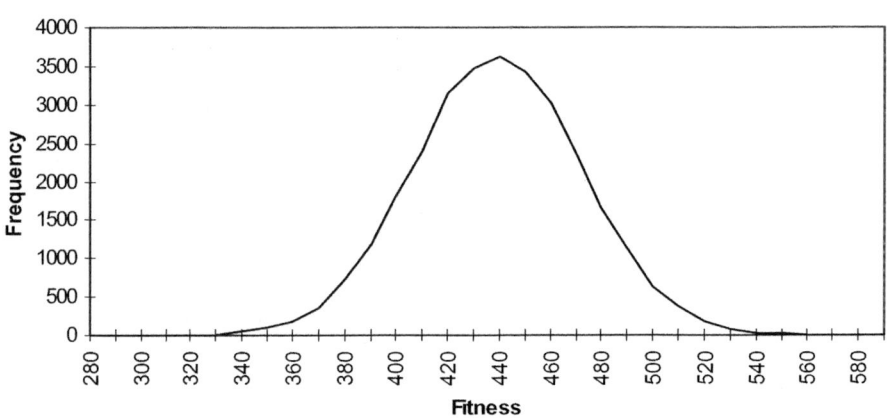

Figure 1: Distribution of fitness for random sample using 'unconstrained' representation.

This sample indicates that the average randomly generated individual breaches about four hundred and forty constraint requirements, and the bulk of the population is in this vicinity. It is difficult to believe that an acceptable solution will breach more than about twenty constraints, so that leaves a considerable challenge for the search mechanism.

In order to progress the search, some sort of move operator must be employed. It is useful to have an indication of the likely effectiveness of operators. Random mutation is one of the operators traditionally used, and the effect of this operator in this representation in this problem is summarised in figure 2. Mutation in this case involves the selection of a single locus on the chromosome, and replacing it with a single randomly generated legal value. The figure is generated from a sample of 30000 individuals, each of which receives one mutation. The resulting change in fitness due to the mutation is noted against the fitness of the original. Rather than keep records of individual fitness numbers, the outcomes are grouped into ranges, and averaged.

The graph shows the phenomenon of above average fitness individuals generally being adversely affected by random mutations, whereas below average fitness individuals are on average improved. The kinks at the extremities can be explained as due to small sample sizes at the tails of the distribution.

No steps were taken to prevent the mutation replicating the original value. As an alphabet of sixteen was used, approximately one in sixteen mutations result in no change. This may be viewed as reducing the slope of the graph slightly.

A graph with a gradients low as this can be taken as indicating that the representation induces severe epistatic effects, or that the problem space has no structure.

A series of runs (with a range of parameter sets) using this approach gave sufficiently poor results to confirm that a better representation should be sought.

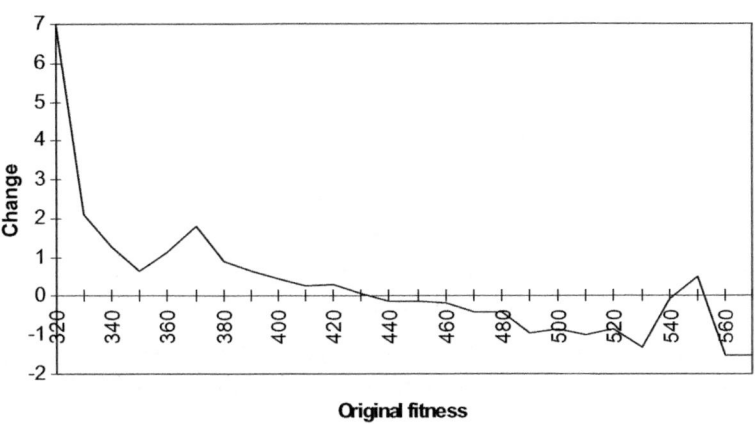

Figure 2: Average change in fitness under mutation.(30000 samples)

Section 3 - Approach 2

Instrumentation of the runs of the first approach seemed to indicate that the major source of penalty points was due to difficulty in maintaining the even spread between the number of times each team was rostered. Clearly, the next step was to reconsider the representation. As a way of reducing the imbalance penalties a permutation representation was employed, with each team appearing a given number of times within a sixteen week roster. This change largely fitted into the framework of the previous system, with relatively small code changes. As before, a sample of randomly generated individuals was evaluated to ascertain if any benefit accrued. The results are noted in figure 3.

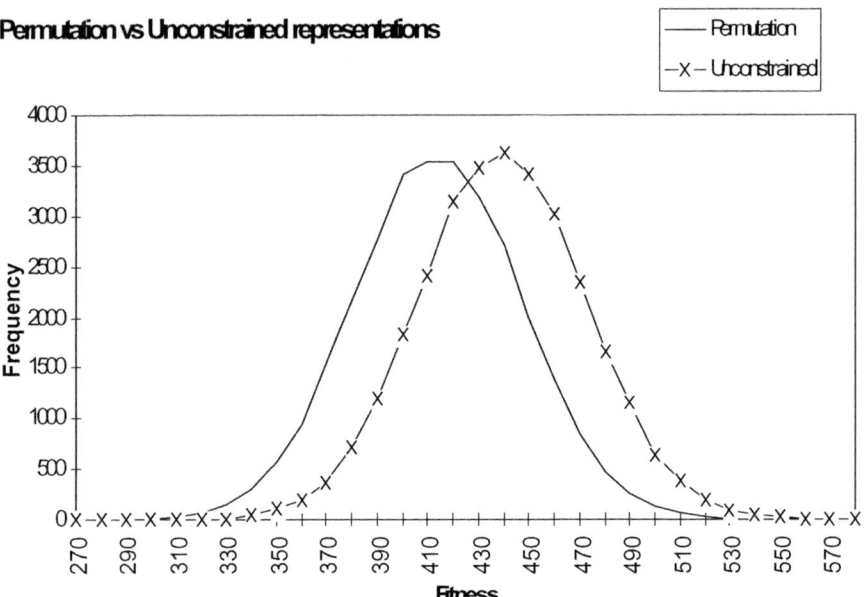

Figure 3: Distribution of fitness for random sample using single permutation chomosome compared to unconstrained chromosome (30000 samples).

Comparing figure 3 with figure 1 we see that the distribution has moved far enough to provide encouragement to code the rest of the system to perform further test runs.

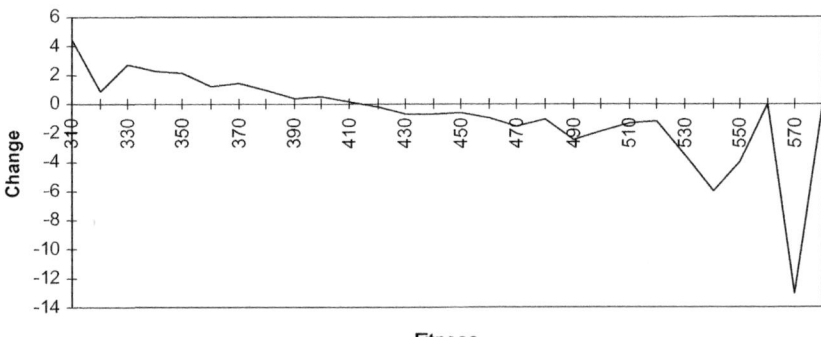

Figure 4: Average change in fitness to permutation representation under mutation.(30000 samples)

Once again performing the change under mutation test, we see that the gradient of the response is very low, again leading us to suspect that there is a large amount of interaction left in the representation.

Using PMX [Goldberg1985] as the crossover method, a number of test runs were performed. Various parameter sets failed to produce even near acceptable results. Again the representation of the candidate solutions permitted far too many constraints to be violated.

It is believed that this in turn led to a highly modal search space that caused difficulties for the search algorithm.

Section 4 - Approach 3

The representation described in the previous section had the effect of forcing all candidate solutions to obey the constraint requiring all team pairings to be equally represented, but the 'equitability' request is left to the evolutionary mechanism. In order to reduce the number of undesirable combinations further, a third encoding was devised. This was made up from four separate components (chromosomes), each encoding a permutation of teams. The first two, each of length sixteen, determine which weekend the team will work a day shift and which weekend they work a night shift. If a team is scheduled to work both day and night shift then this will be detected and accrue the associated penalty. This approach guarantees that not only does each team work an equal number of weekend day and night shifts, but also that the requirement to avoid ruined weekends is avoided. This renders half of the fitness function redundant, thus speeding the evaluation portion of the system by about 25%. Similarly the third and fourth chromosomes, each of length eighty encode five weekday shift and five week night shift appearances for each team.

Figure 5 compares the fitness distribution of this representation, which on the basis of this admittedly small sample seems only a little better. Figure 6 examines the effect of the mutation operation. Examination of the slope of this graph in comparison with the results from the previous two representations reveals that it is more pronounced. This is taken as indication that there is less epistasis evident in the representation. Even so the slope is still relatively shallow, indicating that the problem is still 'hard'.

Preliminary runs with this representation gave a series of results with fitness in the order of ten constraint breaches. Closer examination of the resulting roster however revealed that about three of these were breaches of the 'no double shift' constraint, and were thus not acceptable. Rather than try to recode the representation further to totally avoid the occurrence of double shifts, it was decided to rebalance the weight of penalty accrued by the various breaches. The value for double shift violations was increased to 25, in the hope that this would drive the search into eliminating double shifts. This of course alters the fitness distribution values already produced, but as we were now nearing the ideal solutions

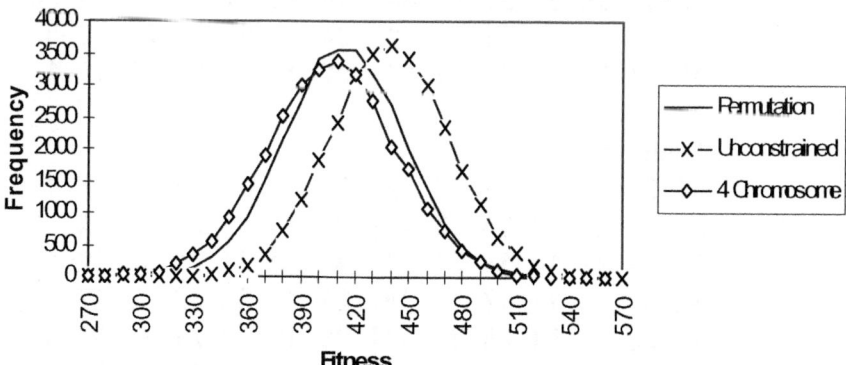

Figure 5: Distribution of fitness for random sample using 4 chromosome arrangement, compared to single permutation chomosome and unconstrained chromosome (30000 samples).

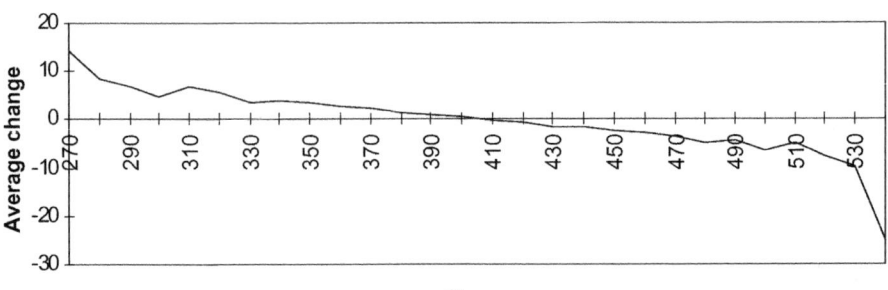

Figure 6: Average change in fitness to 4 Chromosome permutation representation under mutation.(30000 samples)

this was felt to be a small price to pay. It is also possible or even likely that these changes will change the search space slightly, with a resulting change in progress

Section 5 - Results

After making the change described to penalty weightings given in the previous section, the system was run with various parameters. One of the first runs, using a pool size of 20, PMX, binary tournament selection and a 10% mutation rate produced a solution with only 5 minor breaches of the 48 hours per week rule. This was felt to be a practicable solution to the problem, and is shown in figure 7. Despite a certain amount of experimentation with parameters, so far no other solutions have been generated that have such an acceptable fitness value.

```
Paramedic
43 21 43 41 32 34 34|  24 13 12 42 13 43 43|
41 23 14 12 34 24 24|  31 24 13 13 12 42 42|
41 32 42 31 42 31 31|  31 24 34 21 42 13 13|
21 23 43 21 41 24 24|  32 41 32 14 24 31 31|
43 14 32 41 32 13 13|  41 24 32 42 14 12 12|
43 14 34 23 21 34 34|  14 12 14 23 21 23 23|
13 43 42 14 13 21 21|  34 23 13 24 21 42 42|
34 31 43 21 32 41 41|  32 32 13 21 34 12 12|


Technician
41 24 12 32 31 32 32|  41 42 34 21 24 13 13|
42 14 21 43 24 13 13|  23 14 31 24 31 24 24|
34 31 32 14 14 21 21|  43 23 21 34 14 12 12|
34 34 34 12 12 34 34|  23 21 34 12 43 12 12|
13 23 42 41 24 31 31|  23 12 43 21 31 42 42|
43 43 13 21 23 41 41|  42 12 42 13 41 24 24|
12 43 23 14 31 43 43|  14 21 23 42 32 41 41|
41 31 32 43 12 34 34|  12 43 12 31 34 23 23|
```

Figure 7: Screen dump of best result so far.

Section 6 - Conclusions

The system, developed against a practical need and relatively undeveloped as it is has produced a workable roster for a highly constrained problem.

As there are a number of different local optima being found, it is possible to give the 'client' a number of different candidate solutions. This will allow consideration of other 'social' factors not explicitly mentioned in the initial outline of the problem.

This approach of looking at the behaviour of the search space has saved a considerable amount of time in running experiments and analysing the results from them. While the relative shortcomings of the first two representations are to an extent predictable, given the nature of the constraints, it is useful to be able to confirm this without expending large amounts of compute cycles.

Section 7 - Further Work

This document describes the first steps towards solving this problem. The authors are aware that many variants are possible, and even desirable. As mentioned in the body of the text, the length of the cycle is not clear, so an approach allowing a variable length cycle is one possible direction for further study.

It would also be of interest to discover more about the modality of the search space, and it is planned to use a reverse hill-climbing technique, such as described by Jones [Jones1995].

Producing more information on the sensitivity of the problem to different parameters, such as pool size and crossover mechanism remains an item on the agenda.

It has been noted that the addition of heuristics could be used to improve performance, possibly in a similar fashion to that reported by Hart et. al. [Hart1998].

Acknowledgements

The constructive comments of the anonymous referees are noted and appreciated. Thanks also to Colin Wilson for his help.

Bibliography

[Brindle1981]. Brindle, A, "Genetic algorithms for function optimization", Doctoral Dissertation and Technical Report TR81-2, Department of Computer Science, University of Alberta, Edmonton, 1981.

[Chisholm1997]. Chisholm K.J. and Bradbeer P.V.G. "Using a Genetic Algorithm to Optimise a Draughts Program Board Evaluation Function", Proceedings of IEEE ICEC'97, Indianapolis, 1997.

[Eiben1998a] Eiben A.E., Back T,. Schoenauer M, and Schwefel (eds.) "Proceedings of Parallel Problem Solving From Nature - PPSN V", LNCS 1498, Springer Verlag

[Eiben1998b] Eiben A.E., van Hemert J.I., Marchiori E. and Steenbeek A.G, "Solving Binary Constraint Satisfaction Problems Using Evolutionary Algorithms with an Adaptive Fitness Function", in [Eiben1998a]

[Fang1993]. Fang H-L, Ross P. and Corne D. "A promising Genetic Algorithm approach to job-shop Scheduling, rescheduling and open-shop scheduling problems", in [Forrest1993].

[Forrest1993]. Forrest S., (ed.) "Proceedings of the Fifth International Conference on Genetic Algorithms", Morgan Kaufmann, San Mateo, 1993.

[Fogarty1995]. Fogarty T.C (ed.) "Proceedings of the AISB Workshop on Evolutionary Computing", Sheffield, LNCS 993, Springer Verlag, Berlin 1995.

[Goldberg1990]. Goldberg D.E. "Messy Genetic Algorithms: Motivation, Analysis and First Results", Complex Systems, Vol. 3.

[Goldberg1985]. Goldberg D.E. and Lingle, R., "Alleles, loci, and the travelling salesman problem", in [Grefenstette1985].

[Grefenstette1985]. Grefenstette J.J. (ed.) "Proceedings of the First International Conference on Genetic Algorithms", Lawrence Earlbaum, Hillsdale, 1985

[Hart1998] Hart E., Nelson J. and Ross P., "Solving a Real-World Problem Using an Evolving Heuristically Driven Schedule Builder", Evolutionary Computation V6(1):61-80.

[Jones1995]. Jones T., "Evolutionary Algorithms, Fitness Landscapes and Search", PhD Dissertation, University of New Mexico, 1995.

[Jones1994]. Jones P.A. and Bradbeer P.V.G., "Discovery of optimal weights in a concept selection system", in [Leon1994].

[Langdon1995]. Langdon W.B. "Scheduling Planned Maintenance of the National Grid", in [Fogarty1995].

[Leon1994]. Leon R., (ed.) "Proceedings of the 16th Research Colloquium of the BCS Information Retrieval Specialist Group", Taylor Graham, 1994

[Michchalewicz1994] Michchalewicz Z., "Genetic Algorithms + Data Structures = Evolution Programs", (2nd Edition) p256, Springer Verlag, 1994.

[Miller1995]. Miller J.F., Thomson P. and Bradbeer P.V.G. "Ternary Decision Diagram Optimization of Reed-Muller Logic Functions using a Genetic Algorithm for Variable and Simplification Rule Ordering", in [Fogarty1995].

[Paechter1998]. Paechter B., Rankin R.C., Cumming A. and Fogarty T.C., "Timetabling the Classes of an Entire University with an Evolutionary Algorithm." in [Eiben1998a]

[Schaffer1993]. Schaffer J.D. "Proceedings of the Third International Conference on Genetic Algorithms", Morgan Kaufmann, San Mateo, 1993.

[Syswerda1989]. Syswerda G. "Uniform Crossover in Genetic Algorithms", in [Schaffer1993]

[Wren1995]. Wren A. and Wren D. O., "A Genetic algorithm for public transport driver scheduling", Computers in Operations Research, 22(1), 1995.

A Systematic Investigation of GA Performance on Jobshop Scheduling Problems

Emma Hart, Peter Ross

Division of Informatics, University of Edinburgh,
Edinburgh EH1 2QL, Scotland
{emmah,peter}@dai.ed.ac.uk

Abstract. Although there has been a wealth of work reported in the literature on the application of genetic algorithms (GAs) to jobshop scheduling problems, much of it contains some gross over-generalisations, i.e that the observed performance of a GA on a small set of problems can be extrapolated to whole classes of other problems. In this work we present part of an ongoing investigation that aims to explore in depth the performance of one GA across a whole range of classes of jobshop scheduling problems, in order to try and characterise the strengths and weaknesses of the GA approach. To do this, we have designed a configurable problem generator which can generate problems of tunable difficulty, with a number of different features. We conclude that the GA tested is relatively robust over wide range of problems, in that it finds a reasonable solution to most of the problems most of the time, and is capable of finding the optimum solutions when run 3 or 4 times. This is promising for many real world scheduling applications, in which a reasonable solution that can be quickly produced is all that is required. The investigation also throws up some interesting trends in problem difficulty, worthy of further investigation.

1 Introduction

Since the first applications of GAs to scheduling problems, [4], there have been many reported applications of GAs to scheduling problems in general, with the jobshop problem receiving a great deal of attention, for example [2, 8, 9], and more recently [3, 13]. A large variety of representations and operators have been reported, each showing impressive performance of a GA on some small subset of benchmark problems. Often this performance is then extrapolated to claim that the GA in question is a good algorithm for solving job-shop scheduling problems. However, in order to properly evaluate the quality of a method, it is important to show that it works over a wide range of problems. This requires testing the method over an extremely large number of problems – the problems should be chosen such that they exhibit a variety of features, and should vary in difficulty in some configurable way. Rather than use benchmarks problems, it is more useful to use a parameterised problem generator which can generate problem instances at random, and in some tunable manner, so that many different instances of problem classes can be generated.

S. Cagnoni et al. (Eds.): EvoWorkshops 2000, LNCS 1803, pp. 277–286, 2000.

Therefore, in this article we describe a problem generator which we have used to generate instances of jobshop scheduling problems falling into 280 different problem classes. We present the preliminary results from an ongoing investigation, which allows us to come to some general conclusions about the robustness of a GA as a technique for solving scheduling problems. We evaluate our results not only in terms of the capability of the GA to find the optimum solution to the problems we generate (which all have a known optimum value), but also in terms of its accuracy at producing 'reasonable' quality solutions on a reliable basis. For real-world problems, it is seldom necessary to produce *optimum* solutions to problems — "good enough fast enough" will generally suffice. Therefore, it is important to take this into account when making statements about the performance of the GA.

We first describe the features of the problem generator, then present a summary of the results from 28000 experiments on a range of scheduling problems. Some general conclusions about the performance of the GA are then drawn.

2   The Problem Generator

We designed a parameterised, tunable problem generator which generates a *solution* to a jobshop scheduling problem, i.e. a gantt chart, characterised by $P(O, M, D, I, S)$, where O is the number of operations to be scheduled, M is the number of machines, D defines the distribution that the operation sizes are drawn from, (either Gaussian or uniform), I is the total amount of idle time per machine, and S is the amount of slack in the arrival and due dates of each job. The parameters are described in more detail below:

Operations/Jobs/Machines The operations, O, are divided equally between the number of machines, m. A schedule is generated at random, by placing operations into the gantt chart. The operations are then assigned to jobs, j, such that each job is processed only once on each machine, and an operation of a job on a machine cannot begin until its operation on the previous machine has finished. Therefore, although the minimum number of resulting jobs is O/M, the actual number depends on the exact manner in which the tasks are allocated to jobs.

Sizes of Operations The size of each operation in O is drawn from either a uniform distribution between *min* and *max*, $U(min, max)$, or a Gaussian distribution of specified mean and deviation, $G(m, d)$.

Idle Time The amount of idle time on each machine is specified as a percentage of the total processing time of operations on that machine. The idle time is randomly and uniformly distributed between the actual operations.

Slack The maximum amount of slack in the arrival (A) and due-dates D of each job is also specified as a percentage of the total processing time (pt) of the job. If there is no slack, then the arrival date of the job is set to be equal to the time slot at which the job is first processed in the generated schedule, and the due-date is set to the time-slot in which the job finishes. If slack is specified, then the arrival and due dates of each job are altered randomly such that

$$A \rightarrow A - (random(0, S) * pt)$$
$$D \rightarrow D + (random(0, S) * pt)$$

Therefore, no job is ever tardy, and the maximum tardiness objective, T_{max} for each generated problem has an optimum value of 0. For each solution, there is also a known upper bound on the makespan of each problem — if the problem has *no* idle time, then the optimum makespan is known exactly.

3 Experimental Parameters

3.1 The Genetic Algorithm

The GA used is *HGA* which was described by the authors in [6]. This GA outperformed other heuristic combination methods, and compared well to the most recently published results on a number of benchmark problems. It uses an indirect representation, in which each gene on the chromosome encodes a pair (*Method, Heuristic*). The *method* gene denotes the methodology that should be used to calculate a conflicting set of schedulable operations at each iteration of the algorithm. An operation is then chosen using the heuristic denoted by the heuristic gene. The method applied is either the Giffler and Thompson algorithm, [5], which produces an *active* schedule, or a modified G&T algorithm which produces a *non-delay* schedule.

The GA used a population size of 100, uniform crossover and a swap mutation operator. Each experiment is run for a maximum of 500 generations, or until it finds the optimum solution. The maximum tardiness of the best solution is noted, with the number of generations required to find it if the optimum was discovered. The best solutions for each of the 100 trials in each problem class are averaged, and we also record the minimum and maximum solution quality in each class.

3.2 The Problem Classes

In the experiments presented in this article we fix the number of operations, and hence the chromosome length, at 60. This facilitates a thorough and fair investigation of the effects of the other four parameters on problem difficulty. Furthermore, as each GA experiment uses a chromosome population of identical size and length, we can fix the "GA" parameters such as the number of generations to be the same in all experiments.

The other four parameters are varied as shown in table 1, resulting in a total of 280 problem classes, each defined by a tuple $P(O, M, D, I, S)$. (For all experiments in which $S > 0$, then I is fixed at 0, and vice versa). For each of the 280 problem classes, 10 problems are generated from different random number seeds. We then run a GA 10 times on each problem instance, and using T_{max} as the objective function (which we know to have an optimum solution of 0), average the results over the 100 experiments in that class.

In the remainder of this article, the term problem *class* refers to the general class of problems defined by a tuple $P(O, M, D, I, S)$. The term problem *instance* refers to one of the 10 problems generated for a problem *class*.

Parameter	Value
Total Number of Operations	60
Number of Machines	2,3,4,5,6
Task Size Distribution	U(0,10), U(0,30), U(0,50), U(0,100)
	G(50,1), G(50,10), G(50,20), G(50,40)
Idle Time	5%, 10%, 25%
Slack	10%, 25%, 50%

Table 1. Experimental Parameters

4   Results

As space limitations do not allow us to present the results of all 280 experiments here, we attempt to present summaries of the findings and general trends observed, and discuss some individual problem classes in more detail. The complete set of results can be found at [1].

4.1   Overall Performance

Firstly, we note that in every problem class, the GA is able to find the optimum solution for at least 1 instance of the problems in that class. On the whole, better performance is observed in the 140 experiments in which the operation sizes were distributed uniformly, (regardless of distribution size). For example, in 81% of problem classes that had a uniform distribution of task size, then running the GA on problem instances of that class resulted in the optimum solution being found in at least half of all experiments. This compares to a figure of 59% for problem classes with a Gaussian distribution of task size. These statements should be treated with some caution however as they may simply be an artifact of the parameters chosen to define the uniform and Gaussian distributions, and the two series of experiments cannot be directly compared.

Turning our attention to those problem classes which are solved to "reasonable" accuracy, we note that 55% of the problem classes with Gaussian task size distribution are solved to within 10 time units of the optimum solution. This increases to 96% for problem classes with uniformly distributed operation sizes.

Tables 2 and 3 show the problem classes in which all generated instances where solved with 100% accuracy. There is no obvious pattern, except that it

Idle Time	Slack	Std.Dev	Machines
10%	0	10	2
25%	0	1	3
25%	0	1	2
25%	0	10	2
25%	0	20	2
25%	0	40	2
25%	0	40	3
0	50%	40	6

Table 2. Problems with Gaussian Operation Size Distribution which are solved with 100% accuracy

Idle Time	Slack	Max Size	Machines
25%	0	10	2
25%	0	30	2
25%	0	50	2
25%	0	100	2
25%	0	30	4
0	25%	10	5
0	25%	10	6
0	50%	30	6
0	50%	50	6

Table 3. Problems with Uniform Operation Size Distribution which are solved with 100% accuracy

appears that for perfect performance, either a large value of idle time or large slack is required in all cases, and that for Gaussian distributions of operation size, a large value of standard deviation helps.

For problems with a Gaussian distribution of operation size, the worst performance is observed in 2 problem classes which both result in the optimum solution only being in 1 run of 1 of the 10 problem instances. These problems classes are $P(60, 5, G(50, 10), 0, 0)$ and $P(60, 6, G(50, 10), 0, 0)$. In the uniform case, the worst performance, again with only 1 optimum solutions is for a problem class $P(60, 6, U(50), 0, 0)$.

4.2   Number of Machines

Intuitively, it would be expected that as the number of machines increases for a fixed number of operations, O, then the problems would become easier to solve. This is because the number of schedulable operations in the conflict set produced as a result of applying the G&T or non-delay algorithm must reduce as the number of machines increases. Table 4 shows the percentage of problem classes with $m \in 2, 3, 4, 5, 6$ in which the optimum solution was found. For a uniform distribution of jobs, the results are as expected. However, for the Gaussian distributions, we see exactly the opposite — i.e the percentage of optimum solutions decreases as the number of machines is increased.

	Number of Machines				
	2	3	4	5	6
Gaussian	63	56	55	55	58
Uniform	62	65	74	75	73

Table 4. % of problem classes resulting in an optimum solution vs no. machines

	Slack Parameter			
	0.0	0.1	0.25	0.5
Gaussian	33	32	60	82
Uniform	46	50	76	80

Table 5. % of problem classes resulting in an optimum solution vs slack

4.3   Slack

Table 5 shows the percentage of problem classes with $slack \in (0.0, 0.1.0.25.0.5)$ in which the optimum solution was found. For problems with Gaussian distributions of task size, we observe that increasing the slack in due and arrival dates to 10% does *not* increase performance, as would be expected, and in fact results in a slight decrease. Increasing the slack to larger values however increases performance. This trend is not observed with uniform distributions, which tend to become easier to solve as the slack parameter is increased.

Examining the effect of the slack parameter in more detail in the Gaussian experiments, we notice a correlation between the number of machines parameter m and the slack S in determining solution quality. Table 6 shows the minimum value of the slack parameter S that was required for each set of problem classes with machine m before an improvement was observed in solution quality compared to the equivalent experiment with $S = 0.0$. Where no improvement was observed, adding slack had a *detrimental* effect on solution quality.

| Number of Machines | Minimum Percentage | Standard Deviation |
Machines	Slack Required	
2	0.5	20,40
3	0.25	all values of std. dev.
4	0.1	all values of std. ded.
5	0.05	all values of std. dev.
6	0.05	all values of std. dev.

Table 6. Minimum Value of Slack Parameter Required to Improve Solution Quality

In problem classes where operation size was uniformally distributed, then adding slack generally increases performance. The only exceptions where a significant difference occurs are for the problem classes $P(60, 2, U(0,\text{max}), 0.0, 0.1)$, where $max \in (10, 30, 50, 100)$.

4.4   Idle Time

For problems with both uniform and Gaussian task distributions, adding idle time generally improves performance, as is seen in table 7, which shows the percentage of problem classes in which the optimum solution was found for various values of I. This is as expected — inserting idle time into the schedule allows some flexibility in the exact placement of operations in the schedule, without necessarily decreasing schedule quality.

5   A Phase Transition in Problem Classes with Gaussian Distributions

In the majority of problems with a Gaussian distribution of operation sizes, we notice an interesting transition in the difficulty of the problems as the standard

	Idle Time			
	0.0	0.05	0.1	0.25
Gaussian	33	48	55	85
Uniform	46	65	75	89

Table 7. % of problem classes in which the optimum solution was achieved vs idle time

deviation of the distribution is varied.The GA performs best on those problems with large standard deviations. For very small standard deviations, i.e when the all tasks have very similar sizes, then the GA also performs reasonably well. However, for a range of values of standard deviation in the middle, performance decreases considerably. For example, figure 1 shows an expanded graph for the problem class $P(60, 5, G(50, sd), 0.0, 0.0)$, in which some extra experimental points have been added. A clear peak in difficulty is seen, centered around a standard deviation of 7.

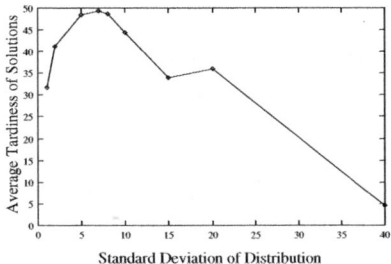

Fig. 1. Solution Quality vs Standard Deviation of Task Size Distribution

This appearance of this phenomenum shows remarkably similar properties to earlier work performed by the authors in the timetabling domain. Work described in [11] showed that there was a clear phase transition in the performance of a GA on a sequence of (solvable) timetabling problems designed to be of increasing difficulty; the GA tested could solve very lightly constrained, and also very highly constrained problems, however, for moderately constrained problems, the GA would often fail to find a solution. The appearance of similar phase-transition regions has also been reported in other classes of constraint satisfaction problems, for example see [7, 10, 12].

In the timetabling case, it was noted that other (non-evolutionary) algorithms also failed on the same subset of problems, suggesting that it was not the GAs "fault", but that the problems were intrinsically difficult. We have not yet investigated the performance of other non-evolutionary methods on these problems, but expect to see a similar pattern in performance.

1. Calculate the set C of all operations that can be scheduled next

2. Calculate the completion time of all operations in C, and let m^* equal the machine on which the minimum completion time t is achieved.

3. Let G denote the conflict set of operations on machine m^* - this is the set of operations in C which take place on m^*, and whose start time is less than t.

4. Select an operation from G to schedule

5. Delete the chosen operation from C and return to step 1.

Fig. 2. Giffler and Thompson Algorithm

A possible reason, still to be looked at in more detail, is the use of the Giffler and Thompson (G&T) algorithm (or the modified non-delay version) in constructing the conflict sets of operations at each iteration. The G&T algorithm is shown in figure 2. For problems in which there is a large deviation in operation sizes, then it is possible that the conflict set is generally smaller at each iteration, and therefore it is more straightforward to choose the 'correct' operation. When all the operations are of similar size, then it seems likely that the size of the conflict set is non-trivial, and hence it is more difficult to choose the 'correct' operation. Figure 3 shows the size of the conflict set at each iteration, averaged over 100 runs of experiments in which the standard deviation of operations sizes was set to 5, and then to 40, using a constant mean of 50. Early in the scheduling process, there is a small region, highlighted on the figure, in which there a fewer items in the conflict set for the case when $sd = 40$. As we know that the placement of operations early in the schedule is crucial to the success of the algorithm, this may provide a clue, however the matter needs further attention.

6   Problem Classes with Uniform Distributions

As noted earlier, better results appear to be obtained when the operation sizes are uniformally distributed, compared to those problems with gaussian distribution of operation sizes. Solution quality tends to decrease as the range of the distribution increases. This information is summarised in table 8 which reports the percentage of optimum solutions obtained for instances of problems tested with each different range value.

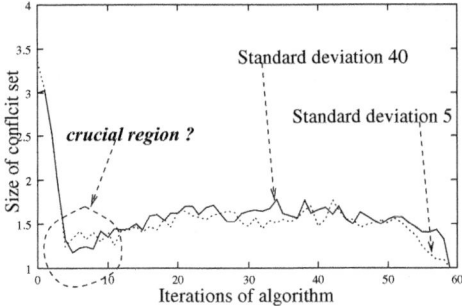

Fig. 3. Size of Conflict Set For Different Distributions of Operation Sizes

	Maximum Task Size			
	10	30	50	100
% optimum solutions	80	76	71	68

Table 8. Percentage of problem instances resulting in an optimum solution vs maximum task size

7 Conclusion

This article has presented some initial observations made whilst attempting to perform a systematic investigation on the performance of a genetic algorithm on a range of job-shop scheduling problems. The investigation involved a total of 280 different problem classes, each containing 10 randomly generated problems.

Initial findings suggest that although the GA does not find the optimum solution for all problem classes, for most instances of all problem classes it is capable of finding the optimum in at least 1 in 10 trials, and that the quality of solution is generally satisfactory. This suggests the GA appears to be a relatively robust method of tackling such problems, and that since runs are reasonably fast, running the GA several times to find a solution is a viable strategy. Some interesting trends have been observed — we now intend to try and understand and explain these trends, and to determine whether they are a feature of the problems themselves or due to the genetic algorithm itself.

This study concentrated on problem classes which all contained a fixed number of operations. Therefore, in order to complete the study, we will also use the generator to test GA performance on a sequence of much larger problems. Although we expect performance to degrade as the problems get bigger, it is fruitful to determine at what point this happens.

Finally, further work is planned to compare the performance of other non-evolutionary algorithms on the same set of problem classes, to see if similar trends are observed, and to compare overall performance. These other methods will include constraint satisfaction techniques, and simple scheduling rules.

Acknowledgements

Emma Hart is supported by EPSRC grant GR/L22232.

References

1. http://www.dai.ed.ac.uk/~emmah/jobshop-expts.html.
2. Sugato Bagchi, Serdar Uckun, Yutaka Miyabe, and Kazuhiko Kawamura. Exploring problem-specific recombination operators for job shop scheduling. In R.K. Belew and L.B. Booker, editors, *Proceedings of the Fourth International Conference on Genetic Algorithms*, pages 10–17. San Mateo: Morgan Kaufmann, 1991.
3. Brizuela. C.A. and N. Sannomiya. A diversity study in genetic algorithms for jobshop. In *Proceedings of the Genetic and Evolutionary Computation Conference*, pages 75–83, 1999.
4. L. Davis. Job shop scheduling with genetic algorithms. In J. J. Grefenstette, editor, *Proceedings of the International Conference on Genetic Algorithms and their Applications*, pages 136–140. San Mateo: Morgan Kaufmann, 1985.
5. B. Giffler and G.L. Thompson. Algorithm for solving production scheduling problems. *Operations Research*, 8(4):487–503, 1960.
6. E. Hart and P. Ross. A heuristic combination method for jobshop scheduling problems. In *Parallel Problem Solving from Nature, PPSN-V*, pages 845–854, 1998.
7. T. Hogg, A. Huberman, and C.P. Williams. Phase transitions and the search problem. *Artificial Intelligence*, 81(1-2):1–15, 1996.
8. S-C. Lin, E.D. Goodman, and W.F. Punch. A genetic algorithm approach to dynamic job-shop scheduling problems. In Thomas Bäck, editor, *Proceedings of the Seventh International Conference on Genetic Algorithms*, pages 481–489. Morgan-Kaufmann, 1997.
9. R. Nakano and T. Yamada. Conventional genetic algorithms for job shop problems. In R.K. Belew and L.B. Booker, editors, *Proceedings of the Fourth International Conference on Genetic Algorithms*, pages 474–479. San Mateo: Morgan Kaufmann, 1991.
10. P. Prosser. An empirical study of phase transitions in binary constraint satisfaction problems. *Artificial Intelligence*, 81(1-2):81–109, 1996.
11. P Ross, E Hart, and D Corne. Some observations about ga-based exam timetabling. In *Practice and Theory of Automated Timetabling*, pages 115–130, 1997.
12. B.M. Smith and M.E. Dyer. Locating the phase transitions in binary constraint satisfaction problems. *Artificial Intelligence*, 81(1-2):155–181, 1996.
13. P. Van Bael, D. Devogelaere, and M. Rijckaert. The job shop problem solved with simple, basis evolutionary search elements. In *Proceedings of the Genetic and Evolutionary Computation Conference*, pages 665–670, 1999.

An Ant Algorithm with a New Pheromone Evaluation Rule for Total Tardiness Problems

Daniel Merkle[1] and Martin Middendorf[2]

Institute for Applied Computer Science and Formal Description Methods,
University of Karlsruhe, Germany
{[1]merkle,[2]middendorf}@aifb.uni-karlsruhe.de

Abstract. Ant Colony Optimization is an evolutionary method that has recently been applied to scheduling problems. We propose an ACO algorithm for the Single Machine Total Weighted Tardiness Problem. Compared to an existing ACO algorithm for the unweighted Total Tardiness Problem our algorithm has several improvements. The main novelty is that in our algorithm the ants are guided on their way to good solutions by sums of pheromone values. This allows the ants to take into account pheromone values that have already been used for making earlier decisions.

1 Introduction

Ant Colony Optimization (ACO) is an evolutionary metaheuristic to solve combinatorial optimization problems by using principles of communicative behaviour found in real ant colonies (for an introduction and overview see [5]). Recently the ACO approach has been applied to scheduling problems, like Job-Shop [2, 7], Flow-Shop [13], and the Single Machine Total Tardiness problem [1]. Bullnheimer et al. [1] have compared an ACO algorithm with several other heuristics to solve the Single Machine Total Tardiness problem (e.g. decomposition heuristics, interchange heuristics and simulated annealing). They have shown that the ACO algorithm found the optimal solution of 125 benchmark problems more often than the other heuristics (these benchmark problems where generated with the same method from [12] as the benchmarks problems used in this paper).

In this paper we propose alternative and improved ways to solve the Single Machine Total Tardiness problem by ACO. Moreover, we also study the weighted version of the total tardiness problem.

In ACO algorithms several generations of artificial ants search for good solutions. Every ant of a generation builds up a solution step by step going through several probabilistic decisions until a solution is found. In general, ants that found a good solution mark their paths through the decision space by putting some amount of pheromone on the edges of the path. The following ants of the next generation are attracted by the pheromone so that they will search in the solution space near good solutions. In addition to the pheromone values the ants will usually be guided by some problem specific heuristic for evaluating the possible decisions.

S. Cagnoni et al. (Eds.): EvoWorkshops 2000, LNCS 1803, pp. 287–296, 2000.

The approach used in [1] and [13] to solve scheduling problems with ACO algorithms is to use a pheromone matrix $T = \{T_{ij}\}$ where pheromone is added to an element T_{ij} of the pheromone matrix when a good solution was found where job j is the ith job on the machine. The following ants of the next generation then directly use the value of T_{ij} to estimate the desirability of placing job j as the ith job on the machine when computing a new solution.

Here we propose a different approach. Instead of using only the value of T_{ij} the ants use $\sum_{k=1}^{i} T_{kj}$ to compute the probability of placing job j as the ith on the machine. A problem with using only T_{ij} can occur when the ant does not chose job j as the ith job in the schedule. Because, if the $T_{i+1,j}, T_{i+2,j}, \ldots$ values are small then job j might be scheduled much later than at the ith place (and possibly long after its due date). It is likely that this will not happen when using $\sum_{k=1}^{i} T_{kj}$. Note, that this approach differs from nearly all other ant algorithms proposed so far, in that we base one possible decision of an ant on several pheromone values. The only other work that uses several pheromone values to estimate the quality of one possible decision is [11]. Moreover, we let the ants make optimal decisions when this is possible and use a heuristic that is a modification of the heuristic used in [1].

This paper is organized as follows. The Single Machine Total Weighted Tardiness Problem is defined in Section 2. In Section 3 we describe an ACO algorithm for the unweighted problem. The pheromone summation rule is introduced in Section 4. Section 5 contains further variants and improvements. The choice of the parameter values of our algorithms used in the test runs and the test instances and are described in Section 6. The results are reported in Section 7. A conclusion is given in Section 8.

2   The Single Machine Total Weighted Tardiness Problem

The Single Machine Total Weighted Tardiness Problem (SMTWTP) is to find for n jobs, where job j, $1 \leq j \leq n$ has a processing time p_j, a due date d_j, and a weight w_j, a non-preemptive one machine schedule that minimizes $T = \sum_{j=1}^{n} w_j \cdot \max\{0, C_j - d_j\}$ where C_j is the completion time of job j. T is called the total weighted tardiness of the schedule. The unweighted case, i.e. $w_j = 1$ for all $j \in \{1, \ldots, n\}$, is the Single Machine Total Tardiness Problem (SMTTP).

It is known that SMTTP is NP-hard in the weak sense [8] and SMTWTP is NP-hard in the strong sense [10]. A pseudopolynomial time algorithm for SMTWTP in case that the weights agree with the processing times (i.e. $p_j < p_h$ implies $w_j \geq w_h$) was given in [10]. Observe, that the last result implies that SMTTP is pseudopolynomial time solvable. For an overview over different heuristics for SMTWTP see [4].

3   ACO Algorithm for SMTTP

The ACO algorithm of Bullnheimer et al. [1] is described in this section. The general idea was to adapt an ACO algorithm called ACS-TSP for the traveling salesperson problem of Dorigo et al. [6] for the SMTTP. In every generation each of m ants constructs one solution. An ant selects the jobs in the order in which they will appear in the schedule. For the selection of a job the ant uses heuristic information as well as pheromone information. The heuristic information, denoted by η_{ij}, and the pheromone information, denoted by τ_{ij}, are an indicator of how good it seems to have job j at place i of the schedule. The heuristic value is generated by some problem dependent heuristic whereas the pheromone information stems from former ants that have found good solutions. With probability q_0, where $0 \leq q_0 < 1$ is a parameter of the algorithm, the ant chooses a job j from the set S of jobs that have not been scheduled so far which maximizes

$$[\tau_{ij}]^\alpha \, [\eta_{ij}]^\beta$$

where α and β are constants that determine the relative influence of the pheromone values and the heuristic values on the decision of the ant. With probability $1 - q_0$ the next job is chosen according to the probability distribution over S determined by

$$p_{ij} = \frac{[\tau_{ij}]^\alpha \, [\eta_{ij}]^\beta}{\sum_{h \in S} [\tau_{ih}]^\alpha \, [\eta_{ih}]^\beta}$$

The heuristic values η_{ij} are computed according the Modified Due Date rule (MDD), i.e.,

$$\eta_{ij} = \frac{1}{\max\{T + p_j, d_j\}} \tag{1}$$

where T is the total processing time of all jobs already scheduled.

After an ant has selected the next job j, a local pheromone update is performed at element (i, j) of the pheromone matrix according to

$$\tau_{ij} = (1 - \rho) \cdot \tau_{ij} + \rho \cdot \tau_0$$

for some constant ρ, $0 \leq \rho < 1$ and where

$$\tau_0 = \frac{1}{m \cdot T_{EDD}}$$

and T_{EDD} is the total tardiness of the schedule that is obtained when the jobs are ordered according to the Earliest Due Date heuristic (EDD), i.e., with falling values of $1/d_j$. The value τ_0 is also used to initialize the elements of the pheromone matrix.

After all m ants have constructed a solution the best of these solutions is further improved with a 2-opt strategy. The 2-opt strategy considers swaps between all pairs of jobs in the sequence. Then it is checked whether the so derived schedule is the new best solution found so far.

The best solution found so far is then used to update the pheromone matrix. But before that some of the old pheromone is evaporated according to

$$\tau_{ij} = (1 - \rho) \cdot \tau_{ij}$$

The reason for this is that old pheromone should not have a too strong influence on the future. Then, for every job j in the schedule of the best solution found so far some amount of pheromone is added to element (ij) of the pheromone matrix where i is the place of job j in the schedule. The amount of pheromone added is ρ/T^* where T^* is the total tardiness of the best found schedule, i.e.,

$$\tau_{ij} = \tau_{ij} + \rho \cdot \frac{1}{T^*}$$

The algorithm stops when some stopping criterion is met, e.g. a certain number of generations has been done or the best found solution has not changed for several generations.

4   The Pheromone Summation Rule

In this section we describe a new approach of using the pheromone values which is used in our ACO algorithm for SMTTP. In general, a high pheromone value τ_{ij} means that it is advantageous to put job j at place i in the schedule. Assume now that by chance an ant chooses to put some job h at place i of the schedule that has a low pheromone value τ_{ih} (instead of a job j that has a high pheromone value τ_{ij}). Then in order to have a high chance to still end up with a good solution it will likely be necessary for the ant to place job j not too late in the schedule when j has a small due date. To some extend the heuristic values η_{lj} for $l > i$ will then force the ant to choose j soon. But a problem occurs when the values τ_{lj} are small (because no good solutions have been found before that have job j at some place $l > i$). Then the product $(\eta_{lj})^\alpha \cdot (\tau_{lj})^\beta$ is small and it is likely that the ant will not choose j soon. In this case the ant will end up with a useless solution having a high total tardiness value.

To handle this problem we propose to let a pheromone value τ_{ij} also influence later decisions when choosing a job for some place $l > i$. A simple way to guaranty this influence is to use the sum of all pheromone values for every job from the first row of the matrix up to row i when deciding about the job for place i. When using this pheromone summation rule we have the following modified decision formulas. An ant chooses as next job for place i in the schedule with probability q_0 the job $j \in S$ that maximizes

$$(\sum_{k=1}^{i} [\tau_{kj}])^\alpha \cdot [\eta_{kj}]^\beta \tag{2}$$

and with probability $1 - q_0$ job $j \in S$ is chosen according to the probability distribution over S determined by

$$p_{ij} = \frac{(\sum_{k=1}^{i} [\tau_{kj}])^{\alpha} \cdot [\eta_{ij}]^{\beta}}{\sum_{h \in S} (\sum_{k=1}^{i} [\tau_{kh}])^{\alpha} \cdot [\eta_{ih}]^{\beta}} \qquad (3)$$

5   Further Variations and Improvements

In this section we describe further variations and improvements that we used in our ACO algorithm.

5.1   Modified Heuristic

A problem when using the heuristic values according to formula (1) is that the values of $\max\{T + p_j, d_j\}$ become much larger — due to T — when deciding about jobs to place at the end of the schedule than they are when placing jobs at the start of the schedule. As a consequence the heuristic differences between the jobs are, in general, small at the end of the schedule. To avoid this effect we used the following modified η values

$$\eta_{ij} = \frac{1}{\max\{T + p_j, d_j\} - T} \qquad (4)$$

For the weighted problem SMTWTP we multiplied every value on the right side of equation (4) with the weight w_j of job j. Note that jobs with a small weighted processing time p_j/w_j have a high heuristic value when $T + p_j \geq d_j$.

5.2   Deterministic Scheduling Between Due Dates

Consider the construction of a schedule for the unweighted problem SMTTP. Assume that some jobs have already been scheduled. Assume further that the sum T of the processing times of all jobs scheduled so far lies between some due date d_j and a due date $d_h > d_j$ and every other due date is smaller than d_j or larger than d_h. For this case it is easy to show that it is optimal to schedule all jobs with a due date $\leq d_j$ before scheduling a job with a due date $\geq d_h$ as long as the sum of the processing times of the scheduled jobs is at most d_h. Moreover when there are several jobs with due date $\leq d_j$ it is optimal to schedule these jobs ordered by increasing processing times. If the ants apply this deterministic rule whenever possible we say that the ants work locally deterministic. Then the ants will switch between probabilistic and deterministic behaviour.

6   Test Instances and Parameters

We tested the different variants of ACO algorithms on 125 benchmark instances for SMTWTP of size 100 jobs that are included in the OR-Library [14]. These benchmark instances were generated as follows: for each job $j \in [1 : 125]$ an integer processing time p_j is taken randomly from the interval $[1 : 100]$, an integer weight w_j is taken randomly from the interval $[1:10]$ and an integer due date d_j is taken randomly from the interval

$$\left[\sum_{j=1}^{125} p_j \cdot (1 - TF - \frac{RDD}{2}), \sum_{j=1}^{125} p_j \cdot (1 - TF + \frac{RDD}{2})\right]$$

The value RDD (relative range of due dates) determines the length of the interval from which the due dates were taken. TF (tardiness factor) determines the relative position of the centre of this interval between 0 and $\sum_{j=1}^{125} p_j$. The values for TF and RDD are chosen from the set $\{0.2, 0.4, 0.6, 0.8, 1.0\}$. The benchmark set contains five instances for each combination of TF and RDD values. For the unweighted problem SMTTP we used the same benchmark instances but ignored the different weights.

Our results for SMTWTP were compared to the best known results for the benchmark instances that are from [3] and can be found in [14].

The parameters used for the test runs are: $\alpha = 1$, $\beta = 1$, $\rho = 0.1$, $q_0 \in \{0, 0.9\}$. The number of ants in every generation was $m = 20$. Every test was performed with 4 runs on every instance. Every run was stopped after 500 generations.

We used a 2-opt strategy to improve the best solution that was found in every generation which differs slightly from the 2-opt strategy used in [1]. For every pair of jobs it was checked exactly once whether a swap of these jobs improves the schedule. A swap that improves the schedule was fixed immediately. Thus we tried exactly 4950 swaps per generation.

In the following ACS-SMTTP (or short ACS) denotes the algorithm of [1] as described in Section 3 but with the new 2-opt strategy described in the last paragraph. Our algorithm ACS-SMTWTP-Σ is similar to ACS but uses the pheromone summation rule as described in Section 4. Algorithm ACS-SMTWTP-H is similar to ACS but uses the new heuristic from Section 5.1. Algorithm ACS-SMTWTP-D is similar to ACS but additionally uses the deterministic strategy from Section 5.2 for scheduling between due dates. Algorithms that use combinations of new features are denoted by ACS-SMTWTP-XYZ where X,Y,Z$\in \{\Sigma$, H, D$\}$ (e.g. ACS-SMTWTP-HΣ uses the new heuristic and the pheromone summation rule). For shortness we write ACS-XYZ for ACS-SMTWTP-XYZ.

7   Experimental Results

The influence of the pheromone summation rule (called Σ-rule in the following) and the modified heuristic was tested on weighted and unweighted problem instances. Since the parameter q_0 has some influence on the results we performed tests with $q_0 = 0$ and $q_0 = 0.9$.

Table 1 shows the results for SMTWTP. The average total tardiness values found by the ACO algorithms for SMTWTP were compared to the average total tardiness of the best known solutions that are from [3]. The average total tardiness per instance of the best solutions from [3] is 217851.34. Table 1 shows that ACS-ΣH performed better than ACS-H and also that ACS-Σ performed better than ACS (this holds for both cases $q_0 = 0$ and $q_0 = 0.9$). In all cases the difference of the total tardiness values compared to the best known solutions are at least 61.1% lower for the ACO algorithm with Σ-rule (79.5 for ACS-ΣH compared to 204.5 for ACS-H with $q_0 = 0.9$). Moreover, the ACO algorithms with Σ-rule found for more instances a better total tardiness than their counterparts without Σ-rule (at least 5.3 times as often). The differences of the total tardiness values compared to the best known values over the first 200 generations are shown in Figure 1. The best solution of ACS-ΣH was found after an average of 80 generations, which was after less than 3.5 seconds on a 450 MHz Pentium-II processor.

Table 1 also shows that the ACO algorithms with modified heuristic performed in all cases better than their counterparts using the heuristic from [1]. For $q_0 = 0.9$ the advantage of the modified heuristic is smaller than for $q_0 = 0$ (e.g. for $q_0 = 0.9$ ACS-ΣH has a 60.2% smaller difference to optimal total tardiness than ACS-Σ compared to a 92.9% smaller difference for $q_0 = 0$).

Table 1. Influence of pheromone summation rule and new heuristic on solution quality for SMTWTP. Total Tardiness: average difference to total tardiness of best found solutions from [3] (average over 500 test runs, 125 instances and 4 runs for each instance); Better: comparisons between ACS-ΣH and ACS-H (respectively ACS-Σ and ACS), number of instances with smaller average total tardiness (average over 125 instances and 4 runs for each instance).

weighted		ACS-ΣH	ACS-H	ACS-Σ	ACS
Total	$q_0 = 0$	191.8	3024.7	946.1	9914.7
Tardiness	$q_0 = 0.9$	79.5	204.5	200.0	1198.6
Better	$q_0 = 0$	97	2	106	0
	$q_0 = 0.9$	86	16	97	3

Table 2 shows the results for the unweighted problem SMTTP. The results are compared with the average of the best total tardiness values we found for the unweighted instances, i.e. 54309.5. Similarly as for the weighted problem in all cases the ACO algorithms with Σ-rule are better than their counterparts without Σ-rule. Also the modified heuristic performed better in all cases than the heuristic from [1].

Since the 2-opt strategy significantly influences of the quality of the solutions we also compared the ACS-ΣH with ACS-H when using no 2-opt strategy. The results can be found in Table 3 for SMTWTP and in Table 4 for SMTTP. The only case where ACS-ΣH performed not significantly better than ACS-H is the unweighted case with $q_0 = 0.9$. In this case ACS-H found a slightly better average total tardiness ACS-ΣH (difference is 331.5 for ACS-H and 332.3 for ACS-ΣH). On the other hand ACS-ΣH found for more instances better solutions than ACS-H (For 65 instances ACS-ΣH found better solutions than ACS-H whereas ACS-H performed better than ACS-ΣH for 33 instances).

Fig. 1. SMTWTP: Average difference to total tardiness of best found solutions from [3] over the first 200 generations.

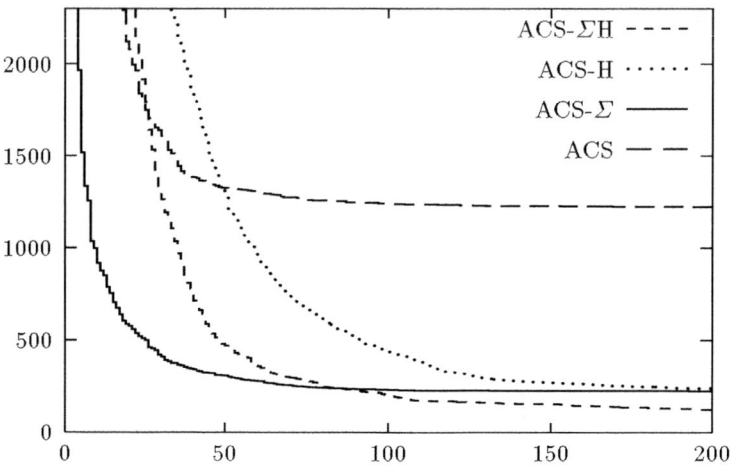

Table 2. Influence of pheromone summation rule and new heuristic on solution quality for SMTTP. Total Tardiness: average difference to total tardiness of best found solutions (average over 500 test runs, 125 instances and 4 runs for each instance); "Better" as in Table 1.

unweighted		ACS-ΣH	ACS-H	ACS-Σ	ACS
Total	$q_0 = 0$	47.9	48.5	112.9	256.4
Tardiness	$q_0 = 0.9$	7.0	19.0	8.7	26.3
Better	$q_0 = 0$	53	32	82	17
	$q_0 = 0.9$	53	22	67	14

Table 3. Influence of pheromone summation rule and new heuristic on solution quality for SMTWTP when using no 2-opt. "Total Tardiness" as in Table 1; "Better" as in Table 1 but comparison between ACS-ΣH and ACS-H.

no 2-0pt, weighted		ACS-ΣH	ACS-H
Total	$q_0 = 0$	11894.4	22046.8
Tardiness	$q_0 = 0.9$	1733.2	1793.5
Better	$q_0 = 0$	76	48
	$q_0 = 0.9$	67	42

Table 4. Influence of pheromone summation rule and new heuristic on solution quality for SMTTP when using no 2-opt. "Total Tardiness" as in Table 2; "Better" as in Table 1 but comparison between ACS-ΣH and ACS-H.

no 2-0pt, unweighted		ACS-ΣH	ACS-H
Total	$q_0 = 0$	3943.7	4515.8
Tardiness	$q_0 = 0.9$	332.3	331.5
Better	$q_0 = 0$	59	50
	$q_0 = 0.9$	65	33

The influence of the deterministic strategy for scheduling between due dates for SMTTP has only a minor influence on the results for the unweighted benchmark instances from the OR-Library. The reason is that these instances have small gaps between the due dates. Thereby, the deterministic strategy does come into play only rarely. Hence, we created new test instances which have two neighboured due dates that have a large gap in between. We changed each of the problem instances from the OR-Library as follows. The jobs were ordered by their due dates and the due dates of jobs 41 to 59 were set to the same due date that job 40 has. The average of the best total tardiness values we found for these modified instances was 56416.3. Table 5 shows for $q_0 = 0$, that ACS-ΣHD performed much better than ACS-ΣH and also that ACS-HD performed much better than ACS-H. For $q_0 = 0.9$ the ACS-ΣHD algorithm could not profit from the deterministic scheduling between due dates.

Table 5. Influence of deterministic strategy between due dates on solution quality for SMTTP and problem instances with modified due dates. "Total Tardiness as in Table 2; "Better" as in Table 1 but comparison between ACS-ΣH and ACS-ΣHD.

unweighted		ACS-ΣH	ACS-ΣHD	ACS-H	ACS-HD
Total	$q_0 = 0$	101.4	45.7	120.1	3.8
Tardiness	$q_0 = 0.9$	2.9	8.7	11.1	9.2
Better	$q_0 = 0$	8	78	1	92
	$q_0 - 0.9$	36	14	29	36

8   Conclusion

We have introduced a new method to use the pheromone values in an Ant Colony Optimization (ACO) algorithm for the Single Machine Total Weighted Tardiness problem. An ACO algorithm using this pheromone summation rule gives better solutions for 125 benchmark than its counterpart that does not use the summation rule. This holds also for the unweighted total tardiness problem. Moreover, we proposed a new heuristic that can be used by the ants when searching for a solution. For the unweighted problem we have shown that the ACO algorithm can profit from ants that switch between a deterministic behaviour (in case that optimal decisions can be made) and the "standard" probabilistic behaviour.

References

1. A. Bauer, B. Bullnheimer, R.F. Hartl, C. Strauss: An Ant Colony Optimization Approach for the Single Machine Total Tardiness Problem; in: Proceedings of the 1999 Congress on Evolutionary Computation (CEC99), 6-9 July Washington D.C., USA, 1445-1450, 1999.
2. A. Colorni, M. Dorigo, V. Maniezzo, M. Trubian: Ant System for Job-Shop Scheduling; *JORBEL - Belgian Journal of Operations Research, Statistics and Computer Science*, 34: 39–53 (1994).
3. R.K. Congram, C.N. Potts, S. L. van de Velde: An iterated dynasearch algorithm for the single-machine total weighted tardiness scheduling problem; submitted to *INFORMS Journal on Computing*.
4. H.A.J. Crauwels, C.N. Potts, L.N. Van Wassenhove: Local Search Heuristics for the Single Machine Total Weighted Tardiness Scheduling Problem; *INFORMS Journal on Computing*, 10: 341–359 (1998).
5. M. Dorigo, G. Di Caro: The ant colony optimization meta-heuristic; in: D. Corne, M. Dorigo, F. Glover (Eds.), *New Ideas in Optimization*, McGraw-Hill, 1999, 11-32.
6. M. Dorigo, L. M. Gambardella: Ant colony system: A cooperative learning approach to the travelling salesman problem; *IEEE Trans. on Evolutionary Comp.*, 1: 53-66 (1997).
7. M. Dorigo, V. Maniezzo, A. Colorni: The Ant System: Optimization by a Colony of Cooperating Agents; *IEEE Trans. Systems, Man, and Cybernetics – Part B*, 26: 29-41 (1996).
8. J. Du, J.Y.-T. Leung: Minimizing the Total Tardiness on One Machine is NP-hard; *Mathematics of Operations Research*, 15: 483–496 (1990).
9. P. Forsyth, A. Wren: An Ant System for Bus Driver Scheduling; Report 97.25, University of Leeds - School of Computer Studies, 1997.
10. E.L. Lawler: A 'pseudopolynomial' algorithm for sequencing jobs to minimize total tardiness; *Annals of Discrete Mathematics*, i: 331–342 (1977).
11. R. Michels, M. Middendorf: An Ant System for the Shortest Common Supersequence Problem; in: D. Corne, M. Dorigo, F. Glover (Eds.), *New Ideas in Optimization*, McGraw-Hill, (1999) 692-701.
12. C.N. Potts, L.N. Van Wessenhove: Single machine tardiness sequencing heuristics; *IEE Transactions*, 23: 346-354 (1991)
13. T. Stützle: An ant approach for the flow shop problem; in *Proc. of the 6th European Congress on Intelligent Techniques & Soft Computing (EUFIT '98)*, Vol. 3, Verlag Mainz, Aachen, 1560-1564, 1998.
14. http://mscmga.ms.ic.ac.uk/jeb/orlib/wtinfo.html.

A New Genetic Representation and

Common Cluster Crossover

for Job Shop Scheduling Problems

TEZUKA Masaru, HIJI Masahiro,
MIYABAYASHI Kazunori, and OKUMURA Keigo

Hitachi Tohoku Software, Ltd.
2-16-10, Honcho, Aoba-ku, Sendai, 980-0014, Miyagi, Japan
{tezuka, hiji, miyabay, okumura}@hitachi-to.co.jp

Abstract. This paper describes a genetic algorithm approach for sequencing problems especially for job shop scheduling. In actual problems, setup time should be optimized. In order to reduce setup time, we developed a new genetic representation and an efficient crossover operator called Common Cluster Crossover (CCX). In our representation, chromosomes represent the shift of the order in a sequence. To preserve sub-sequences in crossover operations, we implemented the process to identify the cluster of the sub-sequences and applied CCX that exchanges common clusters between two parents. The approach was tested on two standard benchmarks and applied to an audio parts manufacturer. CCX achieved remarkable results on the actual job shop scheduling problem.

1 Introduction

Job Shop Scheduling Problems (JSSPs) are solved by assigning jobs to resources. Those assignments are called operations. Jobs are basic tasks. Through a chain of operations, materials are processed into intermediate parts, and the parts are processed into products. Consequently each job completes. The sequence among operations of each job must be restricted. At any time, one resource can process only one operation.

JSSPs are known as the most difficult problems and they cannot be solved in a practically timely manner because they are NP-hard. The genetic algorithm (GA) which is recognized as the effective tool to solve combinatorial problems has been applied to JSSPs in various approaches [1][2][3][4].

S. Cagnoni et al. (Eds.): EvoWorkshops 2000, LNCS 1803, pp. 297–306, 2000.

In this paper, we propose a new genetic sequencing approach and a new crossover operator that works efficiently to minimize setup time on JSSPs. When GAs are used in sequencing approach, several specially designed crossover operators such as the partially mapped crossover (PMX)[5], the order based crossover [6], and C1 [7] are used to eliminate infeasible sequences. Those operators preserve some of sub-sequences but frequently destroy others. For example, the order based and C1 preserve the complete sub-sequences inherited from one parent but just a relative position from the other.

In most practical JSSPs, sub-sequences are very important because setup time is required for resources to change setups from the one for a certain item to the one for a new item. Successive production of the same items reduces the setup time. Thus, we consider that preserving good sub-sequences that reduces the setup time is important to reach the good solution quickly.

In order to allow offspring to inherit sub-sequences from the both parents, we propose a new genetic sequencing representation and the Common Cluster Crossover (CCX). In our framework, chromosomes represent the shift of the order in a sequence. We added the process to identify the cluster of the sub-sequences, and crossover operation is applied to the common clusters between two parents.

In Section 2, we describe the details of the representation and CCX. In Section 3, we report the results of two computational experiments, one is for FT10x10 and FT20x5 and the other is for an audio parts manufacturer's scheduling problem. Then, we conclude with the results in Section 4.

2 Preserving Sub-sequences

2.1 Genetic Representation

In our approach, chromosomes represent the shifting distance of the order in a sequence. One chromosome corresponds to one resource, and one individual consists of a group of chromosomes. The number of chromosomes in one individual equals to the number of resources in a schedule.

Each chromosome is decoded to the sequence of the jobs on the resource. Let us consider the case in which j jobs are processed by a certain resource. The length of the chromosome corresponding to the resource is j-1. The i-th allele of the chromosome

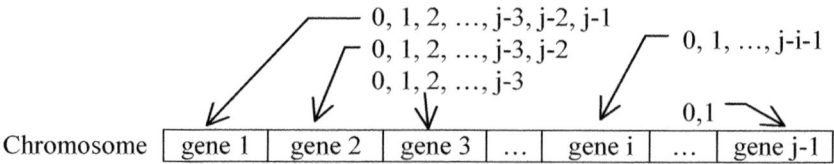

Fig. 1. The alleles.

can take an integer [0, *j-i*-1] as shown in Fig. 1.

The decoding process of the chromosome to the sequence is shown in Fig. 2. In the figure, the chromosome is {5,4,1,0,1}. At first, the sequence is initialized as {1,2,3,4,5,6}. The value of the first gene is five, and this means that the first symbol of the sequence is to be exchanged for the symbol on five steps right. Thus, symbol '1' and '6' are exchanged and the sequence becomes {6,2,3,4,5,1}. The value of the second gene is four, and this means that the second symbol of the sequence is to be exchanged for the symbol on four steps right. Thus, symbol '2' and '1' are exchanged and the sequence becomes {6,1,3,4,5,2}. Repetition of this procedure for the whole genes leads to the sequence {6,1,4,3,2,5}.

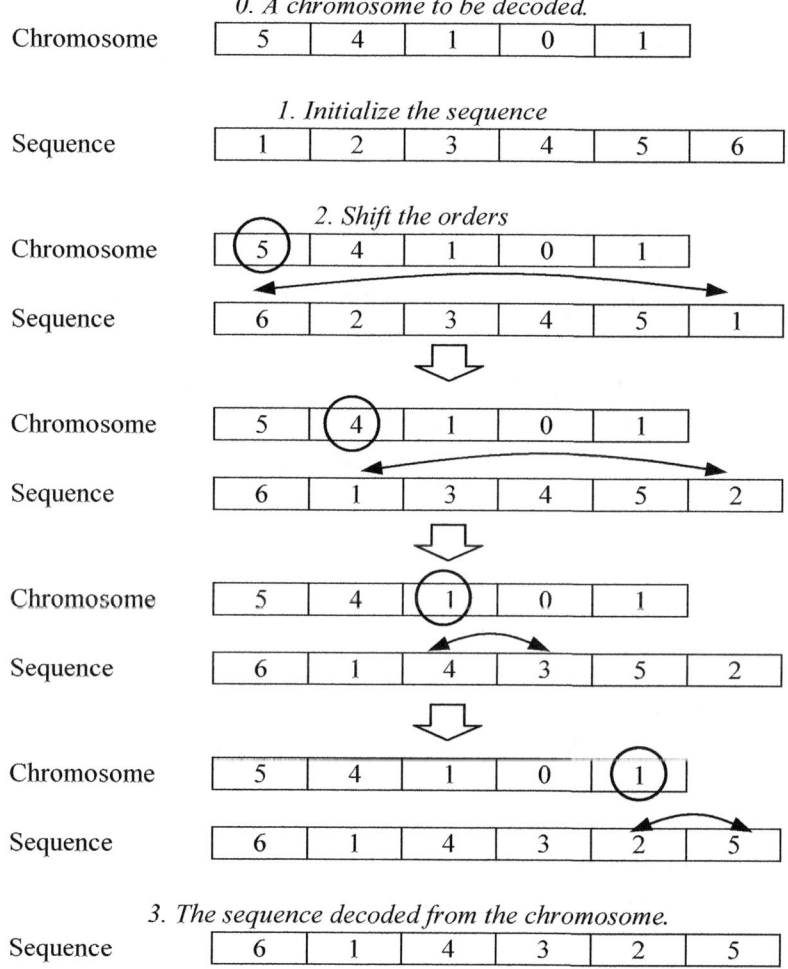

Fig. 2. Decoding process of the chromosome to the sequence.

2.2 Common Cluster Crossover (CCX)

The random selection of the crossover point is typically used in the conventional GA approaches. Even if the new genetic representation we mentioned above is adopted, the random selection can destroy the sub-sequence. In order to select the effective crossover point, we propose the Common Cluster Crossover (CCX). CCX identifies the clusters of sub-sequences and exchanges the common clusters between two parents.

A cluster consists of the genes linked together. Fig. 3 shows the example of clusters. As mentioned in Section 2.1, each value of a gene is the shifting distance. The arrows in the figure indicate the link of the shifts. In this case, the chromosome has two clusters. As shown in the figure, a cluster can consist of several separate segments.

CCX exchanges the common clusters of two parents. Fig. 4 describes CCX process. In this example, Parent 1 has two clusters, A and B. Parent 2 has four clusters, C, D, E, and F. Two dotted arrows indicate the common boundaries of clusters. CCX exchanges the common clusters that are separated by the common boundaries and create two offspring.

Fig. 5 shows the example of CCX. One offspring is reproduced from Parent 1(P1) and Parent 2 (P2) by CCX. P1 has two clusters, A and B. P2 has four clusters C, D, E, and F. Since two boundaries match between P1 and P2, CCX exchanges clusters at

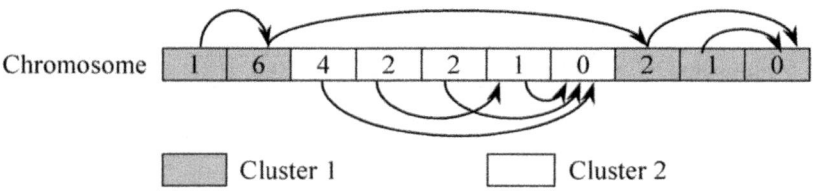

Fig. 3. The cluster of sub-sequences.

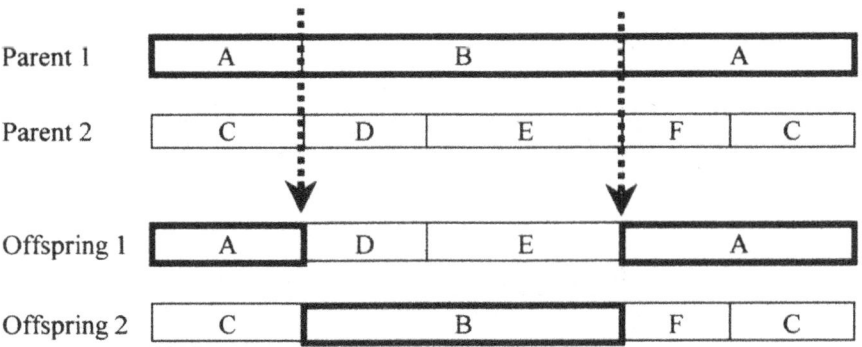

Fig. 4 Common Cluster Crossover (CCX).

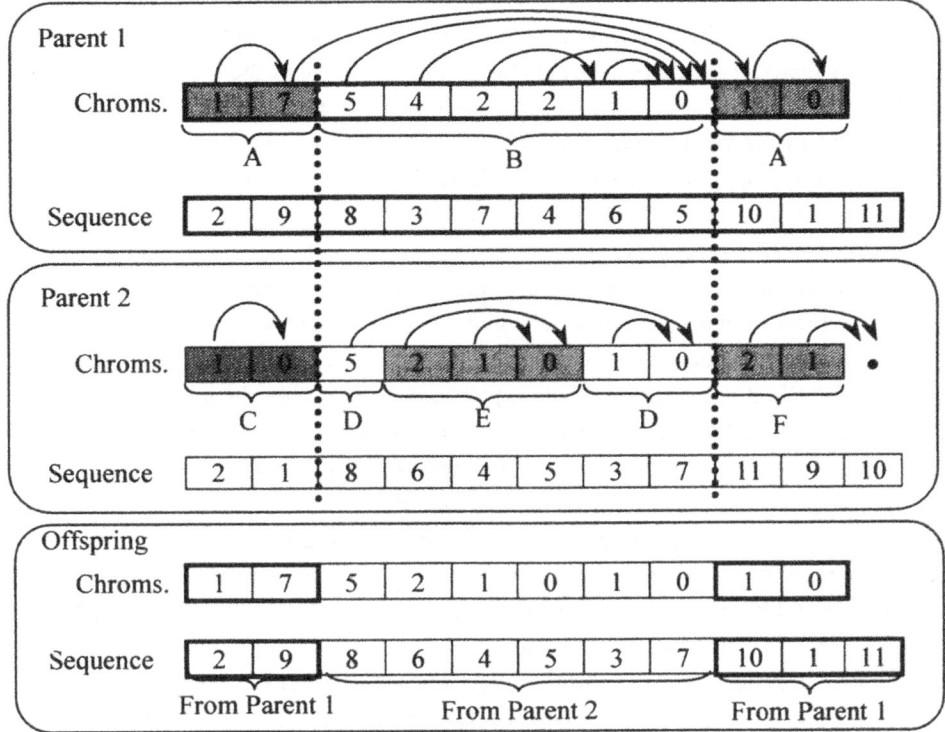

Fig. 5 Common Cluster Crossover.

these points.

I would like to emphasize that the offspring inherits the sub-sequences from both P1 and P2. In this case, the sub-sequence (2, 9, *, *, *, *, *, *, 10, 1,11) was inherited from P1 and (*, *, 8, 6, 4, 5, 3, 7, *, *, *) from P2. We believe that this feature of CCX is very effective to optimize and reduce the setup time in JSSPs.

There are two cases in which CCX cannot be applied. The first case occurs when no position of the boundaries matches between two parents. The second case occurs when both parents consist of one cluster. In other words, there is no boundary. In the first case, CCX is applied according to the clusters of one parent and those of the other parent are ignored. We call this crossover quasi-CCX. In the second case, conventional one-point crossover is applied. Naturally, preservation of sub-sequences is not guaranteed in these cases.

3 Computational Results

3.1 Standard Job Shop Scheduling Problem

The GA based on our representation and CCX has been implemented in Java, JDK1.1.8 on the Pentium 550MHz PC with Windows 98.

The GA was tested on two standard job shop problems, Fisher and Thompson's FT10x10 and FT20x5 [8], as a benchmark. These two problems have been widely used by many researchers as benchmarks. The objective of the benchmark problems is to minimize makespan. The optimal solutions are 930 on FT10x10 and 1165 on FT20x5. In the field of GAs, on FT10x10, Yamada[9] first found the optimum 930 four times among 600 trials in 1992. Kobayashi[10] founds 930 fifty-one times among 100 trials in 1995. On FT20x5, Mattfeld[11] and Lin[12] found optimum 1165 in 1994, 1997 respectively.

It is known that the optimal schedules of makespan are among active schedules. The method to generate active schedules (GT method) was proposed by Giffler and Thompson [13]. Some other approaches use the GT method [4][10][14]. We also used the GT method to interpret the job sequence into an active schedule.

We adopted the ranking selection [15] and the elitist policy [16]. Mutation changes the value of a gene into a random integer following uniform distribution $[0, j-i-1]$ where j is the number of jobs and i is the loci of the gene.

In our experiments, the population size was set to 200 and termination was set to 250 generations. Crossover rate was set to 1.0 and mutation rate was set to 0.01. It took about 400 seconds to run one trial on FT10x10 and 450 seconds on FT20x5.

Fig. 6 and Fig. 7 show the frequency distribution of solutions (makespan) in 100 trials on FT10x10 and FT20x5 respectively. The approach we proposed could not obtain optima but produced highly efficient solutions. I must point out that in practical use it is necessary to create good enough schedules reliably and fast rather than find optimal solutions.

As mentioned in Section 2.2, our crossover operation consists of three kinds of crossover. CCX is applied when there are common clusters between two parents. Quasi-CCX is applied when no boundary matches between two parents. One-point crossover is applied when both parents consist of one cluster. Fig. 8 shows a temporal

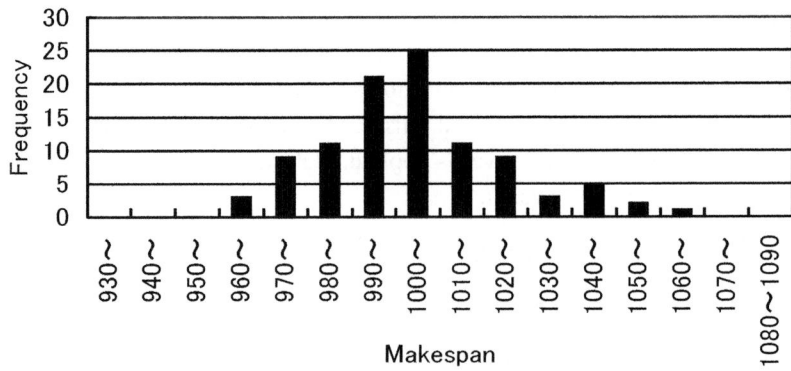

Fig. 6 Frequency distribution of solutions on FT10x10.

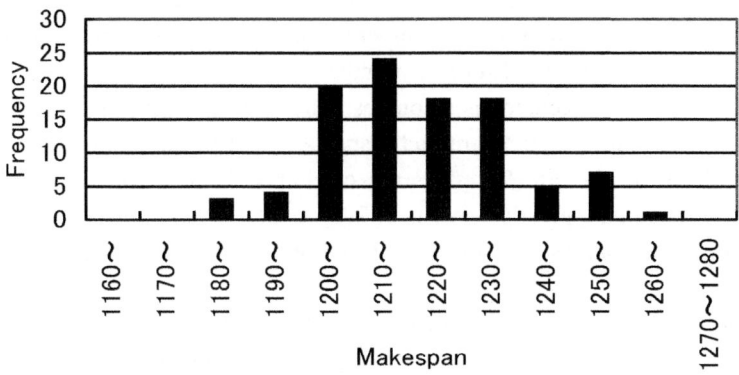

Fig. 7 Frequency distribution of solutions on FT20x5.

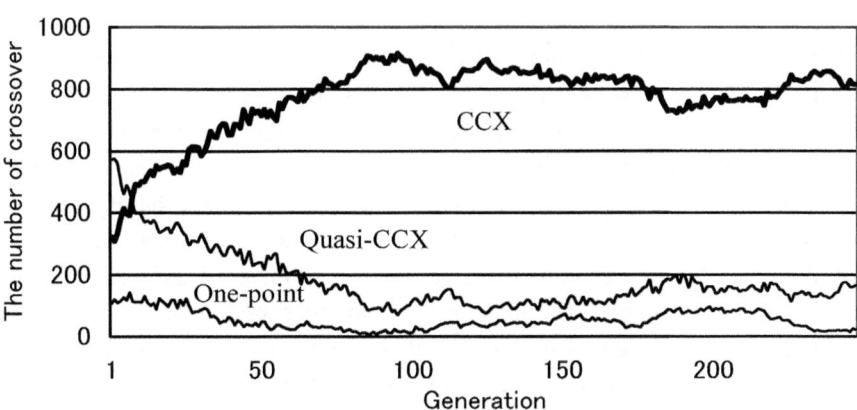

Fig. 8 A typical temporal change of the number of three kinds of crossover.

change of the numbers of three kinds of crossover in a certain trial on FT10x10. In the first generation, about 30% of crossover operations was CCX, 55% was quasi-CCX, and 15% was one-point crossover. The proportion of CCX increased gradually, and in the end of iteration, CCX increased to 80%, quasi-CCX decreased to 15%, and one-point crossover decreased to 5%.

We surmise that the proportion of CCX increased since good sub-sequences as building blocks were created, maintained and combined during iteration.

3.2 Application to an Audio Parts Manufacturer

The GA approach we proposed was applied to the scheduling of an audio parts manu-facturer. Since the scheduling problem involves setup time reduction, we can test the optimization ability of CCX by measuring the reduction in setup time.

The manufacturer is a subcontractor of an electrical appliance manufacturer and receives production orders once a month. According to the production orders, it makes its production schedule once a month.

It has fifteen work centers as resources. One hundred workers are involved in the production. Its main products are audio speakers and headphones. It produces about 7500 products per month. Lot sizes are calculated by a MRP system. Its monthly production schedule typically consists of 300 operations. In the production, setup time is required for resources to convert from the production of a certain item to another item.

Formerly, the manufacturer had used a dispatching-based scheduler. In order to reduce setup time, the worker had to modify the schedule created by the scheduler. It took about three to four hours to modify with a Gantt chart editor. To refine the scheduling process, our GA-based scheduling system was introduced. The objective of the manufacture is to minimize total delivery delay time, the number of delaying jobs, and total setup time.

First, we used total setup time solely as the fitting function in order to test the ability of CCX. The test data consist of 84 jobs and 300 operations. For comparison, we also applied order based operator [6] as the traditional operator to the same problem. Population size was set to 50 and termination was set to 40. To create schedules in a reasonably short time, We set the values smaller than on the benchmarks. It took about 80 seconds to run one trial with both CCX and order based. Fig. 9 shows the frequency distribution of solutions (setup time) in 100 trials. Obviously CCX is superior to the traditional crossover operator in minimizing setup time.

Next, we tested weighted sum of total delivery delay time, the number of delaying jobs, and total setup time as the fitting function on the same data. Population size and termination were also set to 50 and 40 respectively. Table 1 shows the best and the average of CCX and order-based in 100 trials. The result of a legacy dispatching-

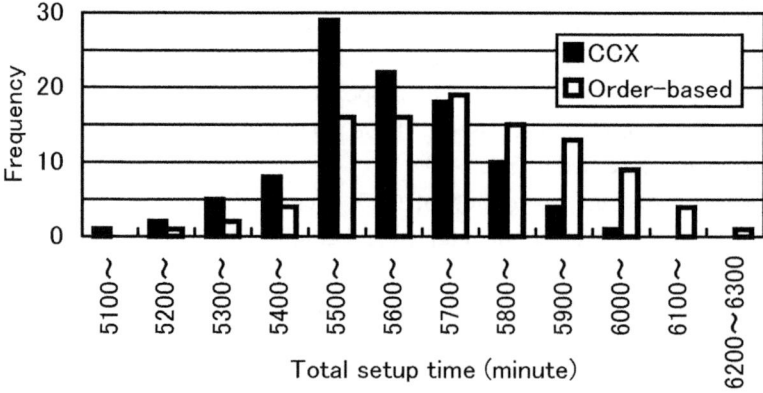

Fig. 9 Frequency distribution of solutions on setup time minimizing.

Table 1. Application to an audio parts manufacturer

		Total setup time (minutes)	Total delivery delay time(days)	The number of delaying jobs
CCX	best	5165	7.16	16
	average	5766	8.40	18
Order-based	best	5575	7.21	15
	average	6085	8.55	18
Legacy scheduler		8425	8.87	25

based scheduler is also shown for comparison. CCX approach successfully optimized the whole indices, especially setup time. What needs to be emphasized here is that in order to minimize total setup time, sub-sequences are quite important because successive production of the same kind of items reduces the total setup time remarkably.

4 Conclusions

In this paper, we proposed a new genetic representation and Common Cluster Crossover (CCX). Chromosomes represent the shifting distance of the order in a sequence and are decoded to the sequence of the jobs. CCX exchanges the common clusters of two parents. Integrated use of our representation and CCX allows offspring to preserve sub-sequences. Our discussion shows that this feature is highly effective in solving JSSPs especially problems to optimize setup time.

We tested our approach on the standard JSSPs FT10x10 and FT20x5 and found the approach was effective.

Since we developed CCX in order to optimize setup time in particular, we also applied our new approach to an audio parts manufacturer. Setup time should be dealt with on the problem. First, only setup time was used as the fitting function to test the ability of CCX, and CCX successfully minimized setup time. Next, the fitting function was set to minimize the weighted sum of the total delivery delay time, the number of delaying jobs, and the total setup time. The proposed approach successfully optimized the JSSP based on the actual data.

References

[1] Lawrence Davis: Job Shop Scheduling with Genetic Algorithms. Proceedings of the First Intl. Conference on Genetic Algorithms and their Applications (1985) 136-140

[2] Sugato Bagchi, Serdar Uckun, Yutaka Miyabe, and Kazuhiko Kawamura: Exploring Problem-Specific Recombination Operators for Job Shop Scheduling. Proceedings of the Forth Intl. Conference on Genetic Algorithms (1991) 10-17

[3] Ryohei Nakano and Takeshi Yamada: Conventional Genetic Algorithm for Job Shop Problems. Proceedings of the Forth Intl. Conference on Genetic Algorithms (1991) 474-479

[4] Christian Bierwirth: A generalized permutation approach to job shop scheduling with genetic algorithms. OR Spektrum Vol. 17 (1995) 87-92

[5] David E. Glodberg and Robert Lingle, Jr.: Alleles, Loci, and Traveling Salesman Problem. Proceedings of the First Intl. Conference on Genetic Algorithms and their Applications (1985) 154-159

[6] Gilbert Syswerda: Schedule Optimization Using Genetic Algorithms. In Handbook of Genetic Algorithms. Lawrence Davis ed. Van Nostrand Reinhold, New York (1991)

[7] Colin R. Reeves: A Genetic Algorithm for Flowshop Sequencing. Computers and Operations Research Vol. 22, No. 1 (1995) 5-13

[8] H. Fisher, G. L. Thompson: Probabilistic learning combinations of local job-shop scheduling rules. In Industrial Scheduling. J.F. Muth, G. L. Thompson ed. Prentice Hall, Englewood Cliffs, New Jersey (1963)

[9] Takeshi Yamada and Ryohei Nakano: A Genetic Algorithm Applicable to Large Scale Job Shop Problems. Parallel Problem Solving from Nature, 2 (1992) 281-290

[10] Shigenobu Kobayashi, Isao Ono, and Masayuki Yamamura: An Efficient Genetic Algorithm for Job Shop Scheduling Problems. Proceedings Of the 6th Intl. Conference On Genetic Algorithms (1995) 506-511

[11] D. C. Mattfeld, H. Kopfer, and C. Bierwirth: Control of Parallel Population Dynamics by Social-like Behavior of GA-Individuals. Parallel Problem Solving from Nature, 3 (1994) 15-24

[12] Shyh-Chang Lin, Erik D. Goodman, and William F. Punch, III: Investigating Parallel Genetic Algorithms on Job Shop Scheduling Problems. Evolutionary Programming VI, Proceedings of Sixth International Conference EP97 (1997), 383-394

[13] B. Giffler and G. L. Thompson: Algorithm for Solving Production Schedule Problems. Operations Research, Vol.8 (1960) 487-503

[14] Robert H. Storer, S. David Wu, and Renzo Vaccari: New Search Spaces for Sequencing Problems with Application to Job Shop Scheduling, Management Science, Vol.38 (1992) 1495-1509

[15] James E. Baker: Adaptive Selection Methods for Genetic Algorithms. Proceedings of the First Intl. Conference on Genetic Algorithms and their Applications (1985) 101-111

[16] Gunar E. Liepins and W. D. Potter: A Genetic Algorithm Approach to Multiple-Fault Diagnosis. In Handbook of Genetic Algorithms. Lawrence Davis ed. Van Nostrand Reinhold, New York (1991)

Optimising an Evolutionary Algorithm for Scheduling

Neil Urquhart, Ken Chisholm and Ben Paechter

School Of Computing
Napier University
219 Colinton Road
Edinburgh
EH14 1DJ
{neilu,ken,benP}@dcs.napier.ac.uk

Abstract. This paper examines two techniques for setting the parameters of an evolutionary Algorithm (EA). The example EA used for test purposes undertakes a simple scheduling problem. An initial version of the EA was tested utilising a set of parameters that were decided by basic experimentation. Two subsequent versions were compared with the initial version, the first of these adjusted the parameters at run time, the second used a set of parameters decided on by running a meta-EA. The authors have been able to conclude that the usage of a meta-EA allows an efficient set of parameters to be derived for the problem EA.

1. A Description of the Problem

The use of Evolutionary Algorithms (EAs) for solving timetabling and scheduling problems has become commonplace in recent years [1], [9]. This paper examines two methods for optimising the parameters used by the algorithm. Although the concepts and methods used for developing and optimising the evolutionary algorithm are intended to be general, for the purposes of presentation within this paper they will be applied to a simple scheduling problem. The scheduling problem under consideration requires a number of jobs to be processed through a factory. Each job has start and end times that form its time window. Within this window the job must "visit" all the resources required for this job, within a specified order. Each resource is mutually exclusive (ie only one job may make use of a resource at any one time).

The various conflicting constraints for the scheduling task can be divided into two categories hard constraints and soft constraints, as shown in table 1. A hard constraint is one that must be satisfied in order to produce a feasible schedule, soft constraints alter the quality of the schedule.

2. A Description of the EA

The general concepts of evolutionary algorithms are well understood. Each individual requires a representation (a genotype) that may be decoded into a specific schedule (a

S. Cagnoni et al. (Eds.): EvoWorkshops 2000, LNCS 1803, pp. 307–318, 2000.

phenotype). Each gene within the genotype represents the start time of the event (stored as a real number). Within each genotype the event genes make up a chromosome, the event genes are grouped so that all the genes for a particular resource are stored together in order of start time. This representation may be considered a direct representation, in that the genes within the chromosome are a direct representation of the finished schedule. The mutation and crossover operators act directly on the items in the schedule. With an indirect representation the genes would represent a set of instructions for building the schedule [10], thus further processing would be required to convert the genotype into a schedule.

Table 1. Hard and soft constraints

Constraint	Type
Event must start after the previous event on this resource has finished.	Hard
Event must start after the previous event in this job has finished.	Hard
The first event must start after the jobs "available" time	Hard
The last event must finish after the jobs "due" time	Soft[1]
The event should take place in normal working hours	Soft
The event should take place on week days	Soft

In the course of these experiments 3 data sets were utilised, they may be summarised as:

Data set 1: 35 Events, within 16 Jobs using 4 Resources, over a 9 day period
Data set 2: 41 Events, within 15 Jobs using 8 Resources, over a 4 day period
Data set 3: 50 Events, within 18 Jobs using 8 Resources, over a 4 day period

Each resource has a 'queue' of events, each event has a start time set to the nearest minute. Thus for an event being placed within a schedule being built for a 4 day period there are *1440*4* possible values for the start time (there are 1440 minutes in a day). Thus if there were 10 events scheduled to use a resource over a 2 day period the number of potential queues for that resource would be *(48*1440)*10 = 619200* different queues. The total number of potential solutions to the scheduling problem is the product of the size of the potential queues for each resource. The vast majority of these schedules are infeasible, and the restricted mutation operator discourages their production.

2.1 Basic Operations

Within the system two-point crossover, based on the Davis system is used to create children, based on exchanging start time genes. The child replaces one of the randomly selected parents. The EA uses steady state population to reduce complexity.

---

[1] A job which is not completed until after the end of its time window may not be desirable, but will still result in a schedule that is feasible.

Use is made of elitism, this ensures that the best individuals of each generation are preserved for inclusion in the next generation.

The mutation operator considers each member of the population for possible mutation, the chance of a specific member being selected is based on the mutation rate which is expressed as a percentage. For instance with a mutation rate of 35% there is a 35% chance of any individual being selected. When an individual is selected, one gene is selected for mutation with a selection pressure of 1/L (where L is the length of the chromosome).

The fitness function is based on the constraints outlined in section 1, with the satisfaction of hard constraints being given greater rewards than the satisfaction of soft constraints. The fitness for each individual event is calculated using the values shown in table 2.

Table 2. Initial weight values used in the fitness function

Constraint	Reward
Current event is "placed" (does not conflict with any other event)	120
All events that make up the current job placed	100
Current event takes place before the job is due for completion	50
Current event takes place after the job is "available"	50
Event takes place between Monday and Friday	25
Event takes place before "night time"	4
Event takes place after "day time"	4

The EA was allowed to run until all the events were placed (ie all the hard constraints are satisfied). There may arise situations where there is no feasible schedule for a given data set (ie if 25 hours of work are scheduled to take place in one day) therefore a limit is set on the total number of generations that may be calculated. This limit is calculated as *no of events *150*. Through initial experimentation it was found that allowing the EA to continue running for a short period after a feasible solution had evolved allowed further improvement with regards to the soft constraints. The number of generations used for this run-on period was calculated as *total number of events * 5.*

2.2 Improving the Quality of Schedules using Lamarckism

To improve the quality of schedules constructed by the system, the authors implemented Lamarckian writeback [11],[5]. Lamarckian theory suggests that knowledge acquired by individuals during their lifetime may be passed genetically onto their descendants. This theory has since been discredited by biologists who maintain that crossover and random mutation are the main influences on evolutionary development. The type of improvement made to a schedule is illustrated in figure 1, it may be accepted that any solution where events within a job "overlap" is not feasible and thus of no practical value. The work of [10] outlines the use of Lamarckianism within an EA used for solving the Napier University timetabling problem.

2.3 Targeted Mutation

In [10] use is made of what is known as "Targeted mutation". This modifies the mutation operator to increase the chances of genes with specific characteristics being selected for mutation. The actual characteristics that are used to select a gene for an increased chance of mutation will vary depending on the application. In most cases the genes that will have the extra bias in favor of mutation will be those that exhibit some defect requiring improvement. An extra weight value will be added to the chances of genes exhibiting that defect. In the case of the problem under consideration the authors decided to implement the bias so that genes relating to events breaking a hard constraint have a higher chance of being selected for mutation. The mutation function was been modified so that if a gene selected for mutation represents an event that does not break a hard constraint there is a chance (expressed as a percentage) that another gene may be selected. Table 3 shows the results of using a 40%, 50%, and 60% bias against the mutation of placed genes.

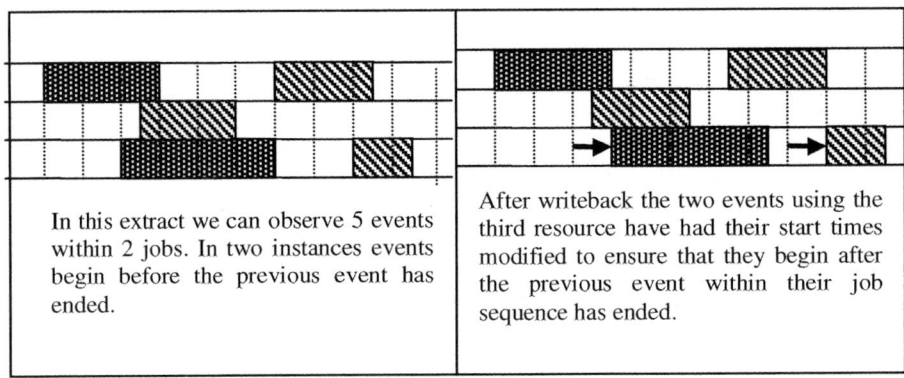

In this extract we can observe 5 events within 2 jobs. In two instances events begin before the previous event has ended.

After writeback the two events using the third resource have had their start times modified to ensure that they begin after the previous event within their job sequence has ended.

Fig. 1. An example of part of a schedule before write back(left) and after (right) writeback

2.4 Premature Convergence and Regeneration

Within the EA it is desirable to maintain a diverse population. When a good solution is found, this individual will begin to dominate and thus reduce the bio-diversity within the population. The convergence of many population members to a similar point within the search space is feature of all EAs. It is desirable though, to try to prevent convergence until the EA beings to converge on the optimum solution. One method of eliminating the premature convergence effect is to implement a regeneration operator. Within the regeneration operator a number of individuals have their genes randomly reset. Thus in the case of the scheduling algorithm under consideration a number of members of the population will have the times of their all events randomly mutated. Consideration must be given to the frequency of the usage of the operator and the percentage of the population involved. If the regeneration operator is utilised too frequently then the population will not have time to evolve, if it affects too many individuals with a high fitness then the EA may loose useful genetic material.

The operator as implemented initially, reinitialises the lower 50% of the population randomly every 50 generations. The regeneration operator in practice applies the mutation operator to all of the genes in the schedule, thus creating a new, random individual. The times taken to produce a feasible schedule using regeneration are shown in table3. It may be seen therefore that using regeneration offers a substantial (approximately 40%) saving in time.

3. Setting the EA Parameters

The parameters used within the EA to specify population size, crossover rate, mutation rate and regeneration have the biggest bearing on the success or failure of an EA. Conventionally the EA designer usually sets these parameters to values decided with reference to previous experimentation. The authors decided to investigate two alternative methods of arriving at a set of parameters. A version of the EA using the initial parameters was constructed and will be referred to as version CP.

Table 3. Time in seconds to produce a feasible schedule using data set 2

Run	1	2	3	4	5	Average
No Bias	570	118	19	502	1457	768.5
50% Bias	110	780	1617	280	1599	563.3
40% Bias	238	353	534	1614	163	716.3
60% Bias	51	1619	137	934	108	646.9
Regeneration and 50% bias	57	181	472	228	1631	343.6

3.1 Modifying the Parameters at Run Time

The first method investigated is based on a simplified version of the system described in [13] and [5]. Each of these parameters starts off set to a low value, when the individual at the head of the population (ie the most fit individual) changes the parameter that invokes the operator responsible for creating that individual is increased slightly. This is simplified from the original system proposed by Davis [13] which not only took into account the most fit individual, but also the operators used to create the ancestors of that individual for several generations previous to the current generation.

A new version of the EA incorporating the ability to modify its parameters at run time was constructed. Whenever the most fit individual changes the parameter controlling the creating operation is modified to increase the usage of the operator that caused the improvement. For instance if a new most-fit individual is created by the mutation operator then the mutation rate is increased by 0.1%, in the event of the cross over operation creating a new most-fit individual then the cross-over rate is increased by 0.1%. In the event of the regeneration operator being responsible then the frequency of the regeneration operator is increased by 1 generation. A new version of the EA implementing this operator was constructed, it will be referred to as VP1. Initial values for the mutation rate and crossover rate are 75% and 50% with the regeneration interval being set at 50 generations, as used in CP. The results obtained when utilising data set2 may be see in table 4.

In order to attempt to decrease the time taken to build a schedule, it was decided to modify the starting values, thus allowing the EA itself to make one or the other become the dominant operator. A second version (VP2) with starting mutation and crossover rates were set to 10% and the starting regeneration interval was set to 1000 generations, thus no operator is allowed to dominate at the start.

Table 4. Fitness of schedules generated using data set 2. (CP – Constant parameters; VP1 – Variable parameters, starts with 75% crossover and 50% mutation rates; VP2 – Variable paramters starts with 10% crossover and mutation rates)

Run	1	2	3	4
CP.	34162.9	34019.4	34193.2	34062.4
VP1	34598.1	34487.4	34497.6	34352.3
VP2	34214.1	34382.8	34424.7	34372.8

From the figures in table 4 it may be seen that version VP1 gives the highest average fitness when building schedules based on data set 2. The results would suggest that in order to be effective the run time parameters must have starting values that are proven to provide an efficient result when used without modification. It was found that when starting with low initial values the EA is unable to progress due to the lack or crossover, mutation or regeneration.

Based on the results obtained from the testing of versions VP1 and VP2, optimising the parameters at run-time does not appear to give any advantages over using parameters that the user has arrived at using "trial and error" methods. The reasons for failure may be connected with the initial parameter values. If the parameters are initially set to small values (as in version VP2), there is a lack of activity within the GA, and the parameters cannot increase, and thus increase the activity within the algorithm until new most-fit individuals are being created.

3.2   Using a Meta-Level EA to Optimise the Parameters

The parameters of an EA may themselves be optimised by using another EA [13], [4], known as a meta-EA, to attempt to evolve a set of efficient parameters. Within this meta-level EA the genes represent the parameters of the EA that we wish to optimise (hereinafter referred to as the problem EA). The fitness function of the meta-EA is based on the results of running the problem EA with the parameters stored in each individual within the meta-EA population.

The parameters of the problem EA to be optimised are listed in table 5. The fitness of each individual within the meta-EA was calculated by running the scheduling-EA using the parameters encoded within that individual, the fitness was calculated as the fitness of the final schedule less a time value.

Table 5. Parameters used for the meta-EA

Parameter	Value
Population size	10
Elite size	1
Mutation Rate	40%
Crossover Rate	50%

During initialisation the first 2 members of the population were set to equal the parameters used with version CP and the remaining members were initialised randomly.

In order to ascertain the ability of the meta-EA to evolve an efficient set of parameters the following experiment was carried out over 6 days. Six identical workstations were allowed to run the meta-EA for approximately 136 hours, each of the EAs was running completely independently. In total 3961 different configurations were evaluated, this would mean each PC evaluated on average 660 different configurations, which would be 12.5 mins per evaluation.

Table 6. Parameters requiring optimisation by the meta-EA

Chromosome	Min	Max	Note
Pop size	1	200	
Elite size	0	10	If 0 then elitism is switched off.
Mutation rate	0	100	% chance of any one individual being mutated.
Crossover rate	0	100	% chance of any individual being selected for crossover.
Bias mutation in favour of fitness	False	True	If true provide increase chances of being mutated to the most fit examples
Regenerate time	0	100	No of generations between re-generation
Regenerate PC	0	100	Percentage of population to re-generate.

In all of the runs the meta EA fitness initially increased rapidly, before slowing. It should be remembered that the graph covers a period of approximately 6 days. If it had been practical to have allowed the meta-EA experiment to continue the graph suggests that some, if not all of the fitness rates would have continued to grow.

After the meta EA experiment had finished the parameter set was extracted from the individual with the highest fitness. This is shown in table 7. The percentage of the population to be regenerated has been decreased and the interval between regeneration increased from the settings used in version CP of the GA. The population and the elite sizes have both been increased. These parameters have been incorporated in a new version of the EA known as MP. It is now possible to compare version MP with version VP1 and version CP, the results are shown in table 7.

Table 7. :The parameters evolved using the meta-GA

Parameter	Value
Crossover Rate	68
Mutation Rate	30
Regeneration Interval	74
Regeneration % of population	29
Population size	95
Elite size	4
Bias mutation in favour of fitness	Yes

Table 8. A comparison of the fitness value of schedules

Run	1	2	3	4
CP	34500.08	34715.8	34162	32891
VP1	34485.08	34598.1	34487.4	34518
MP	34545.05	34454.8	34435.8	34611

4. Conclusions and Future Work

4.1 Some General Conclusions on the EA

The representation used is a basic direct representation. Although many EA scheduling techniques, make extensive use of memetic algorithms, the basic scheduling problem outlined in this report was solved without recourse to a purely memetic algorithm. A number of changes from the biological model were necessary to allow the system to produce schedules with a high fitness in a reasonable time. These changes were the use of a steady-state population, the use of restricted mutation and the use of Lamarkian writeback.

The mutation operator was extensively modified and it was restricted to only mutating event start times to within the start and end time for that job, rather than anywhere on the schedule. The mutation operator was also given a bias to discourage the mutation of genes that represented already placed events. It may be concluded that the mutation operator works most effectively when it is restricted from producing schedules that are obviously infeasible. The mutation operator's power stems from its ability to create changes that are not influenced by existing genetic material within the population. This ability to make "radical" changes can cause a bigger decrease than increase in fitness, if the changes are detrimental to the overall problem that the EA is attempting to solve.

The directed mutation operator showed a significant improvement, the average times to build a feasible schedule decreased from 768.5 to 563.3 seconds for versions with no bias in favour of badly placed events, and 50% bias respectively. The standard deviation in time taken without bias was 632.3 seconds, but adding a 50% bias, this decreases to 584 seconds. To further examine the effects of targeted mutation two more versions of the EA were constructed with 40% and 60% bias against the mutation of placed events. A more detailed comparison of typical runs is shown in figure 2.

The fitness function is of paramount importance to the EA, and a subtle alteration to the fitness function will radically affect the output obtained from the EA. The exact values used as weights within the fitness function were established by "trial and error". It was not possible to allow the meta-EA to optimise the fitness function values during the meta-EA experiment. If this had taken place it is likely that the meta-EA would have altered the fitness function values to give a high fitness regardless of the actual fitness of the schedule. It is acknowledged that a more efficient set of weights may exist. Some research into establishing an optimum set of weights for a fitness function within a drafts program [4] has been carried out with considerable success

The use of the re-generation operator provided a marked improvement. Initially this operator appeared to be very disruptive and could possibly cause the loss of genetic material. The re-generation operator requires skill in its usage, if too many individuals are re-initialised or the operator invoked too frequently then useful genetic material will be destroyed. Further research to find if there is an optimum percentage of population to re-initialise and an optimum frequency of use would be desirable. The meta-EA lowered the percentage of the population to be re-initialised to 29%. The re-generation interval was increased to 74 generations.

4.2 Some Conclusions on Parameter Optimisation

Figure 3 shows the fitness growth rates for typical runs using versions CP, VP1 and MP, when running data set 2. Version CP represents the EA running with parameters decided by the user using "trial and error", version VP1 represents the best of the versions that allowed the parameters to be optimised at run time, finally version MP represents the result of the meta-EA.

When trying to pick a "best" version of the EA it is important to notice that different versions display different qualities. For instance version CP produced schedules with a slightly higher average fitness (for data set 3, see table 10) than MG, but the time required to produce the schedules is significantly less (see table 11). It is the conclusion of the authors that version MP appears to represent the best "trade-off" in terms of achieving an acceptable level of feasible schedules, achieving a higher average fitness than the versions and completing the runs in significantly less time. Versions VP1 and VP2 were unable to produce schedules that were efficient when compared with the results obtained using CP and MP (see table 9). The main reason for this appears to be due to the subtle nature of the links between the parameters. Adjusting the parameters based only on the most fit individual in the population is not necessarily productive. The entire population contributes to each other fitness through crossover therefore although one operator may appear to be dominant in terms of creating the individual with the highest fitness and this operator may not be dominant throughout the population.

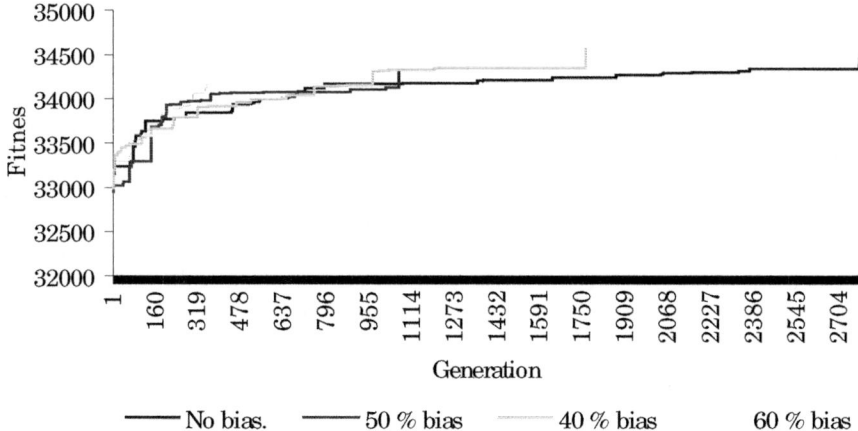

Fig. 2. Fitness Growth with 0,40,50 and 60% mutation bias using data set 2

A major conclusion that may be drawn from the work is that the optimisation and selection of the EA parameters may be best carried out by a meta-EA. In this case the meta-EA parameters recorded a small but significant increase in performance from those obtained using the more traditional "trial and error" methods. Within version CP, the version that used the "trial and error" parameters 86% of all the EA runs resulted in feasible schedules, using version MP that increased to 93%.

It may be seen that using parameters derived by the meta-EA it is possible to achieve a faster and higher growth rate. Optimising with the meta-EA gives superior results to those achieved using parameters decided by the user (version CP) or by modifying the parameters at run time (VP1 and VP2, see figure 4)

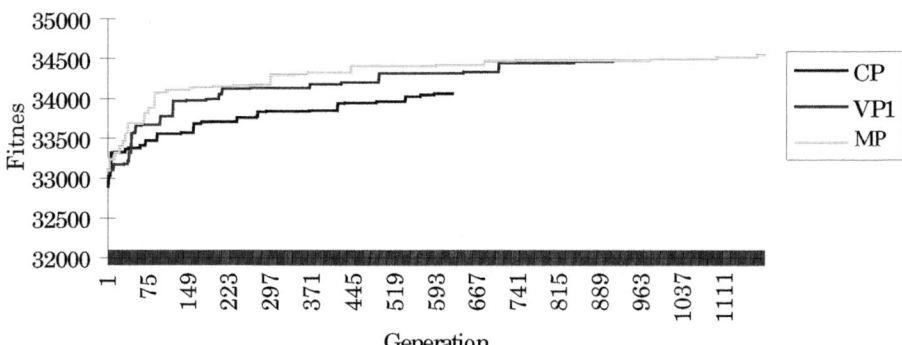

Fig. 3. Fitness growth comparison for typical runs of versions CP, VP1 and MP

Table 9. Total jobs placed (averaged over 10 runs)

	Data set 1 (16 jobs)	Data set 2 (15 Jobs)	Data set 3 (18 jobs)
CP	16	14.7	17.7
VP1	15.8	14.4	17
VP2	15.9	12.1	13.6
MP	16	15	17.8

Table 10. Schedule fitness (Averaged over 10 runs)

	Data set 1	Data set 2	Data set 3
CP	74626.5	34433.383	53698.60
VP1	74500.13	34260.041	53628.33
VP2	74592.81	33950.86	53009.92
MP	74600.81	34478.634	53681.26

Table 11. Time taken for run (seconds averaged over 10 runs)

	Data set 1	Data set 2	Data set 3
CP	207.2	805	1388.3
VP1	214.5	678.4	1260.9
VP2	412.1	1657.6	2431
MP	154.3	367.5	338

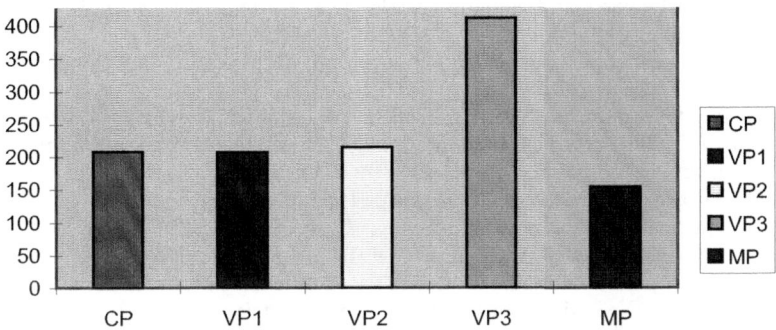

Fig. 4. Average time (seconds) to build a feasible schedule

4.3 Future Work

The authors plan to apply meta-EA principles to the optimising of an EA based system used to solve a routing problem. Recent work into differential evolution has raised the possibility of using differential evolution to optimise the weighting values used within the fitness function [2]. It may also be possible to use differential evolution to optimise the parameters for the whole EA. A major problem with the meta-EA is the processing time required, differential evolution may achieve similar results to the meta-EA, but with fewer CPU cycles being required.

References

[1] Production Scheduling and Rescheduling with genetic algorithms. Bierwirth C, Mattfeld D. *Evolutionary Computation volume 7*, No 1. MIT Press 1999.

[2] Co-Evolving Draughts Strategies with Differential Evolution, Chap 9 pp 147-158 in *NewIdeas in Optimiszation, Corne* D, Dorigo M, Glover F Eds. McGraw-Hill 1999.

[3] Building and Optimising a Scheduling GA. BSc Honours dissertation. Urquhart, N Chisholm, K.(supervisor). *Napier University, Edinburgh* 1998.

[4] Machine Learning Using a Genetic Algorithm to Optimise a Draughts Program Board Evaluation Function. Chisholm K.J, Bradbeer P.V.G.. *Proceedings of IEEE ICEC'97, Indianapolis, USA*, 1997.

[5] An introduction to Genetic Algorithms, Mitchell, M. MIT press 1996.

[6]. Extensions to a Memetic Timetabling System. Paechter B, Norman M, Luchian H. *Practice and theory of Automated Timetabling*, Burke and Ross Eds. Springer Verlag 1996.

[7] Evolutionary Computation, Fogel D B. IEEE Press 1995.

[8] Specialised Recombinative Operators for Timetabling Problems, Burke E, Elliman D, Weare R. *Proceeding of Evolutionary Computing AISB Workshop Sheffield UK* April 1995 ed Fogarty, T. Springer-Verlag 1995.

[9] Optimising a Presentation Timetable Using Evolutionary Algorithms. Paechter B . *Lecture Notes In Computer Science No 864*, Springer-Verlag 1994.

[10] Two solutions to the General Timetable Problem Using Evolutionary Methods, Paechter B, Cumming A, Luchian H, Petriuc, M. *Proceedings of the IEEE World Congress on Computational Intelligence*, June 1994.

[11] A Case for Lamarckian Evolution, Ackley D H, Littman M L. *Artificial Life III*, Langton C ed. Addison-Wesley 1994.

[12] Genetic Algorithms in Search Optimisation and Machine Learning, Goldberg D. Addison-Wesley 1989.

[13] Adapting Operator Probabilities in Genetic Algorithms, Davis L. *Proceedings of the third International Conference on Genetic Algorithms.* Schaffer J. ed. Morgan Kaufmann 1989.

[14] Optimisation & Control Parameters for Genetic Algorithms, Grefenstette J. *IEEE Transactions on Systems, Man and Cybernetics* 1986.

[15] Fast Practical Evolutionary Timetabling.. Corne D, Ross P, Fang H.. *Lecture Notes In Computer Science No 864*, Springer-Verlag 1994.

[16] Adapting operator settings in Genetic Algorithms. Tuson A, Ross P. *Evolutionary Computation Vol 6 No2*. Massachusetts Institute of Technology.

On-line Evolution of Control for a Four-Legged Robot Using Genetic Programming

Björn Andersson, Per Svensson, Mats Nordahl and Peter Nordin

Complex Systems, Chalmers University of Technology,
SE-412 96 Gothenburg, Sweden
{tfemn,nordin}@fy.chalmers.se

Abstract. We evolve a robotic controller for a four-legged real robot enabling it to walk dynamically. Evolution is performed on-line by a linear machine code GP system. The robot has eight degrees of freedom and is built from standard R/C servos. Different walking strategies are shown by the robot during evolution and the evolving system is robust against mechanical failures.

1   Introduction

Robots on legs constitute both one of the largest potentials and one of the largest challenges for intelligent robotic control. Applications are numerous in all environments accessible to humans and animals but inaccessible to wheeled autonomous agents. In general the flexibility of the legged robot increases with decreasing number of legs, unfortunately so does also control complexity. It is less complicated to control a robot with six legs or more, since the robot can have four legs on the ground all the time providing stable static balance [2,4,9]. Already the four-legged case becomes more difficult. A four-legged creature can crawl like a turtle with partial support by its body or by a tail. In this way it is possible to walk on four legs with static balance. Static balance means that the agent is in balance at all moments scarifying some efficiency and flexibility by dragging their body over the ground. However, it is more advantageous in general to use a *dynamic walk* where the agent will fall if interrupted in the middle of a movement.

We have evolved the first controller for a four-legged robot, learn using a genetic programming system and a real robot. The evolution of behavior passes through several stages starting with simple paddling behavior, continuing through crawling, "camel walk", and finally galloping with dynamic elements in the walk.

2   Experiments

Since we are using a real robot the experiment consists of one hardware part, the robot, as well as a software part, the GP system. There is also a communication part of

S. Cagnoni et al. (Eds.): EvoWorkshops 2000, LNCS 1803, pp. 319–326, 2000.

the system, sending control command to the actuators and receiving feedback from the sensors. Most experiments were run on a long time scale of about 20 hours and for this reason we needed a special automatic system that monitors the experiment and pulled the robot back if it had advanced too far.

2.1 The four legged robot

The four-legged robot is built with eight standard R/C servos as actuators, see Figure 1. The servos are of one of the smallest available but still have a momentum of 11 Ncm. The servos are controlled by a standard servo controller card, which allows a PC with a serial port to control up to eight servos. No input sensors on the dog itself are needed for this experiment.

Figure 1: The four-legged robot dog.

Fitness is generated through a computer mouse, which the robot drags behind itself. Positive reinforcement is given by dragging the mouse in the forward direction. Dragging backwards gives a negative fitness signal. Just moving the robot gives a small positive contribution to fitness. The experiment set-up consists of a rectangular box (0.3mX1.2m) where the agent can learn forward walking. If it reaches the end of the box then it is automatically pulled backwards to the other end by an electric motor. Single experiments ran automatically with this set-up for more than 18 hours.

2.2 Control system

The control system is built around a linear genetic programming system evolving binary machine code [1,5]. The evolved structure is a linear string of instructions operating on a set of registers of a register machine. The control system is executed on a PC and normal mouse communication is used for feedback while the servos are controlled through serial communication with a servo control card. The inputs to the evolved programs are placed into the registers of the register machine (processor) while the output vector is whatever remains in the registers upon termination of the program. Each individual that is evaluated is iterated eight times. In each iteration the eight outputs are sent as angles to the servos and the previous output vector is sent as input to the next iteration enabling a sequence to be evolved. Fitness is measured after the eighth iterations (and basically consists of the number of "ticks" that the mouse has moved forward). Some individuals are removed before evaluation, such as those, which do not move the servos at all.

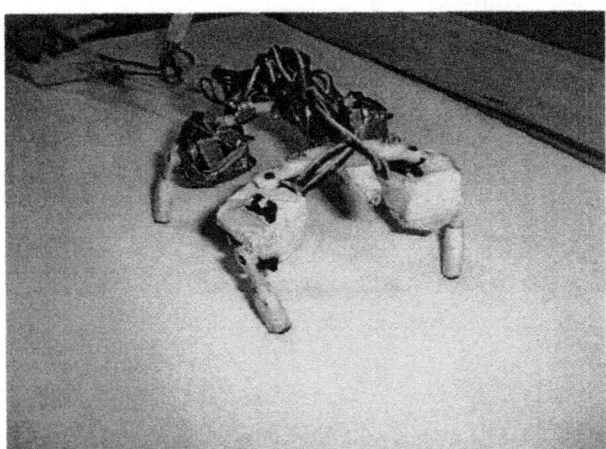

Figure 2: Feedback through computer mouse

Figure 3: Cables to PC and power supply

The population consisted of 100 individuals and the genetic operators where crossover and mutation. Crossover was performed both as two-point string crossover and as homologous linear crossover [3,7]. The system used a small tournament for selection and operates under steady state, see Figure 4. The function set consisted of arithmetic operators and square root. A single constant of 0.5 was used in the terminal set.

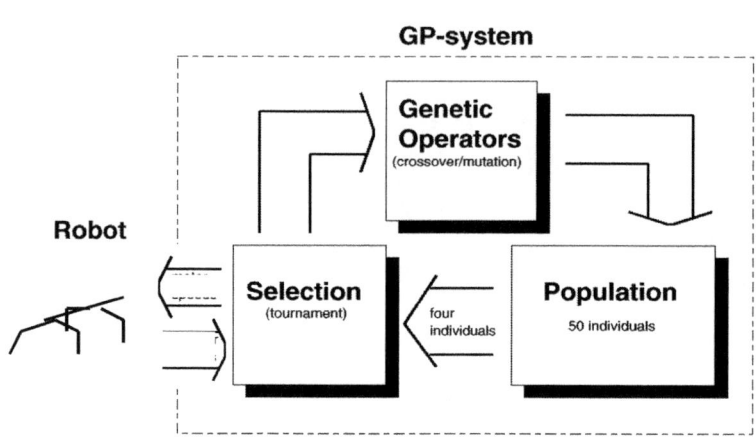

Figure 4: On-line GP system

Our method for using GP with a real-time application is based on a probabilistic sampling of the environment [6]. Different solution candidates (programs) are evaluated in different situations. This is unfair because a good individual dealing with a hard situation can be rejected in favor of a bad individual dealing with a very easy situation. For instance, an individual that gets stuck near the wall of the box and does a good job of moving away from the wall but advances little in the forward direction might get a low score while a poor individual in the middle of the box might perform better. Our experience is, however, that a good overall individual tends to survive and reproduce in the long term. The somewhat paradoxical fact is that sparse training data sets or probabilistic sampling in evolutionary algorithms often both increase speed toward the goal and keep the diversity high enough to escape local optima during search.

3   Results

Evolving walking behavior, through coordination of eight servos and without any domain knowledge or guidance and where the only feedback is from a computer mouse on a rod, is not a trivial problem. The first emerging behavior is usually some kind of chaotic paddling, which slowly moves the robot forward even though it is barely visible. A common local optimum after the paddling is a strange strategy, where the robot stands upright on all legs carefully balancing and moving quickly back and forth without really advancing the robot. It is unclear how this strategy emerges but it could be the result of some lack of symmetry in the hardware and measurement system. The next strategy is often some kind of crawling similar to that of turtle. Other observable behaviors are a "camel walk" where the legs move in parallel and pair wise on each side. The most efficient strategy for walking that has evolved is also the most difficult to learn and *galloping* only appears after many hours of training. Here both the front legs and back legs are parallel and the robot often stands up on its back legs before pushing forward. It uses the dynamic interaction with the heavy mouse lode to achieve maximum speed forward.

Emergence of behavior does not always pass through all of these behaviors but usually most of them appear in order.

Summary of common evolution of behavior over time could look like:
1.   chaotic paddling behavior
2.   crawling
3.   "camel walk"
4.   galloping with dynamic elements in the walk

Like many evolutionary control systems we observe some robustness against mechanical failure. Figure 5 shows the evolution of speed over 111 generations and 15 hours. During evolution a tree servos broke down but the control system adapted and managed to increase the speed after the failures, see Figure 5. Servo breakdowns occurred in generation 20, 75 and 100. The figure clearly indicates how fitness at first decreases rapidly but the system manages to relearn and adapt to the changed (damaged) hardware configuration.

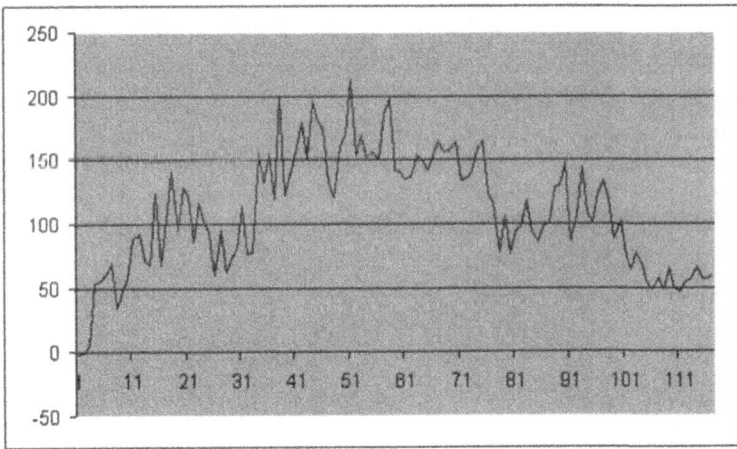

Figure 5: The velocity (in m/h) of the dog when controlled by the (currently) best individual during 117 generations. Servo failure occurred in generation 20, 75 and 100.

4   Future Work

Our intention is to use the results obtained with the four-legged robot for a larger humanoid project, see Figure 6. Here the aim is bi-pedal walking, which is harder than four legged walking and demands much more emphasis on balance. Preliminary work on the ELVIS humanoid confirms the feasibility of the approach [8].

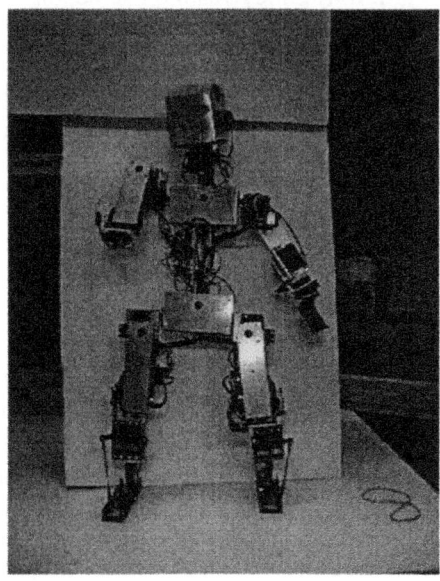

Figure 6: ELVIS humanoid a bipedal walking robot

Making the robot fully autonomous is another ambition for the future. That would mean that the experiments will not be hindered by all the cables and total autonomy would also be interesting from a more philosophical standpoint. Figure 7 shows a chaotic robot made of "garbage" which has learned to move by the same mechanism as the four-legged robot but which is fully autonomous. The on-board GP system runs on a small, embedded PIC-chip and evolves binary code for this tiny processor. Our plan is to move this system to the walking experiments.

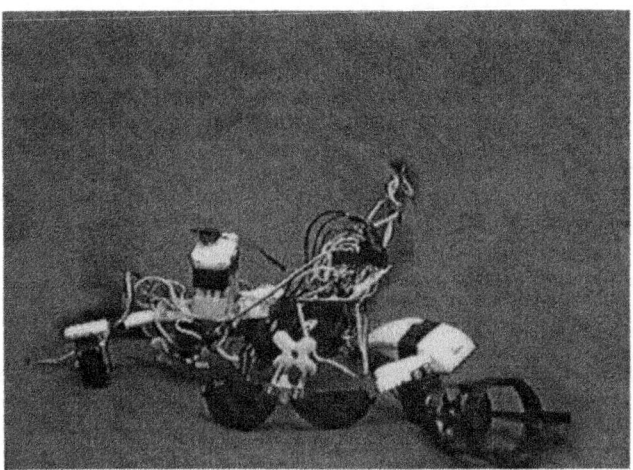

Figure 7: An autonomous chaotic robot made from "trash" with an on-board GP system in an embedded chip.

5   Summary and Conclusions

We have evolved the first controller for a four-legged robot, which learn on-line using a genetic programming system and a real robot. The evolution is of behavior passes through several stages starting with simple paddling behavior, continuing through crawling, "camel walk", and finally galloping with dynamic elements in the walk. A high degree of robustness is seen for mechanical failure - the system is observed to adapt to changes in the mechanical configuration due to component failure.

Acknowledgement

Peter Nordin gratefully acknowledges support from TFR and NUTEK.

References

1.   Banzhaf, W., Nordin, P. Keller, R. E., and Francone, F. D.(1998) *Genetic Programming: An Introduction on the Automatic Evolution of Computer Programs and Its Applications*. Morgan Kaufmann, Germany.

2.   T. Broughton, P. S. Coates, and H. Jackson. (1999) Exploring three-dimensional design worlds using Lindenmeyer systems and genetic programming. In Peter Bentley, editor, Evolutionary Design Using Computers , chapter 14. Academic press, London, UK,.

3. Francone F.D., Conrads M., Nordin J.P.and Banzhaf W.(1999) Homologous Crossover in Genetic Programming, In Proceedings of: Genetic and Evolutionary Computation Conference (GECCO 99) Morgan-Kaufmann

4. M. Anthony Lewis, Andrew H. Fagg, and Alan Solidum. (1992) Genetic programming approach to the construction of a neural network control of a walking robot. In Proceedings of the 1992 IEEE InternationalConference on Robotics and Automation , pages 2618-2623, Nice, France, May 1992. Electronica Bks.

5. Nordin, J.P. (1997) Evolutionary Program Induction of Binary Machine Code and its Applications. Krehl Verlag, Muenster, Germany

6. Nordin, J.P., Banzhaf W.(1997) An On-line Method to Evolve Behavior and to control a Miniature Robot in Real Time with Genetic Programming: The International Journal of Adaptive Behavior, (5) pp 107 - 140 MIT Press, USA.

7. Nordin J. P., Banzhaf W., and Francone F. (1999) Efficient Evolution of Machine Code for CISC Architectures using Blocks and Homologous Crossover. To appear in Advances in Genetic Programming III, (Eds) Langdon, O'Reilly, Angeline, Spector, MIT-Press, USA

8. Nordin J. P., Nordahl M. (1999): An Evolutionary Architecture For A Humanoid Robot, In Proceeding of: The Fourth International Symposium on Artificial Life and Robotics (AROB 4th 99) Oita Japan

9. Graham F. Spencer. Automatic generation of programs for crawling and walking (1994). In Kenneth E. Kinnear, Jr., editor, Advances in Genetic Programming , chapter 15, pages 335-353. MIT Press, 1994.

Optimized Collision Free Robot Move Statement Generation by the Evolutionary Software GLEAM

Christian Blume

Fachhochschule Köln, Abteilung Gummersbach
Am Sandberg 1, D-51643 Gummersbach, Germany
Tel. +49/2261/8196-296, -330 or -332, Fax: +49/2261/819615
email: blume@gm.fh-koeln.de

Abstract. The GLEAM algorithm and its implementation are a new evolutionary method application in the field of robotics. The GLEAM software generates control code for real industrial robots. Therefore GLEAM allows a time related description of the robot movement (not only a static description of robot arm configurations). This internal representation of primitive move commands is mapped to a representation of move statements of an industrial robot language, which can be loaded at the robot control and executed.

Introduction

There are many different ideas and procedures about Genetic Algorithms and Evolutionary Programming, but only some of these lead to implementations which solve „real world problems" for industrial applications. **GLEAM** (**G**enetic **L**e*arning* **A**l*gorithm* and **M**et*hods)* is an algorithm based on evolutionary strategy, which was used to implement a software tool for solving useful and not only academic problems, see [1] and [2].

The aim of the GLEAM application to industrial robots was an implementation with respect to practical requirements and restrictions to proof the GLEAM method. Therefore a first prototype implementation was done at ABB in Västeras (Sweden) and a second one with improved facilities at Daimler Chrysler in Berlin (Germany). Both implementations include the output of robot language code for the specific industrial robot (ABB IRB 2400 and KUKA KR 6).

The Evolutionary Algorithm GLEAM

The principles of GLEAM including its genetic operators and genotype representation are formulated with close connection to the biological evolution (see detailed description in [3]). The genotype of GLEAM is called an „action", which represents one step of a plan to be executed. The plan execution is done by simulation as part of

S. Cagnoni et al. (Eds.): EvoWorkshops 2000, LNCS 1803, pp. 327-338, 2000.
© Springer-Verlag Berlin Heidelberg 2000

the evolution algorithm to get a value for the „fitness" of the plan. In particular GLEAM generates a sequence of basic actions or statements, which are the elements of the plan, which is a member of the evolution population. A plan represents directly the genetic information, see [1]. The purpose of the plan is not of interest for the evolution itself, therefore the kernel of GLEAM including the evolution algorithm was applied to different problems with minor changes. For example, these actions can be the basic commands of a simple robot controller or allocation steps to reserve a machine in a production plan.

The GLEAM method was implemented for several applications:
- planning collision free moves for industrial robots, see [4] and [11]
- generating production plans solving the job-shop-problem, see [5]
- process scheduling in chemical industry, see [6]

GLEAM is implemented as well in Pascal as in C on different hardware platforms, e.g. on PC, work station, and a parallel computer system. It is portable, because the implementation is structured modular and consists of the following main parts:

Basic machine: E.g. initialization, data structure construction and management, error handling

Optimization kernel: Evolution functions for mutation, recombination, population management

Simulation and evaluation: Execution of plans, criterion's check, restrictions, fitness calculation

Overall control: Management of parameter input, display of results, interrupting simulation or optimization and others

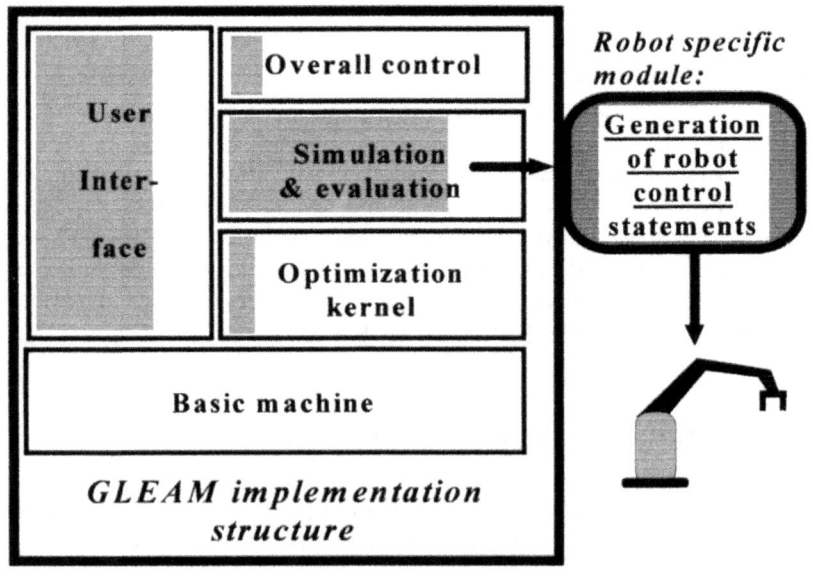

GLEAM implementation structure

Implementation effort to apply GLEAM to another application

<u>User interface</u>: Criterion's priority and fitness definition, target definition, simulation visualization, plan description and others

<u>Robot code generation module</u>, generates a robot program part including the move statements

Simulation and evaluation need the main amount of implementation effort to apply the GLEAM implementation to other applications. It means on the other side, that due to the generality of the method a new application requires less implementation costs because the basic machine, optimization kernel and overall control are not much effected.

Application of GLEAM to Robotics

The first application of GLEAM to robotics was performed because of two reasons:

1. The author is an expert in industrial robot programming and control (Convenor of an ISO working group), therefore a praxis oriented work was performed, see [7]
2. The calculation for controlling an industrial robot to move on a predicted trajectory without collision is very complex but the result is easy to evaluate: everybody can see, if the movement is collision free and takes a shorter time.

The application of GLEAM was something like a benchmark especially to show, that GLEAM is a powerful planning and optimization tool for the control of dynamic processes like the movement of industrial robots.

The GLEAM method applied to robot collision free moves doesn't perform an explicit search of the configuration space, because the search parameters don't include configurations, like in [8] or [9]. With GLEAM the move statements for performing the collision free move are generated directly. The GLEAM method avoids the problem of calculating a smooth path between the configurations and of building up

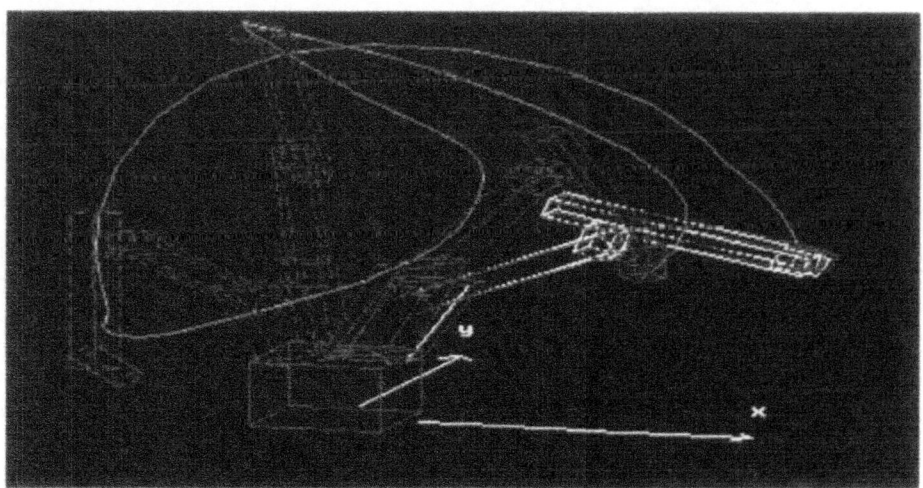

and storing the configuration space, see [10]. The collision free path is optimized by criterion's selected by the user, e.g. the criterion could be a short Cartesian path. Other criterion's like energy or move execution time are integrated into the optimization process. The criterion's could be contradictionary. Therefore every criterion has a priority and the results of the optimization reflect the different priorities.

The evolution performed by GLEAM starts by generating action res. primitive statement sequences for controlling the robot movement. These basic actions for the robot moves look like follows:

- *move robot axis k with a velocity of r degrees per second with an acceleration of*

  b degrees per second^2 (Evolution parameters are k, r, and b)

- *stop robot axis   m   with a slow down of   n   degrees per second^2* (Evolution parameters are m and n)

The fitness of a sequence can be calculated as a function of the following criterion's:

- position precision, i.e. the distance of the end point of the movement and the given target point
- orientation precision, i.e. the difference between the planned and reached orientation of the gripper
- orientation changes, i.e. the orientation moves of the gripper to reach the target orientation
- quality of the move trajectory, i.e. the difference between the move path and a straight line
- length of the move trajectory
- fastness of move execution, i.e. the duration of movement
- program length, i.e. the number of actions res. statements required to perform the robot move
- Intermediate points, i.e. the robot move passes specified intermediate points
- economy, i.e. the energy needed for the move execution

Before the evaluation, the statement sequence is proofed for its plausibility, for example: a stop-axis-statement before a start-move-axis-statement is not meaningful (and therefore canceled). After the plausibility-check the evolution continuous by the mutation of the evolution parameters of the statements and by recombination. As the sequence of elementary moves is essential for the resulting overall movement there are some mutations altering only the sequence of actions.

The complete sequence of actions controls the robot move to a point near the target position without collisions. It is assumed, that the robot control is based on a fixed control cycle time, e.g. of 50 ms. Every control statement is executed by the simulation tool with respect to this cycle time, which can be defined as a control characteristic. The simulation performs the robot move and stores the axis values for every control cycle. To measure the distance from the target position and the path length, the general forward transformation from robot to Cartesian coordinates is applied.

The GLEAM method is able to develop a statement sequence for any kind of robot. A software tool ROBMODEF (**Rob**ot **Mo**del **Def**inition) was implemented providing a graphic supported robot definition. The user can build a robot with up to 16 rotational robot axes, which can be directed into the x- or z-coordinate direction (in zero position), and which can rotate about the x-, y- or z-coordinate axis.

The user can define obstacles in the robot movement area to show collision avoidance. The defined robot model, the obstacle definitions and the (predefined) action definition is called the "action model". It can easily changed or extended by the user and stored for further experiments.

Genotype Representation and Data Structure

The action of GLEAM is one gene and consists of the action code and a number of parameters. The parameters can be of integer, real or character type. The action code

and parameter definition is stored in an (readable) file and can be change very easily. The number of actions, which build up a member of the evolution population, is not limited. The length of such an action chain is flexible and an optimization criteria of the evolution.

GLEAM was designed for the optimization of dynamic processes with respect to the natural evolution. Therefore the evolution process operates about genes res. plans of variable length. Due to this attribute the task of recombination is more complicated than in evolution algorithms of other authors. The most important improvement of GLEAM is the introduction of the concept of so called „sections". A section is formed by a (variable) number of genes res. actions as a substructure of the plan, i.e. the action chain is partitioned into segments. A segment can be regarded as a chromosome of the biological genetic information. So, one member, that means the action chain of the plan, consists of segments, and each segment of a number of actions.

There are new defined evolution operators (for mutation and recombination) to be applied to the sections, like delete a segment or move a segment to another place in the action chain. This concept enables the recombination of „good" sub-structures of a problem solving process to speed up the evolution. If two plans are merged by the recombination, the combination of a segment of plan A with a segment of plan B can be treated as a combination of two sub-solutions. If the two sub-structures res. segments are good in single, they both together in the resulting plan will give a much higher fitness value than the fitness of the both parent plans. To come to such a good solution only by mutation of the genes takes probably much longer time. The concept of segments (the „chromosomes") and actions (the „genes") was very successful for planning and optimization tasks of dynamic processes like robot moves.

Similar to the „traditional" artificial intelligence language LISP, the data structure of GLEAM is based on dynamic lists. The action chain starts with a header node storing all necessary information about the action chain, e.g. chain length (e.i. number of actions and segments), fitness value, and others. The header is followed by actions, all linked by pointers. An action chain can grow or be reduced, it depends on the evolution. Of course, if an action chain (that means a problem solution) gets a good fitness value, and another chain gets the same fitness rank but consists of a lower number of actions, the last one is better and will survive.

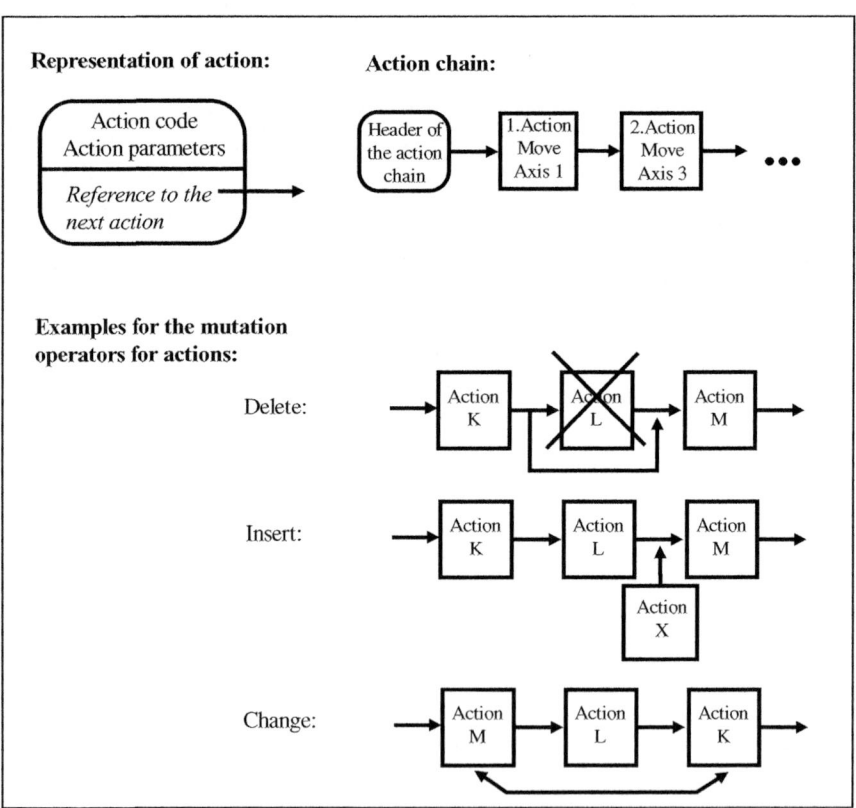

Genotype Oriented Code and Robot Code Generation

The genotype oriented code of GLEAM for robots consists of primitive move commands. The code is simulated during the evolution (without any output, only the fitness value is calculated during the simulation), and if the user wants to see the result of the planning and optimization process, the code of successful action plans is simulated with graphical display of the robot movements.

As mentioned above, the robot control simulation calculates a move step by step, see [11]. Between every calculated via position of the trajectory a user defined cycle time has passed. The move description by primitive commands (used internally by GLEAM) has to be transferred to code formulated in a robot programming language, but programming the same robot move. A new module „robot code generator" was implemented, which calculates the parameter for a move programming by robot move statements.

After every control cycle the values for the robot axes and the Cartesian coordinates for the tool center point TCP are calculated and transferred to the robot code generator (if the user has marked „code generation" with the help of menu input). The first step of the code generation is the declaration of position data and an assignment of values. This position data consists of the intermediate points of the move and the target point. The target point is given by the user in Cartesian coordinates, the code generator calculates the values for the parameters to describe the robot specific orientation and configuration data. After the data definition, a simple main function is generated consisting of move commands and referring to the position data generated before. The robot code can be downloaded to the robot control and executed.

Example to program the ABB IRB 2400 robot:

```
! COMMENT    Position list
! COMMENT    with the move position for every cycle
! COMMENT    (representing the trajectory):

MODULE demo1
    VAR robtarget target:= [[-1647.0,-604.2,1374.7],
        [0.12367,0.69621,0.12367,-0.69621],[-2,0,0,0],
                           [9E9,9E9,9E9,9E9,9E9,9E9]];
    CONST jointtarget P1:= [[   0.5,   45.3,   45.3,
            0.0,  -45.3,  0.0], [9E9,  .... ,9E9]];
    CONST jointtarget P2:= [[   0.7,   45.3,   45.3,
            0.0,  -45.3,  0.0], [9E9,  .... ,9E9]];
    CONST jointtarget P3:= [[   0.7,   45.3,   45.3,
            0.0,  -44.6,  0.0], [9E9,  .... ,9E9]];
    ........
    CONST jointtarget P74:= [[-160.7,   45.2,    0.6,
            16.9,  4.1,-117.0], [9E9,  .... ,9E9]];

! COMMENT   Move statements for the RAPID program:
    PROC main()
        MoveAbsJ P1,v400,z20,toolx;
```

```
    MoveAbsJ P2,v400,z20,toolx;
    MoveAbsJ P3,v400,z20,toolx;
  . . . .
      MoveAbsJ P74,v400,z20,toolx;
      MoveL target,v100,z20,toolx;
  ENDPROC
  ENDMODULE
```

Executing these robot statements, the ABB control performs only one robot move to the target point using the many generated points to perform a „smooth path".

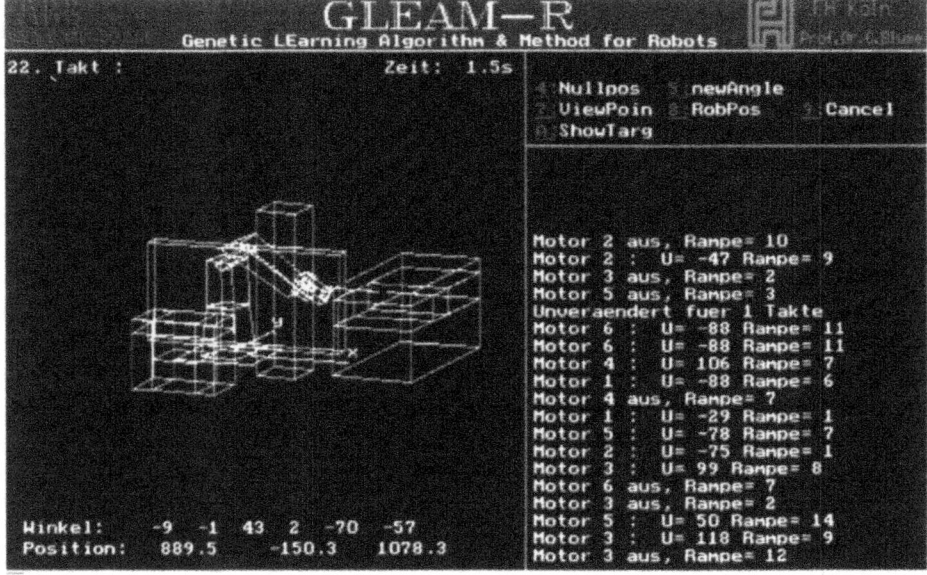

Therefore the resulting movement looks much more „natural", i.e. like a movement of a human arm. The reason is, that the programmer usually „divides" a robot move in only a few (typically 3 to 10) intermediate points with a greater distance between them, and the robot move looks then more „machine-like" or „cornered". A side effect of the new method is the generation of moves, which avoid sudden change of speed or direction. This leads to a control behavior more avoiding wear and tear of the axis motors and gears. Also the generated move was checked to avoid running against axis stops.

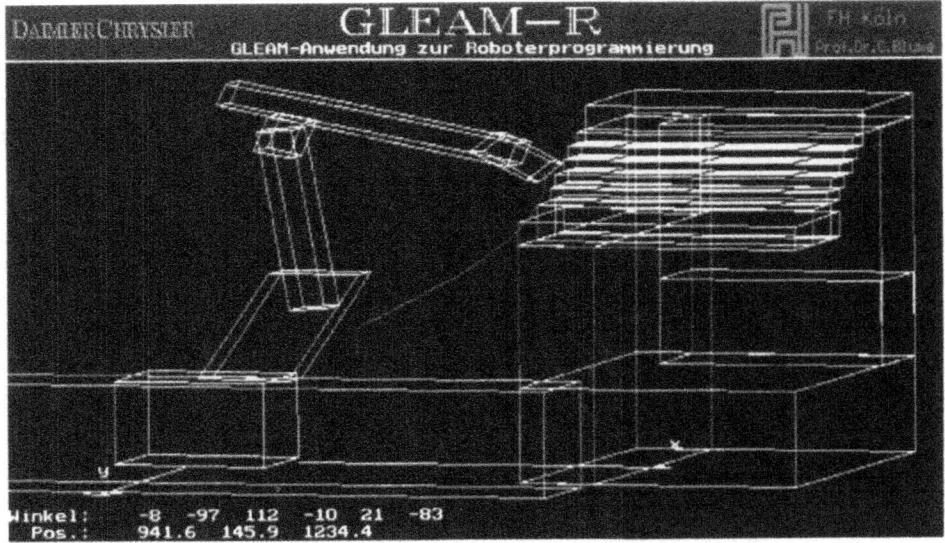

Results of the Implementations for Industrial Robots

The implementations of GLEAM to industrial robots have been done to demonstrate, that GLEAM is able to produce robot code in an industrial environment. The software for the ABB robot programming facility generates robot programs with collision free robot moves. The collision free path is optimized by criterion's of different priorities selected by the user. E.g. the criterion could be a short Cartesian path of the TCP or a short distance of one ore more robot axes. The process of programming optimized robot moves can be done by the software tool, which reduces the needed man power for the user.

Another result is the demonstration, that the method GLEAM is applicable to an industrial robot control of the user. The advantages are controlling robots without complicated mathematics calculations and easy adaptation of different environments and robot models. The result can be used for off-line path planning including a multi-criteria optimization. The generated move trajectory is an optimization of all criterion's, avoiding problems of defining the configuration space or transfer positions.

The implementation for DaimlerChrysler includes several improvements, e.g. the definition of intermediate points by the user. These intermediate points have to be passed by the robot trajectory (more or less), the user can influence the robot path and help the system to find a collision free path in a shorter time.

The implementations have been performed on a PC, they allow a low cost solution and will be accepted also for smaller companies. The software for the GLEAM modules are written in C. The companies ABB and DaimlerChrysler can test the new method and decide, if it will be integrated into their programming software tools, the implementations are the base of feasibility studies to analyze the new method.

Outlook

The GLEAM concept was realized by implementations for different robots: the Mitsubishi RV-M2 robot, the ABB IRB 2400 industrial robot , and the KUKA KR 6 industrial robot. They generate statements for the robot controls to move the robot on an optimized collision free trajectory to a given move target.

In future an implementation for a professional simulation tool for robots will be of interest, e.g. for the IGRIP system of the Deneb company. Such an implementation will demonstrate the application of the GLEAM method to a high level language simulation command language like GSL, and a complex modeled environment of obstacles and robots. The following picture shows the information flow between the system components. The user works with the simulation tool as before and models its machines, work cell, obstacles, and others. He can program the robot moves in a simulation language and simulate the moves.

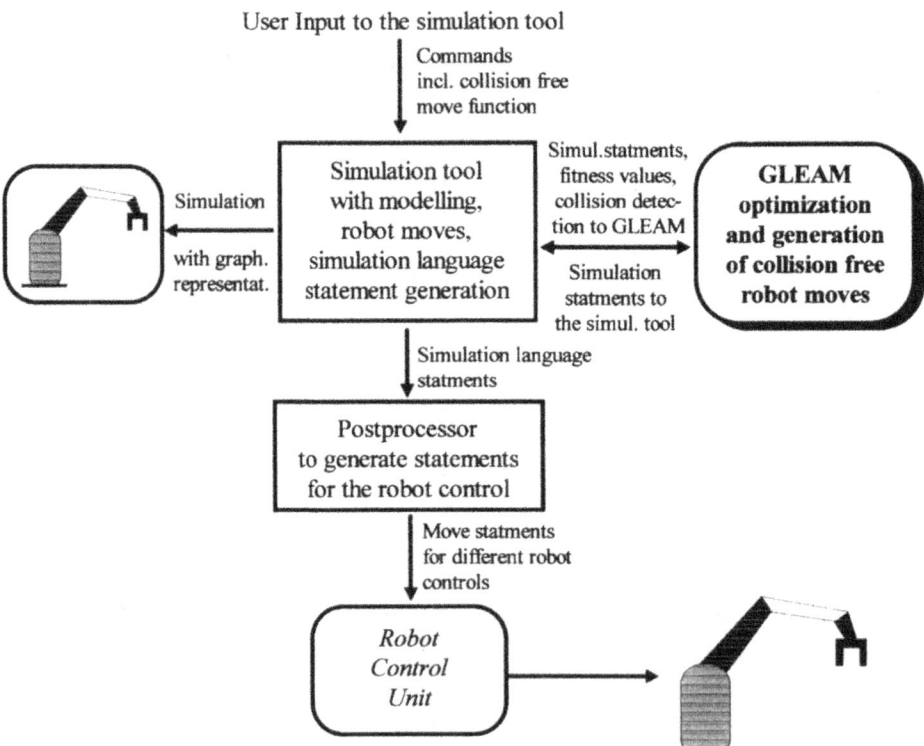

A new function „optimize moves without collision" will be introduced into simulation system. The simulation tool sends the simulation statements already programmed by the user to the GLEAM component. GLEAM tries to optimize the robot moves (shorter execution time) with respect to collision avoidance. During the optimization process GLEAM sends changed statements in simulation language code to the simulation system, which will execute the statements in a „silent" background mode without graphical output. The simulation tool sends back to GLEAM only the information about the „fitness" of the simulated move statements, especially if there was a collision or not. After the optimization process GLEAM sends the result to the simulation tool: the optimized collision free move statements in the simulation language. The user can simulate this result and test, if it is better than his own statements. At the end, the simulation tool generates program code for different robot control units.

It is important, that GLEAM can use solutions already programmed by the user as an input. GLEAM tries to find a better solution (if there exists one), and it generates a trajectory without collisions. The user can visualize the result by simulate it, and he can use the generated statements for other programs or to program any robot.

References

1. Blume, C., Jakob, W.: Closing the Optimization Gap in Production by Genetic Algorithms. Proc. of the *European Congress on Intelligent Techniques and Soft Computing (EUFIT 93)*, 1993, Aachen
2. Blume, C.: Industrielle Anwendungen Evolutionärer Algorithmen (Industrial Applications of evolutionary algorithms). *Contribution:* Planung kollisionsfreier Bewegungen für Industrieroboter (Planning of collision free movements of industrial robots). Ed.: S. Hafner, R. Oldenbourg Verlag, München Wien 1998
3. Blume, C.: GLEAM - A System for Simulated „Intuitive Learning". *Proceedings of the 1st International Workshop on Problem Solving from Nature*, Dortmund, Germany, October 1-3, 1990
4. Blume, C., Jakob, W., Krisch, S.: Robot Trajectory Planning with Collision Avoidance using Genetic Algorithms and Simulation. *Proc. of the 25th International Symposium on Industrial Robots*, 25.-27. April 1994, Hannover, pp. 169-175
5. Planning and Optimization of Scheduling in Industrial Production by Genetic Algorithms and Evolutionary Strategy. *Proc. of the Second Biennial European Joint Conference on Engineering Systems Design and Analysis (ESDA)*, July 4-7, 1994, London, England
6. Blume, C., Gerbe, M.: Deutliche Senkung der Produktionskosten durch Optimierung des Ressourceneinsatzes. (Reduction of production costs by optimizing the resource planning). *atp - Automatisierungstechnische Praxis*, 36 (1994) pp. 5 - 9
7. Blume, C., Früauf, P.: Standardization of Programming Methods and languages for Manipulating Industrial Robots. *27th International Symposium on Industrial Robots*, Oktober 1996, Mailand
8. Dai, F.: Collision-Free Motion of an Articulated Kinematic Chain in a Dynamic Environment, *IEEE Computer Graphics & Application*, January 1989, pp. 70-74

9. Goldberg, D., Parker, J., Khoogar, A.: Inverse Kinematics of Redundant Robots using Genetic Algorithms. *IEEE International Conference on Robotics and Automation 7*, (1989). pp. 271-276

10. Heine, R., Schnare, T.: Kollisionsfreie Bahnplanung für Roboter (Collision free path planning for robots). *Robotersysteme 7*, (1991). pp. 17-22

11. Generation of Optimized Collision Free Robot Move Statements Based on Genetic Algorithms. *Proceedings of the World Automation Congress (WAC '96) Volume 3: Robotic and Manufacturing Systems, May 28-30, 1996, Montpellier, France.* pp. 89 - 94

Self-Adaptive Mutation in ZCS Controllers

Larry Bull & Jacob Hurst

Intelligent Computer Systems Centre
Faculty of Computer Studies & Mathematics
University of the West of England
Bristol BS16 1QY, U.K.
{Larry.Bull,Jacob3.Hurst}@uwe.ac.uk

Abstract. The use and benefits of self-adaptive mutation operators are well-known within evolutionary computing. In this paper we examine the use of self-adaptive mutation in Michigan-style Classifier Systems with the aim of improving their performance as controllers for autonomous mobile robots. Initially, we implement the operator in the ZCS classifier and examine its performance in two "animat" environments. It is shown that, although no significant increase in performance is seen over results presented in the literature using a fixed rate of mutation, the operator adapts to approximately this rate *regardless* of the initial range.

1 Introduction

Within Genetic Algorithms (GAs) [Holland 1975] and Genetic Programming [Koza 1991] the mutation rate is traditionally a global parameter which is constant over time. However, in Evolutionary Strategies [Rechenberg 1973] and later forms of Evolutionary Programming (Meta-EP) [Fogel 1992], the mutation rate is a locally evolving entity in itself, i.e. it adapts during the search process. This "self-adaptive" form of mutation not only reduces the number of hand-tunable parameters of the evolutionary algorithm, it has also been shown to improve performance (e.g. see [Bäck 1992] for results with a self-adaptive GA). In this paper we examine the use of a self-adaptive mutation operator within Michigan-style Classifier Systems (CSs) [Holland et al. 1986], more specifically in Wilson's ZCS [Wilson 1994] system.

The performance of the new operator within ZCS is examined using two simulated autonomous entity/robot - "animat" [Wilson 1985] - tasks: Woods 1 and Woods 7, each of which were originally used by Wilson to introduce and investigate ZCS. In both cases it is found that no benefits in performance are found over the results presented by Wilson using a fixed mutation rate. However, regardless of the initial range for the adapting mutation rates, the final CS animat controllers have roughly the same mutation rate as that used by Wilson. That is, we show the principle of self-adaptation works within the Classifier System framework.

The paper is arranged as follows: the next section introduces ZCS. Section 3 describes how self-adaptive mutation is implemented and Section 4 describes the tasks and examines the effects of the operator. Finally, all results are discussed.

S. Cagnoni et al. (Eds.): EvoWorkshops 2000, LNCS 1803, pp. 339–346, 2000.
© Springer-Verlag Berlin Heidelberg 2000

2 ZCS

ZCS is a "Zeroth-level" Michigan-style Classifier System without internal memory, where the rule-base consists of a number (N) of condition/action rules in which the condition is a string of characters from the usual ternary alphabet {0,1,#} and the action is represented by a binary string. Associated with each rule is a strength scalar which acts as an indication of the perceived utility of that rule within the system. This strength of each rule is initialised to a predetermined value termed S_0.

Reinforcement in ZCS consists of redistributing strength between subsequent "action sets", or the matched rules from the previous time step which asserted the chosen output or "action". A fixed fraction (β) of the strength of each member of the action set ([A]) at each time-step is placed in a "common bucket". A record is kept of the previous action set [A]$_{-1}$ and if this is not empty then the members of this action set each receive an equal share of the contents of the current bucket, once this has been reduced by a pre-determined discount factor (γ). If a reward is received from the environment then a fixed fraction (β) of this value is distributed evenly amongst the members of [A]. Finally, a tax (τ) is imposed on all matched rules that do not belong to [A] on each time-step in order to encourage exploitation of the stronger classifiers. Hence this is different from the traditional "Bucket-brigade" algorithm [Holland et al. 1986] and is known [Wilson 1994] to be similar to Watkin's Q-learning [1989] reinforcement algorithm.

ZCS employs two discovery mechanisms, a panmictic GA and a covering operator. On each time-step there is a probability p of GA invocation. When called, the GA uses roulette wheel selection to determine two parent rules based on strength. Two offspring are produced via mutation (probability μ) and crossover (single point with probability χ). The parents then donate half of their strengths to their offspring who replace existing members of the rule-base. The deleted rules are chosen using roulette wheel selection based on the reciprocal of rule strength. If on some time-step, no rules match or all matched rules have a combined strength of less than ϕ times the rule-base average, then a covering operator is invoked.

The default parameters presented for ZCS, and unless otherwise stated for this paper, are: $N = 400$, $S_0=20$, $\beta = 0.2$, $\gamma = 0.71$, $\tau = 0.1$, $\chi = 0.5$, $\mu = 0.002$, $p = 0.25$, $\phi = 0.5$

Thus ZCS represents a "basic classifier system for reinforcement learning that retains much of Holland's original framework while simplifying it so as to increase understandability and performance" [Wilson 1994]. For this reason the ZCS architecture has been chosen to examine the basic behaviour of classifier systems with self-adaptive mutation rates. The reader is referred to [Wilson 1994] for full details of ZCS.

3 Self-Adaptive Classifier System Controllers

3.1 Self-Adaptation

In this paper we use the same form of self-adaptive mutation as in Meta-EP. That is, each rule has its own mutation rate μ, stored as a real number, which is passed to its offspring, either under recombination or directly (depending upon the satisfaction of χ). The offspring then applies its mutation rate to itself using a Gaussian distribution, i.e. $\mu_i' = \mu_i + N(0,\mu_i)$, before mutating the rest of the rule at the resulting rate. It is noted that this form of self-adaptation is simpler than that typically used in Evolutionary Strategies, where a Lognormal is applied to μ, however the simpler form is shown to be adequate here and has been suggested to work better in noisy environments [Angeline et al. 1996].

We also note that this is in contrast to the adaptive form of crossover introduced by Wilson (1987) for CS, under which a system entropy measure was used to alter the operator rate; Wilson showed benefits from increasing crossover as entropy dropped using predetermined rules of change.

3.2 Classifier Systems in Evolutionary Robotics

A number of investigators have examined the use of Classifier Systems in evolutionary robotics. Dorigo, in conjunction with many others (see [Dorigo & Colombetti 1999] for a comprehensive overview), has used multiple CSs in a hierarchy to control an autonomous robot in a variety of environments. To our knowledge this remains the only hardware implementation to date. A large body of work exists on the use of CSs to control simulated robots however, e.g. [Riolo 1991], [Cliff & Bullock 1993], [Donnart & Meyer 1994], [Stolzmann 1999], etc. The reader is referred to [Lanzi et al. 2000] for a full CS bibliography.

The performance of a Michigan-style classifier system - ZCS - with self-adaptive mutation in simulated evolutionary robotics tasks is now examined, with the aim of determining ways to improve their use in real environments.

4 Results in Woods 1 and Woods 7

4.1 The Tasks

Wilson [1994] introduced two multi-step "woods" environments with which to examine the performance of ZCS. Woods 1 is a two dimensional rectilinear 5x5 toroidal grid. Sixteen cells are blank, eight contain rocks and one contains food. ZCS is used to develop the controller of a robot/animat which must traverse the map in search of food. It is positioned randomly in one of the blank cells and can move into any one of the surrounding eight cells on each discrete time step, unless occupied by a rock. If the animat moves into the food cell the system receives a reward from the environment (1000), and the task is reset, i.e. food is replaced and the animat randomly relocated (Figure 1).

On each time step the animat receives a sensory message which describes the eight surrounding cells. The message is encoded as a 16-bit binary string with two bits representing each cardinal direction. A blank cell is represented by 00, food (F) by 11

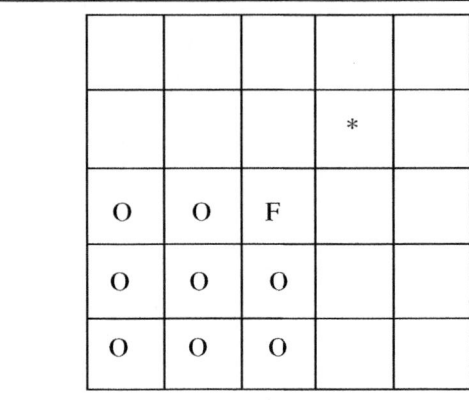

			*	
O	O	F		
O	O	O		
O	O	O		

Fig. 1: The Woods 1 environment showing the CS controlled animat * adjacent to the food goal.

and rocks (O) by 10 (01 has no meaning). The message is ordered with the cell directly above the animat represented by the first bit-pair, and then proceeding clockwise around the animat.

The trial is repeated 10,000 times and a record is kept of a moving average (over the previous 50 trials) of how many steps it takes for the animat to move into a food cell on each trial. If it moved randomly Wilson calculates performance at 27 steps per trial, whilst the optimum is said to be 1.7 steps.

Woods 7 is a more complex and non-Markov version of Woods 1 on a 58x18 grid. Fifty-seven cells evenly scattered around the map contain food. Each of these has rocks positioned randomly in two of the eight surrounding cells. The rest of the map is blank. Wilson states that random search will take 41 steps to food, whilst the optimum is 2.2 steps per trial (not shown - see [Wilson 1994]). It is noted that, since ZCS has no temporary memory (see [Cliff & Ross 1995][Tomlinson & Bull 1998]), it cannot be expected to solve Woods 7 optimally, however "it is still of interest to see how well [it] can do" [Wilson 1994]. All inputs and other task details are the same in Woods 7 as Woods 1.

All results in this paper are the average of ten runs.

4.2 Results

Figure 2 shows the performance of the self-adaptive mutation operator within ZCS on Woods 1. Here the initial possible range of mutation rates was centred around the fixed rate of 0.002 used by Wilson, i.e. $0<\mu<0.004$. It can be seen that the use of self-adaptation has had no real beneficial/detrimental effects on performance, if anything learning is a little quicker (Fig. 2a). Examination of the average mutation rate in the

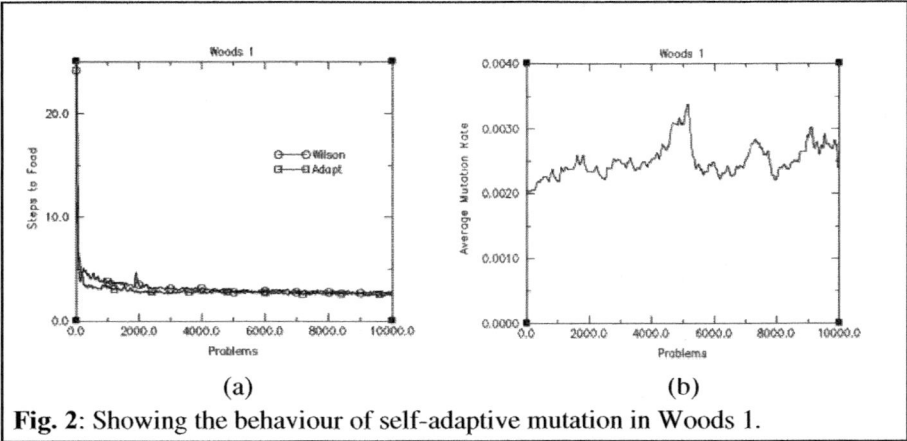

Fig. 2: Showing the behaviour of self-adaptive mutation in Woods 1.

rule-base (Fig. 2b) shows that it rises slightly from the mean of 0.0020 up to around 0.0025. That is, although no significant improvements in performance are seen here, there is obviously a slight selective pressure for a higher mutation rate than that used by Wilson.

The resulting rule-bases at the end of the 10,000 trials have been examined. Although the average mutation rate appears roughly equal to the fixed rate used by Wilson, analysis shows a wide range of individual rates. Typically, a large proportion of the

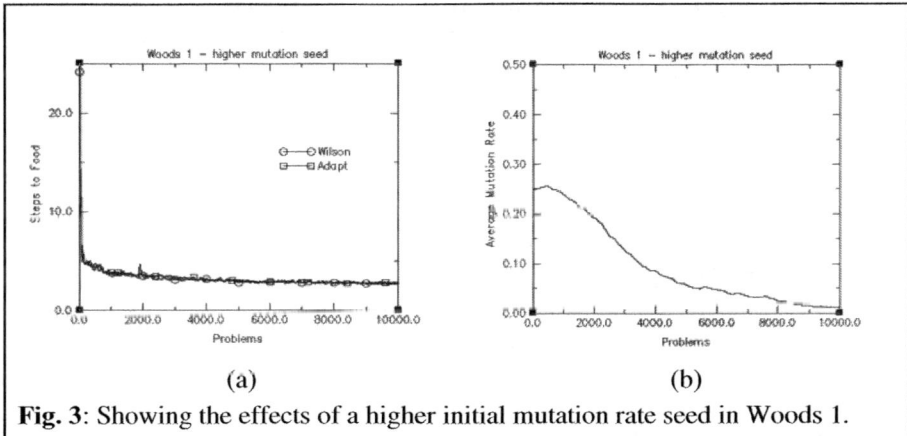

Fig. 3: Showing the effects of a higher initial mutation rate seed in Woods 1.

rules have no mutation, i.e. $\mu_i = 0$, whilst others have mutation rates up to and over 1. The former of these is associated with rules for cells closest to the food goal in the environment, whilst the latter are associated with rules for cells furthest from the goal. *Hence it appears that the degree of evolutionary search on the rules of a particular situation/cell, during the learning process, is directly correlated with the position of the rules in the inductive/reinforcement chain.* Runs over a much larger number of trials show the average mutation rate eventually going to zero (not shown).

We have also examined the robustness of the self-adaptive approach to starting the system with an "inappropriate" mean mutation rate. That is, we were interested in whether self-adaptation can be used to remove the mutation parameter from the

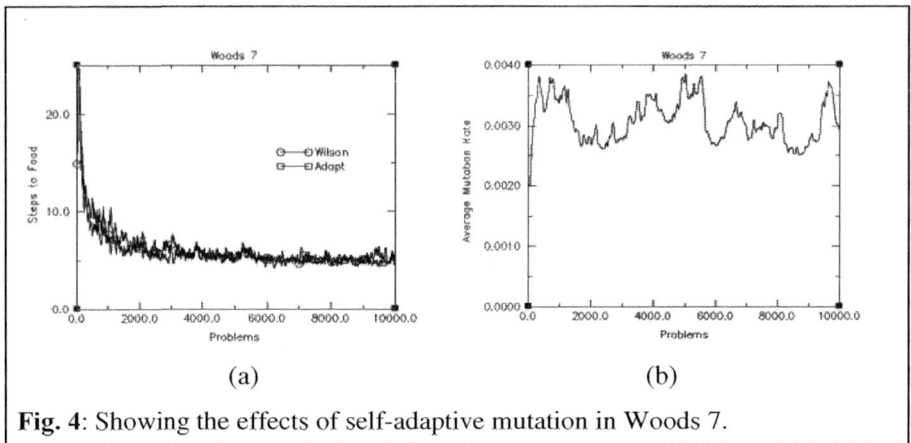

(a)   (b)

Fig. 4: Showing the effects of self-adaptive mutation in Woods 7.

designer's control somewhat and hence ease the use of Classifiers in complex tasks, particularly evolutionary robotics. Figure 3 shows results from an initial range $0<\mu<0.5$. It can again be seen that no real benefit/detriment is found over the use of a fixed mutation rate (Fig. 3a), but that now the average mutation rate continually falls towards 0.002 after a slight delay (Fig. 3b).

Results from the more complex and non-Markov Woods 7 task were very similar to those above. Figure 4 shows that there is no significant change in performance when

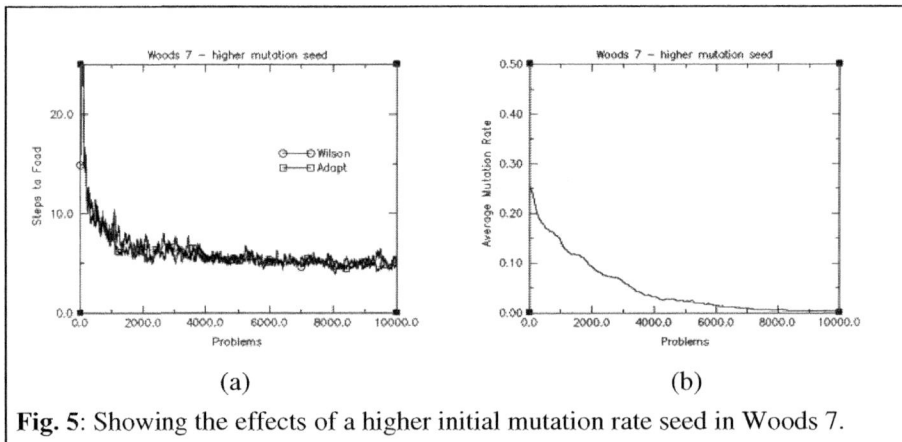

(a)   (b)

Fig. 5: Showing the effects of a higher initial mutation rate seed in Woods 7.

the initial mutation rates are seeded around Wilson's fixed rate of 0.002 (Fig. 4a). There is a rapid increase in the average mutation rate, up to around 0.004 (Fig. 4b), which then deviates around 0.003, a slightly higher rate than seen in Woods 1. That is, in a different environment we see a different form of self-adaptation occurring in the mutation rate. Tests for robustness with much higher initial mutation rates (e.g. $0<\mu<0.5$) also showed no change in performance and that the average mutation rate falls toward 0.002, although more quickly than in Woods 1 (Figure 5).

Due to the complexity of the environment analysis of the resulting rule-bases is more difficult here, but the same general correlated effect in terms of mutation/strength convergence appears to occur as described above in Woods 1.

5 Conclusions

In this paper it has been shown that it is possible to use a self-adaptive mutation operator within Michigan-style Classifier Systems - specifically Wilson's ZCS - with the aim of improving their performance as controllers for autonomous mobile robots. We are now moving these experiments onto a real robot platform in conjunction with the Intelligent Autonomous Systems Laboratory, Faculty of Engineering at UWE. Further enhancements to the self-adaptive mechanism are also currently being investigated, as well as implementing them in the more sophisticated XCS system [Wilson 1995].

Acknowledgements

Thanks to Andy Tomlinson for a number of useful discussions during this work.

References

Angeline, P.J., Fogel, D.B., Fogel, L.J. (1996) A Comparison of Self-Adaptation Methods for Finite State Machines in a Dynamic Environment. In L.J. Fogel, P.J. Angeline, & T. Bäck (eds.) *Evolutionary Programming V*, MIT Press, pp. 441-449.

Bäck, T. (1992) Self-Adaptation in Genetic Algorithms. In F.J. Varela & P. Bourgine (eds.) *Toward a Practice of Autonomous Systems: Proceedings of the First European Conference on Artificial Life*, MIT Press, pp263-271.

Cliff, D. & Bullock, S. (1993) Adding `Foveal Vision' to Wilson's Animat. *Adaptive Behavior* 2(1):47-70.

Cliff, D. & Ross, S. (1995) Adding Temporary Memory to ZCS. *Adaptive Behavior* 3(2): 101-150.

Donnart, J-Y. & Meyer, J-A. (1994) Spatial Exploration, Map Learning, and Self-Positioning with MonaLysa. In P. Maes, M. Mataric, J-A. Meyer, J. Pollack & S.W. Wilson (eds.) *From Animals to Animats 4: Proceedings of the Fourth International Conference on Simulation of Adaptive Behaviour,* MIT Press, pp 204-213.

Dorigo, M. & Colombetti, M. (1999) *Robot Shaping: An Experiment in Behavior Engineering*. MIT Press.

Fogel, D.B. (1992) *Evolving Artificial Intelligence*. PhD dissertation, University of California.

Holland, J.H. (1975) *Adaptation in Natural and Artificial Systems*. University of Michigan Press.

Holland, J.H., Holyoak, K.J., Nisbett, R.E. & Thagard, P.R. (1986) *Induction: Processes of Inference, Learning and Discovery*. MIT Press.

Koza, J.R. (1991) *Genetic Programming*. MIT Press.

Lanzi, P-L., Stolzmann, W. & Wilson, S.W. (eds.)(2000) *Proceedings of the Second International Workshop on Learning Classifier Systems*, Springer-Verlag.

Rechenberg, I. (1973) *Evolutionsstrategie; Optimierung technischer Systeme nach Prinzipen der biologischen Evolution*. Frommann-Holzboog Verlag.

Riolo, R. (1991) Lookahead Planning and Latent Learning in a Classifier System. In J-A. Meyer & S.W. Wilson (eds.) *From Animals to Animats: Proceedings of the First International Conference on Simulation of Adaptive Behaviour*, MIT Press, pp316-326.

Stolzmann, W. (1999) Latent Learning in Khepra Robots with Anticipatory Classifier Systems. In A.S. Wu (ed.) *Proceedings of the 1999 Genetic and Evolutionary Computation Conference Workshop Program*, Morgan Kauffman, pp290-297.

Tomlinson, A. & Bull, L. (1998) A Corporate Classifier System. In A.E. Eiben, T. Bäck, M. Schoenauer & H-P. Schwefel (eds.) *Parallel Problem Solving from Nature - PPSN V*, Springer, pp. 550-559.

Watkins, C. (1989) *Learning from Delayed Rewards*. PhD dissertation, University of Cambridge.

Wilson, S.W. (1985) Knowledge Growth in an Artificial Animal. In J.J. Grefenstette (ed.) *Proceedings of the First International Conference on Genetic Algorithms and their Applications*, Lawrence Erlbaum Associates, pp 16-23.

Wilson, S.W. (1987) Classifier Systems and the Animat Problem. *Machine Learning* 2:199-228.

Wilson, S.W. (1994) ZCS: A Zeroth-level Classifier System. *Evolutionary Computation* 2(1):1-18.

Wilson, S.W. (1995) Classifier Fitness Based on Accuracy. *Evolutionary Computation* 3(2):149-177.

Using a Hybrid Evolutionary-A* Approach for Learning Reactive Behaviours

Carlos Cotta and José M. Troya

Dept. of Lenguajes y CC.CC., University of Málaga,
Complejo Tecnológico (3.2.49), Campus de Teatinos,
E-29071, Málaga, Spain
{ccottap, troya}@lcc.uma.es

Abstract. A hybrid approach for learning reactive behaviours is presented in this work. This approach is based on combining evolutionary algorithms (EAs) with the A* algorithm. Such combination is done within the framework of Dynastically Optimal Forma Recombination, and tries to exploit the positive features of EAs and A* (e.g., implicit parallelism, accuracy and use of domain knowledge) while avoiding their potential drawbacks (e.g., premature convergence and combinatorial explosion). The resulting hybrid algorithm is shown to provide better results, both in terms of quality and in terms of generalisation.

1 Introduction

The control of autonomous mobile agents is a complex task to which great efforts are devoted due to its practical applications. In general, such control is achieved by means of both *planning* and *reactive* components [13]. Each of these components has it own particularities, and can be examined in combination (e.g., [1, 2, 7]) or in isolation (e.g., [11, 12]). In line with the latter, this work focuses on the acquisition of reactive behaviours in mobile agents.

Reactive behaviours are driven by a stimulus-to-response mapping, i.e., the agent receives some information about its local environment and decides the action(s) to carry on exclusively on the basis of such information. This kind of behaviour has usually the advantage of not requiring any underlying global model of the world in which the agent is located. The obvious drawback of reactive systems is the fact that they can get stuck into dead-ends, situations in which the correct action does not only depend on the locally available information but also the structure of the world at a higher-level (hence the necessity of long-term planning capabilities). Nevertheless, reactive systems have been shown to provide a very good performance in a wide variety of scenarios and remain a very suitable option when response-time is critical.

There exist several techniques for designing reactive systems. These can be typically classified into reinforcement learning and optimisation techniques. Algorithms such as Holland's bucket brigade, Sutton's temporal difference learning or Watkins's Q-learning lie in the first class. Within the second class, evolutionary algorithms deserve special attention because of their power and flexibility.

S. Cagnoni et al. (Eds.): EvoWorkshops 2000, LNCS 1803, pp. 347–356, 2000.

Regarding the use of these techniques for this purpose (e.g., [8, 9, 11]), a critical point is the use of as much domain knowledge as possible. Otherwise, the user would be relying on a fortuitous matching between her algorithm and the problem under consideration [14]. Such specialised algorithms are usually termed *hybrid* evolutionary algorithms [6].

This work presents a hybrid evolutionary algorithm for acquiring reactive behaviours. In the proposed algorithm, domain knowledge is included by using a specialised technique (the A* algorithm) as an internal operator. The remainder of the article is organised as follows. First, the agent and the worlds used in the experiments are described (Sect. 2). Next, the classical A* approach for solving the posed problem is shown (Sect. 3). Then, the hybrid algorithm is introduced (Sect. 4). Subsequently, experimental results are presented (Sect. 5). Finally, some conclusions are extracted and future work is outlined (Sect. 6).

2   The Agent and its World

The agent used in this work is located in a two-dimensional toroidal grid-world in which several obstacles are distributed. The purpose of the agent is to reach a certain target point from its initial location within an allowed time. To do so, the agent is capable of making some elementary actions such as moving straight ahead a single grid square, turning 90° to its left, or turning 90° to its right. Obviously, the agent must avoid obstacles while navigating through its world. For this purpose, it is equipped with proximate sensors that can inform of the presence or absence of obstacles in front of the agent, 90° to its left, or 90° to its right (see Fig. 1, left). In addition, these sensors can also detect whether the target point is in any of these three locations or not.

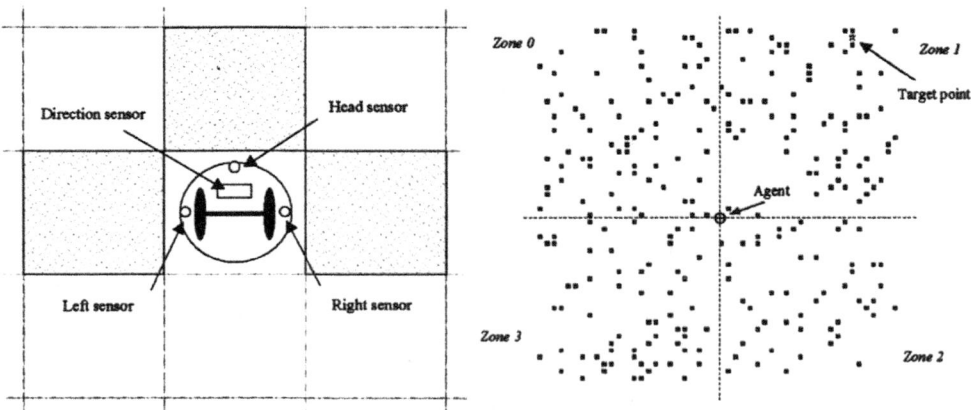

Fig. 1. (Left) Structure of the agent used in experiments. (Right) Example world and regions into which it is divided according to the location of the target point.

The agent is equipped with a direction sensor as well. This sensor allows determining in which of four imaginary regions of the world the target point is located. These regions are illustrated in Fig. 1 (right). It must be noted that these regions are not absolute but relative to the agent's actual orientation. For example, the agent is facing North in Fig. 1 and hence the target point is in zone 1. Now, if the agent turned 90° to its right, the target would be in zone 0. Notice also that these regions are determined taking into account the toroidal shape of the world. Thus, if the agent were a few positions South from the location shown in the previous example, the target point might happen to be in zone 2.

According to this description, the goal is to design a reactive behaviour allowing the agent reaching its target in as many situations as possible. Such reactive behaviour can be defined in a variety of ways, e.g., using a neural network [15], a fuzzy rule-base [8], a cellular automata [3], etc. This work is in line with the latter approach. To be precise, a lookup-table is sought relating every possible sensorial input with a primitive action. At each time-step, the agent must look up the action that corresponds to the current inputs and carry it out. Since each proximate sensor can provide three different inputs (OBSTACLE, NO-OBSTACLE, TARGET), and the direction sensor can return four values, the resulting table has $3^3 \cdot 4 = 108$ entries. Since three primitive actions – MOVE-AHEAD, TURN-LEFT, TURN-RIGHT – are available, this implies a search space of $3^{108} \approx 3 \cdot 10^{51}$ tables.

3   A Classical Approach: A*

A classical approach for finding the lookup-table mentioned above is the utilisation of the A* algorithm. Based on incrementally constructing solutions in an *intelligent* fashion, this technique constitutes a powerful tool for solving search problems to optimality. Before getting into the application of this technique to the design of reactive behaviours, some notation details must be given.

Let W be the current world, and let α be the configuration of the agent (position and orientation). Now, let $\mathcal{I}(W, \alpha)$ be the sensorial input of the agent when configured according to α. Let $\hat{M}$ be a (possibly underspecified) function relating sensorial inputs with actions, and let $\hat{M}_1 \supset \hat{M}_2$ whenever $\hat{M}_2$ provides the same outputs that $\hat{M}_1$ does and $\hat{M}_2$ is defined in at least one case in which $\hat{M}_1$ is not. Finally, let τ be the maximum allowed time for reaching the target and let $\varXi$ be a function such that $\varXi(W, \alpha, \hat{M}) = (\langle \alpha_1, \cdots, \alpha_k \rangle, \eta)$. This function provides a trace of the agent trajectory across configuration space when behaving according to $\hat{M}$. The value η is an indication of the final status of the agent: AT-TARGET, COLLISION, TIMED-OUT or UNKNOWN. This latter value is returned whenever no action is specified in $\hat{M}$ for the current input.

Now, the application of the A* algorithm requires the availability of an optimistic evaluation function ψ such that $\psi(W, \alpha, \hat{M})$ provides a lower bound on the number of steps necessary for reaching the target when the agent is configured as α and behaves according to any $\hat{M}', \hat{M} \supset \hat{M}'$. It is easy to see that making ψ return the Manhattan distance from the agent's current location to the target point fulfils this requirement. Having defined this function, the whole process is as follows:

1. Let $P_0^0 = (\alpha_0, \hat{M}_0, 0, t_0)$, where α_0 is the initial configuration of the agent, $\hat{M}_0$ is a fully underspecified function, and $t_0 = \psi(\mathcal{W}, \alpha_0, \hat{M}_0)$. Let $P^* = (\hat{M}_0, \infty)$ be the current best solution. Insert P_0^0 in the node queue.
2. If the node queue is empty, go to 3. Otherwise let $P_j^i = (\alpha, \hat{M}, h, t)$ be the first element in the queue.
   (a) Let $\Xi(\mathcal{W}, \alpha, \hat{M}) = (\langle \alpha_1, \cdots, \alpha_k \rangle, \eta)$.
   (b) if $\eta \neq$ UNKNOWN then
       i. if $\eta =$ COLLISION or $\eta =$ TIMED-OUT then $\bar{P} = (\hat{M}, t' + \tau)$, where $t' = h + k + \psi(\mathcal{W}, \alpha_k, \hat{M})$.
       ii. if $\eta =$ AT-TARGET then $\bar{P} = (\hat{M}, h + k)$.
       iii. If $\bar{P}$ is better than P^*, update the latter and purge nodes in the queue.
       iv. Go to 2.
   (c) Create three nodes P_{j+1}^{3i+1}, P_{j+1}^{3i+2}, and P_{j+1}^{3i+3} from P_j^i. Each node is $P_{j+1}^{3i+r} = (\alpha_k, \hat{M}_r, h', t_r)$, where $\hat{M}_r$ is obtained by extending $\hat{M}$ to return the rth possible action when the input is $\mathcal{I}(\mathcal{W}, \alpha_k)$, $h' = h + k$, and $t_r = \psi(\mathcal{W}, \alpha_k, \hat{M}_r)$. Insert these nodes in the queue keeping it ordered according to the sum of the last two components of each node.
   (d) Go to 2.
3. Return P^*

This algorithm will thus return the lookup-table allowing the agent reach the target in minimal time from the given starting point. Since the problem has been posed with the goal of obtaining a generalisable reactive behaviour, the process must be slightly modified. To be precise, a training set is selected and the A* algorithm tries to find the table that minimises the sum of the times required to reach the target in each training case or, if such a solution is not possible, a table that firstly maximises the number of training cases solved and secondly minimises the total time. Notice that no global model of the world (i.e., high-level knowledge about the distribution of obstacles) is required. All information used for finding the optimal solution is locally obtained through simulation.

This algorithm has been evaluated on a set of nine different worlds. These worlds are named as Wxy, where $x \in \{10, 25, 50\}$ indicates the dimension of the world (each world is a $x \times x$ grid), and $y \in \{a, b, c\}$ indicates the density of obstacles (5%, 10% and 20% respectively). For each world, a training set of five cases has been selected. Subsequently, the best solution found has been tested for generality on a test set whose size depends on the dimension of the world (50, 400 and 2000 cases respectively). The results are shown in Table 1.

These results are very indicative of the two main drawbacks of the A* algorithm. On the one hand, it is very sensitive to the size of the task to be solved. As it can be seen, the algorithm expended a high computational effort for solving W25b and ran out of memory in three cases (W25c, W50b, and W50c). Moreover, it did not find any fully satisfactory solution for all training cases in W50b

Table 1. Results of the A* algorithm on nine different worlds. The cost values are measured as the number of single simulation steps carried out.

World	Timeout	Optimal solution	Iterations	Cost	Performance on test set
W10a		13.20	11615	289857	74%
W10b	25	13.80	19664	300391	88%
W10c		14.00	26429	281091	58%
W25a		32.40	75818	10352732	63%
W25b	150	31.60	222788	29083614	48%
W25c		[26.40, 51.00]	>300000	>25000000	36%
W50a		54.60	119864	36861614	66%
W50b	400	[45.60, 479.40]	>150000	>37000000	14%
W50c		[34.80, 1629.80]	>110000	>16000000	2%

and W50c (1 and 4 training cases were left unsolved[1]). On the other hand, the solutions found are not very generalisable. This is a direct consequence of the internal functioning of the A* algorithm. Assume that the final solution is found when evaluating node P_j^i. This node was obtained as successive extensions of $P_{j-1}^{i\backslash 3}, P_{j-2}^{i\backslash 9}, \cdots, P_0^0$. Hence, it contains information regarding the best decisions to be taken only in the situations found during this *optimal* path, i.e., the path from the root node of the implicitly defined search tree to the optimal leaf node. All that may have been learnt in solving other situations is discarded since these situations do not take place in this optimal path.

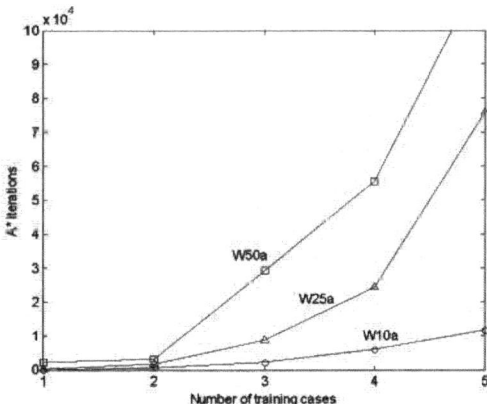

Fig. 2. Growth of the computational cost of the A* algorithm when the number of training cases is increased.

---

[1] These results were not bad a priori since there might exist no better solution. However, further experimentation with the hybrid EA showed that this was not the case.

This generalisation problem could be solved by considering a larger training set whose optimal solution covered all possible situations. However, the subsequent combinatorial explosion makes this approach unrealistic. This is illustrated in Fig. 2. As it can be seen, the computational cost of the algorithm grows very fast when the size of the training set is increased. For this reason, it is clear that alternative approaches must be found. These will be discussed in next section.

4   The Hybrid EA-A* Approach

Evolutionary algorithms constitute a very suitable alternative to A* for finding the lookup-table. A naïve approach for applying EAs to this problem would firstly consist of defining an encoding function for storing the lookup-table into an individual, e.g., a linear chromosome in which the rows of the table are consecutively arranged. Since this is an orthogonal representation [10] (i.e., all combinations of genes are feasible), the next step would simply involve selecting any of the standard genetic operators that can be found in the literature (e.g., single-point crossover – SPX –, uniform crossover – UX –, etc.).

However, such a simple approach is likely to provide very poor results. Recall that this is highly epistatic problem in which the value of each gene (i.e., a specific action to be carried out when a certain sensorial input is received) does not contribute with a fixed amount to the fitness of an individual. On the contrary, the goodness of the reactive behaviour defined is determined by the interplay between all genes. For this reason, a blind recombination operator that randomly shuffles the genetic material of recombined solutions will provably produce solutions with a phenotype (reactive behaviour) completely unrelated to the parents, even when the latter are genotypically similar. In an extreme situation, it may even reduce to macromutation.

The algorithm would be largely more effective if it were able to extract positive behavioural patterns from existing solutions and transmit them to the offspring. This can be achieved within the framework of Dynastically Optimal Forma Recombination [4] (DOR). This framework comprises a family of recombination operators of the form

$$\text{DOR} : \mathcal{S} \times \mathcal{S} \times \mathcal{S} \to [0, 1], \tag{1}$$

where $\mathcal{S}$ is the search space and $\text{DOR}(x, y, z)$ is the probability of generating z when recombining x and y. Besides the obvious $\sum_{z \in \mathcal{S}} \text{DOR}(x, y, z) = 1$, the probability distribution induced by these operators verify that

$$\text{DOR}(x, y, z) > 0 \Rightarrow \{[z \in \Gamma(\{x, y\})] \wedge [\forall w \in \Gamma(\{x, y\}) : \phi(w) \geq \phi(z)]\} \tag{2}$$

where ϕ is the fitness function (to be minimised without loss of generality) and $\Gamma(\{x, y\})$ is the dynastic potential [10] of x and y, i.e., the set of solutions that can be built using nothing but the information contained in x and y.

Thus, the solutions created by DOR are the best that can be constructed using the genetic material of the parents. On the one hand, this implies that

DOR is a fully transmitting operator, i.e., no implicit mutation (genetic information not present in any of the parents) is introduced in the offspring. On the other hand, the fitness-oriented functioning of DOR makes valuable portions of solutions be transmitted to offspring only if they contribute to a good resulting behaviour. In other words, DOR is capable of identifying valuable high-order formae, preventing their disruption. This intelligent combination of information has provided very good results on epistatic problems [5].

In order to implement DOR, it is required to use an embedded A*-like mechanism so as to find the best solution in the dynastic potential of the parents. In this case, the algorithm described in Sect. 3 can be used. It is only necessary to modify step 2c by considering that the possible actions to be taken in a given situation $\mathcal{I}$ are just those present in any of the parents for $\mathcal{I}$. Notice that the search carried out by this subordinate A* algorithm is thus restricted to small portions of the search space and hence its computational cost is largely reduced with respect to the original unrestricted version. Moreover, individuals in the population tend to be more similar as the EA converges and, subsequently, the dynastic potential of selected solutions tends to be smaller and DOR is less computationally expensive.

This combination of EAs and A* has an additional advantage. Each individual carries an information that reflects its past evolution (in fact, the evolution of its ancestors). This way, things that were learnt in the past are retained as long as they do not negatively affect the present behaviour. This "accumulated history" effect is also present in a simple EA, but the learning capabilities of the hybrid algorithm are larger. For this reason, solutions obtained with the hybrid EA are expected to be more general than either the EA or the A* algorithm by themselves. This will be studied in next section.

5   Experimental Results

Experiments have been done with a steady-state EA (*popsize* = 100, p_c = .9, p_m=1/chromosomeLength) using ranking selection ($\eta^+ = 2.0, \eta^- = 0.0$). This algorithm has been run 40 times for each operator and test world. In order to make a fair comparison between DOR and the other simpler operators, each run is terminated when a fixed number of simulation steps ($\tau \cdot 10^5$ in these experiments, where τ is the timeout value) is reached. Thus, the internal calculations performed by DOR are effectively accounted. As in Sect. 3, a training set of five cases is used in the fitness function.

First of all, Fig. 3 shows how the hybrid EA is much more successful in solving the training cases. As it can be seen, while standard operators only provide an acceptable performance on the smallest instances and with the lowest obstacle density, DOR consistently yields satisfactory results: above a 70% of the runs provide a fully successful solution for the training set (the percentage is 100% for 5 out of 9 test worlds). The exception is world W50c for which none of the operators could find a full solution (it must be noted that such a solution may

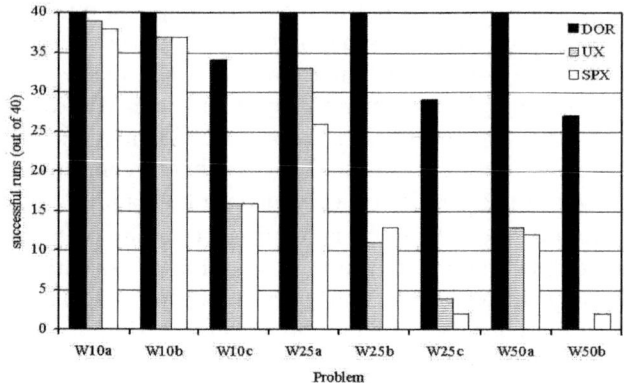

Fig. 3. Number of runs in which each operator provided a fully satisfactory solution for the training set.

not exist). Nevertheless, DOR was capable of solving 3 out of the 5 training cases while SPX could only solve one and UX could not solve any of them.

Table 2 shows a more detailed summary of the results. Notice that DOR is not only more effective in finding satisfactory solutions, but also provides higher-quality results. By comparing the median values[2] provided by DOR with the optimal/best-known solutions (see Table 1), it can be seen that DOR yields near-optimal solutions. Moreover, the lower variance of DOR results with respect to SPX and UX indicates a more stable algorithm.

Table 2. Comparison of different genetic operators on nine different environments. All results correspond to series of forty runs.

World	Timeout	SPX			UX			DOR		
		mean	σ	median	mean	σ	median	mean	σ	median
W10a		17.36	12.06	16.00	15.88	7.47	15.20	13.85	0.87	13.80
W10b	25	22.15	25.96	16.20	18.10	9.26	16.20	14.35	0.74	14.20
W10c		51.49	38.39	59.60	53.85	38.81	59.60	18.75	10.75	14.00
W25a		176.70	210.00	45.60	121.63	193.32	40.00	33.17	2.14	32.40
W25b	150	376.08	285.43	399.30	380.96	272.08	468.40	36.65	6.67	34.00
W25c		520.32	179.07	616.80	502.23	218.40	616.40	99.92	107.70	38.80
W50a		442.39	348.27	438.00	423.87	335.45	438.00	61.17	11.72	58.80
W50b	400	1532.52	554.56	1636.60	457.65	457.65	1636.60	337.32	506.03	103.8
W50c		2021.45	60.77	2031.20	2031.20	0.00	2031.20	1638.78	538.75	1650.60

---

[2] The median value seems to be a more representative measure of the quality of the results than the mean value since the former is much less sensitive to outliers. Furthermore, it provides an reasonable alternative to averaging the fitness of solutions that solve the whole training set with solutions that do not solve any training case.

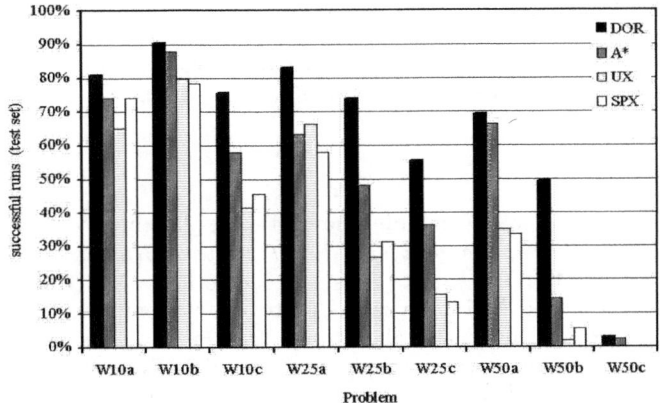

Fig. 4. Percentage of the test set solved for each of the techniques considered.

Finally, the results obtained with the EA are tested for generality. Fig. 4 shows the results. Firstly, notice the poor results of standard EAs. The solutions provided by UX and SPX do not reach 50% success in 6 out of 9 worlds. The A* algorithm performs better than standard EAs, but its performance quickly drops when the density of obstacles is increased. The hybrid EA provide the overall best results, outperforming both A* and standard EAs on all worlds. Moreover, this improvement is larger on instances with higher obstacle densities. It must be noted that the results on W50c are not satisfactory for any algorithm (although the hybrid algorithm remains the best). This is a really hard instance as mentioned before, and may require longer evolution times and/or a larger training set to cope with such a tough environment.

6 Conclusions

This work has presented a hybrid approach for learning reactive rule-bases. By combining EAs with the A* algorithm, a synergetic system has been achieved. This hybrid algorithm has been shown to provide higher-quality results than standard EAs. These results are also better than those of the A* algorithm in terms of their generalisation to previously unseen test cases. Furthermore, the hybrid EA is capable of tackling instances in which the A* algorithm would suffer the effects of the combinatorial explosion.

Future work will try to extend these results to more sophisticated agents. In this sense, notice that most details of the agent are encapsulated within the simulation function Ξ and hence they do not affect the presented algorithm qualitatively. Nevertheless, it is clear that issues regarding simulations of higher computational cost are worth studying. Work is in progress in this area. Additionally, new environments and tasks to be solved will be tackled as well.

Acknowledgement

This work is supported by the Spanish *Comisión Interministerial de Ciencia y Tecnología* (CICYT) under grant TIC99-0754-C03-03.

References

1. K. Ali and A. Goel. Combining navigational planning and reactive control. In *Theories of Action, Planning, and Robot Control. Bridging the Gap: Proceedings of the 1996 AAAI Workshop*, pages 1–9, Menlo Park, CA, 1996. AAAI Press.

2. C.T.C. Arsene and A.M.S. Zalzala. Control of autonomous robots using fuzzy logic controllers tuned by genetic algorithms. In *Proceedings of the 1999 Congress on Evolutionary Computation*, pages 428–435. IEEE NNC - EP Society - IEE, 1999.

3. T.D. Barfoot and D'Eleuterio G.M.T. An evolutionary approach to multiagent heap formation. In *Proceedings of the 1999 Congress on Evolutionary Computation*, pages 420–427. IEEE NNC - EP Society - IEE, 1999.

4. C. Cotta, E. Alba, and J.M. Troya. Utilising dynastically optimal forma recombination in hybrid genetic algorithms. In A.E. Eiben, Th. Bäck, M. Schoenauer, and H.-P. Schwefel, editors, *Parallel Problem Solving From Nature V*, volume 1498 of *Lecture Notes in Computer Science*, pages 305–314. Springer-Verlag, Berlin, 1998.

5. C. Cotta and J.M. Troya. Tackling epistatic problems using dynastically optimal recombination. In B. Reusch, editor, *Computational Intelligence. Theory and Applications*, volume 1625 of *Lecture Notes in Computer Science*, pages 197–205. Springer-Verlag, Berlin Heidelberg, 1999.

6. L. Davis. *Handbook of Genetic Algorithms*. Van Nostrand Reinhold Computer Library, New York, 1991.

7. J.-Y. Donnart and J.-A. Meyer. Learning reactive and planning rules in a motivationally autonomous animat. *IEEE Transactions on Systems, Man, and Cybernetics*, 26(3):381–195, 1996.

8. F. Hoffmann and G. Pfister. Learning of a fuzzy control rule base using messy genetic algorithms. In F. Herrera and J.L. Verdegay, editors, *Genetic Algorithms and Soft Computing*, pages 279–305. Physica-Verlag, Heidelberg, 1996.

9. J.R. Koza. *Genetic Programming*. MIT Press, Cambridge MA, 1992.

10. N.J. Radcliffe. The algebra of genetic algorithms. *Annals of Mathematics and Artificial Intelligence*, 10:339–384, 1994.

11. A.C. Schultz and J.J. Grefenstette. Using a genetic algorithm to learn behaviours for autonomous vehicles. In *Proceedings of the AIAA Guidance, Navigation and Control Conference*, pages 739–749, Hilton Head SC, 1992.

12. C. Thornton. Learning where to go without knowing where that is: the acquisition of a non-reactive mobot behaviour by explicitation. Technical Report CSRP-361, School of Cognitive and Computing Sciences, University of Sussex, 1994.

13. G. Weiss. *Multiagents Systems: a Modern Approach to Distributed Artificial Intelligence*. The MIT Press, Cambridge MA, 1999.

14. D.H. Wolpert and W.G. Macready. No free lunch theorems for search. Technical Report SFI-TR-95-02-010, Santa Fe Institute, 1995.

15. B. Yamauchi and R. Beer. Integrating reactive, sequential and learning behaviour using dynamical neural networks. In D. Cliff, P. Husbands, J.-A. Meyer, and S. Wilson, editors, *From Animals to Animats 3: Proceedings of the Third International Conference on Simulation of Adaptive Behaviour*, pages 382–391, Cambridge MA, 1994. MIT Press/Bradford Books.

Supervised Evolutionary Methods in Aerodynamic Design Optimisation

D.J. Doorly, S Spooner and J. Peiró

Aeronautics Department, Imperial College, London SW7 2BY, UK

Abstract. This paper outlines the application of evolutionary search methods to problems in aeronautical design optimisation. The procedures described are based on the genetic algorithm (GA) and may be applied to other areas. Although easy to implement, a simple genetic algorithm is often found in applications to be of low efficiency and to suffer from premature convergence. To improve performance, two alternative strategies are investigated. In the first, a learning classifier scheme is used to tune the GA for a particular class of problems. The second strategy uses a parallel distributed genetic algorithm supervised by single or competing agents. The implementation of each procedure, and results for typical design problems are outlined. The agent supervised distributed genetic algorithm is found to provide a model with a very high degree of adaptibility, and to lead to considerably improved efficiency.

1 Introduction

The principles of evolutionary computation are well established and are described in texts such as [1, 3, 2, 4]. These methods continue to grow in diversity, and are becoming more commonly used for many problems in engineering design. In aerospace vehicle design, the genetic algorithm (GA) has been applied to a range of problems, as described for example in [5, 6, 7, 8, 9]. The apparent robustness of the GA, the ease with which it can be applied, and its ability to handle discontinuous or even discrete data make it attractive as a search procedure.

There are other search procedures however (including various gradient search methods) which are more efficient than the GA, albeit that they may only work for a more restricted class of problems. Furthermore, although evolutionary methods have been found to be quite adept at locating global optima in highly multimodal problems, there is usually no guarantee that they will, and their rate of convergence to the optimum solution can be very slow. Ideally one would like an optimisation procedure to work well across a broad range of problems. Bearing in mind the no free lunch 'rule' however, some degree of matching of technique to problem seems inevitable.

Three simple ways of improving an evolutionary search procedure for a particular type of application are:

1. tuning the parameters to suit the class of problem,
2. improving its ability to adapt to the problem,
3. hybridisation with other techniques appropriate to the class of problem.

S. Cagnoni et al. (Eds.): EvoWorkshops 2000, LNCS 1803, pp. 357–366, 2000.

We discuss only the first two of these here, and will use the GA as our basic evolutionary procedure. The construction of a suitable hybrid is very problem dependent, and we consider that an improved GA may either outperform a hybrid routine or may at worst lead to more effective hybrids. (Hybrid routines will in any case be discussed elsewhere). The rest of the paper is organised as follows. The common design problem of shape or form optimisation is introduced in the context of aerodynamic or aeroelastic design. The application of a real encoded GA to a typical problem is also briefly outlined. After this, we examine the use of a learning classifier system to tune the GA parameters, for a particular class of such design problems. The distributed genetic algorithm or DGA is then introduced, and finally the use of agent and multi-agent supervision to improve the adaptibility of the DGA is described.

2   Shape optimisation & outline of basic GA

A frequent task in aeronautical design is to find the 'best' aerodynamic shape of an airfoil (in 2D) or a wing (in 3D), subject to certain constraints. The basic GA we use for this task is constructed as follows.

- The *encoding* specifies the shape of a trial solution, though it may also specify structural parameters such as type of material, weight, rigidity etc. For simplicity, let us consider the case of a 2D wing section or airfoil. Then an array $c[j]$ may be used to specify the ordinates of a B-spline control polygon, which in turn describes an airfoil shape, as shown in fig. 1.

  We use a real array of 20 control point ordinates to encode the shape. (For

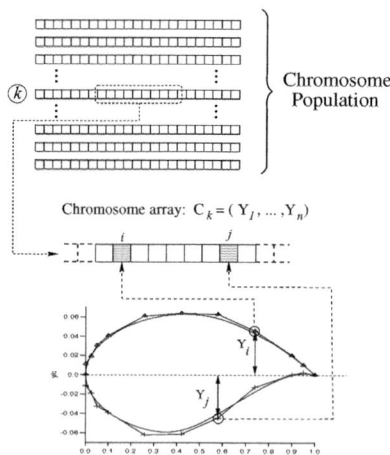

Fig. 1. Representation of airfoil geometry.

aeroelastic optimisation of wings instead [13], we encode the design as a surface interpolated between a series of spanwise sections, with the encoding specifying a standard basic profile, together with the chord, thickness, twist and structural parameters for each section).

- An *initial population* of random airfoils is generated, (each initial shape is ensured to be valid, i.e. encodings which produce surface crossings or violate thickness constraints are discarded).
- The *fitness* is evaluated using a CFD flow solver (or a CFD + structural solver). (Either an unstructured mesh based solution of the Euler equations, or a viscous/inviscid panel flow solver is used, [11]. In a *direct* optimisation, the search is for the shape which best meets the requirements (e.g. high lift/drag (L/D)), whereas in an *inverse* optimisation, the shape which best matches a given pressure coefficient distribution (C_p) is sought.
- Fitness values are *range-scaled* and remapped to the interval $[0, 1]$, with individuals below the mean fitness assigned a base value (typically $O(0.4)$), and those above the mean assigned a value scaled quadratically up to 1.
- *Roulette wheel* is used for selection in the basic GA and DGA operations, though the learning GA allows other methods (binary tournament etc.)
- *Two point crossover* is again the basic crossover type, (with one point, uniform etc. allowed in the learning GA). Mutation and crossover are applied as separate operators. The probability of mutation of a given gene is low, (typically 0.005), but is commonly increased at later stages.
- *Elitist population replacement* is applied, where the best of each generation is automatically carried through, (here together with a slightly mutated copy of the best). The population replacement routine ensures that no over-replication (i.e. excess identical or almost identical duplication) occurs.

3   Classifier Learning Directed GA

The classifier learning system is described in standard texts on machine intelligence; much of the research in this area also follows from ideas put forward by Holland [14]. The classifier learning system adds a layer on top of the GA. Rules to control operators (mutation/crossover, type of crossover, selection scheme etc.), and parameter values (e.g. mutation rate) are prescribed, and their effectiveness when implemented in the GA are assessed. In a static mode, the entire set of rules can undergo genetic operations to evolve better rules and rule combinations. In a dynamic mode, a system of reward paybacks can be used to determine rule selection.

The objective of the procedure here was to train the GA in a static mode for a particular problem type. The efficiency of the trained and basic GAs were then compared; firstly when given problems of a very similar nature to those used in training, and secondly for slightly different problems. In the learning GA outlined in fig. 2, an initial population of 30 rules was used, which were randomly initialised. The rules were encoded as 15 genes in an IF-THEN-AND configuration, and were designed to work in groups, with the final 2 genes being

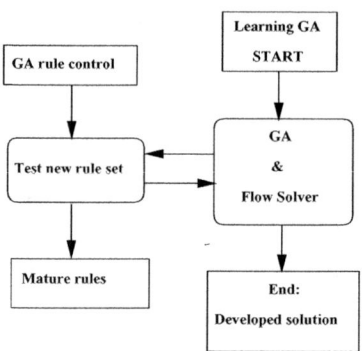

Fig. 2. Classifier Learning System/GA

index references to other rules to be used in combination with the current one. The rule gene string comprised:

- two IF operators, with parameter values determining the generation number, and average gene diversity respectively which triggered operation of the rule,
- the following 11 gene values (type THEN) set operator type and parameters,
- the final two (type AND) determined the combination of rules to be used.

The procedure was repetitively applied to the inverse design of respectively *symmetric* and *non-symmetric* airfoils. (To recall, in inverse design, the problem is to find the shape which matches a specified pressure coefficient (C_p) distribution; it is often required in real applications, and for comparative tests of search methods purposes it is preferrable to direct design, as the target is given). Each rule was tested twice, and the top 10 peforming rules were then isolated and repetitively tested over 15 separate runs.

The GA was trained for inverse optimisation of symmetric and non-symmetric airfoil sections, (respectively using NACA airfoil types -0012,-0022,-0024 for symmetric training, and -23015, -4421 for non-symmetric training). Figure 3 shows results for the inverse design of a non-symmetric airfoil (23015), comparing the performance of the GA using: rules developed for this type of airfoil, rules developed for a different (symmetric type), and the basic (untrained) GA. It can be seen that the trained GA outperforms the baseline GA for inverse optimisation whether it is trained on a class of similar airfoils, or a dissimilar class. However, the difference in performance resulting from the class of problem used for training shows how very specific training may be needed to obtain the highest gains.

At present, the cost of CFD evaluations is generally so high, that the benefits of training do not appear worthwhile. However other possibilities, such as using a simpler approximate evaluation method for training only, or reducing the complexity of the scheme, (and hence degree of training), may yet render it more practical in this area. Also, the classifier system may still prove useful within the context of an agent supervised DGA.

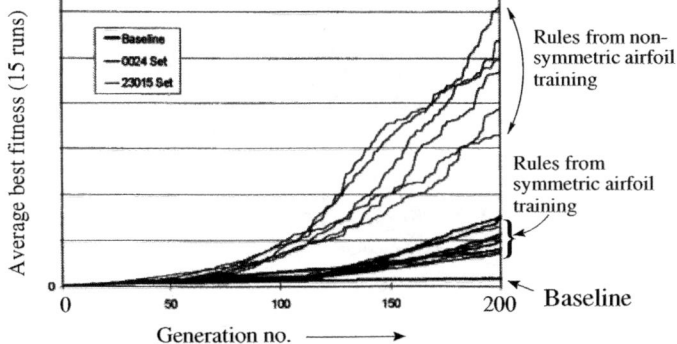

Fig. 3. Application of learning GA to inverse design of NACA-23015 airfoil. Comparison of performance with rule sets developed for similar airfoils, for different airfoils, and basic GA performance.

4   Distributed Genetic Algorithm (DGA)

Previous work has already shown that the DGA outperforms the GA on many test problems and in design optimisation, [10], [16]. Applications of the DGA to aeronautical design problems are also described in [11, 13]. Briefly the DGA differs from the standard GA in that the population of trial solutions is split into semi-isolated subpopulations or 'demes'. (The demes are considered analogous to island populations, where geography acts as a barrier to exchange). Restrictions on the recombination and genetic exchange between subpopulations are imposed; the exchange is limited to the migration of a few individuals (often only the best one or two) from one neighbouring deme to another every m generations, with typically $m = 5$ or $m = 10$ in our implementations.

Thus the parameter set for the basic GA (operator probabilities, selection mode, etc.) is enlarged to include the number of individuals migrating, barriers to acceptance of immigrants, exchange frequency, geographical exchange radius and topology of the demes. With limited exchanges between demes, the DGA is then ideally suited for coarse grain parallelisation (fig.4) whether on a parallel supercomputer or network of workstations. Nang [12] surveys the parallel GA, of which the 'stepping stone' connected DGA shown in fig. 4 is one type.

For a workstation network [11], provided the flow solution can be run on a single workstation, the only communication required between processors involves the exchange of a limited number of chromosomes, at intervals of several generations. The communication requirements are extremeley low, given that by far the bulk of the computational effort is devoted to the flow solution which performs the evaluation. Each processor may be responsible for a number of demes. For a heterogeneous network of processors of different speeds or loadings, load balancing can be achieved by varying the mapping of demes to processors, or by altering the number of individuals treated by a given processor.

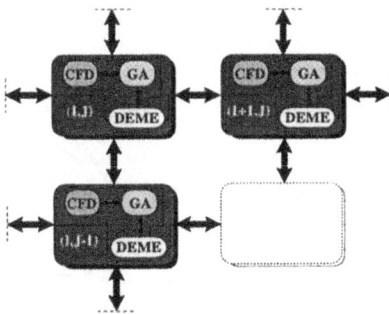

Fig. 4. Distributed GA (DGA) mapped to processor array.

4.1 Application to aeronautical design

Inverse design Comparison of the convergence behaviour obtained using a distributed [13] and a conventional GA, with the same total population and number of evaluations, for the viscous inverse design of a NLF(1)-0115 airfoil is shown in fig. 5. For the DGA, the population of 180 was distributed onto 9 subpopulations of 20, each residing on a different processor, as in fig.4 above. The migration between the islands occurred in a stepping stone fashion, ie. migration occurred only between immediate neighbours every five generations. As the results indicate, the DGA greatly outperforms the conventional GA algorithmically. This gain is then further multiplied almost ninefold with the distributed processor implementation. (The implementation was done using the MPI standard on a network of workstations; it has also been implemented on a multiprocessor machine).

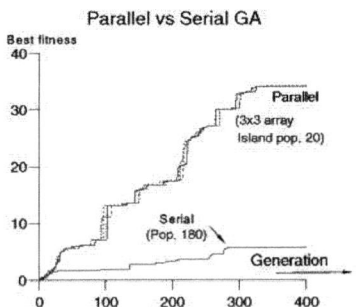

Fig. 5. Comparison of convergence rate of parallel DGA and single population version. (If computing speed up rate (nearly 9) were applied, parallel DGA performance gain would appear even more dramatic)

Direct Airfoil Optimisation Results Application of the DGA to problems of direct airfoil optimisation for low speed and transonic cases are described in [6], [11]. For both inverse and direct optimisation problems however, the use of an agent to supervise the operation of the DGA has been found to improve the performance, as described next.

5   Agent Supervision of DGA

The better ability of the DGA at maintaining population diversity appears to account for its notable gain in performance over the single population GA. Eventually however, the population on each island converges. Adding an agent to supervise the operation of the parallel GA provides a capability whereby the DGA can adapt more generally than is possible with a sequential GA. Although adaptation can be built into a GA (e.g. in the adaptive operator fitness of Davis [4], the use of agents provides a more general framework by decoupling the tasks of higher level supervision from the lower level optimisation. The agent supervised paradigm is very well suited to a distributed computing environment, where agents can direct the operation of the GA on local or global populations, and can additionally direct processing resources. At one extreme, the agent layer may be combined with the DGA software to execute as a single (albeit distributed) entity, or at another, it may run as an entirely separate distributed program, communicating with the DGA by reading external output messages and writing to action inputs.

A simple agent supervision of the DGA is as follows. The agent receives status messages from the DGA, (i.e. generation number, measures of population convergence, fitness changes etc.), and parameter settings (local mutation and crossover rates/types etc.). The agent layer then instructs the DGA to take either global or local action. Examples of such actions could be the introduction of a mechanism to improve diversity between island populations, actions to favour specific local niches within islands, or actions to improve the parallel load balancing by adjusting the deme placing or sizing.

5.1   Infection Agent

A simple implementation of agent supervision of the DGA which was used for direct airfoil design employs a single agent supervisor [15] to act as a vector for infections, with low population diversity encouraging epidemics. On infected islands, individuals close to the global best have greatly reduced fitness, and undergo increased mutation; results [15] shows the pattern of infection changes dynamically. When implemented for the problem of inverse design optimisation, there was a gain in late solution convergence, beginning just beyond the point when the population on all the islands initially converges towards a global 'champion'.

Applying the procedure to direct design (viscous L/D optimisation of a low speed airfoil section, at operating points of 3 and 8 degrees incidence, and a

Reynolds number of 4 million, with a moment constraint ($|Cm| < 0.97$)) produced the shape and C_p distributions shown in fig. 6, (though this is not yet fully a converged solution).

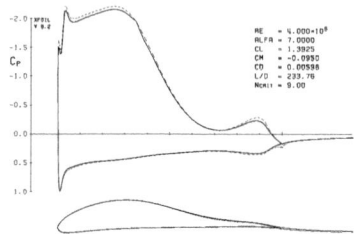

Fig. 6. L/D optimization: Distribution of C_p on the surface of the 'best' airfoil. This computation was performed using XFOIL.

The addition of the infection in these problems was found to show a clear improvement over the solution obtained in [11] for corresponding effort; the improvement in the fitness after island convergence is shown in fig. 8a further below.

5.2 Evolutionary Agent Supervision

In the learning classifier method presented earlier, adaptation occurs through repetitively solving a problem or class of problems. The agent supervision can adapt dynamically however, especially if a large number of generations are to be evaluated.

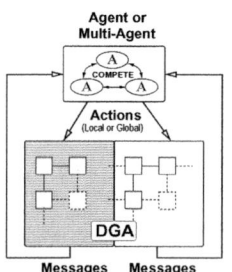

Fig. 7. Competing agent supervised DGA

For example, one may place several islands in a group under the control of one agent, and other groups of islands under the control of other agents, with each agent adopting different strategies. If migration between groups is eliminated, the

relative improvement over a number of generations may be compared. The worse
performing agents then modify either the parameter settings or the rules which
they apply to the DGA under their control, and may also replace their population
partly or fully with that of the best group. An example of the effectiveness of the
approach is shown in fig.8b, where the simple agent infection approach described
previously is supervised by a pair of competing agents. A population of 200 (10
individuals per deme; demes connected in a 4 x 5 array) was used in this case
for the inverse design of a NACA 0012 airfoil. The agents apply different genetic
operators to their respective groups of subpopulations, and their performance
as managers is compared after a certain number of generations. In the example,
only the mutation rate was altered by the supervising agents; at a lower level,
the infection strategy was still implemented. As can be seen from fig. 8b, these
early results are encouraging; further work will consider the effects of controlling
different parameters, and different competition mechanisms.

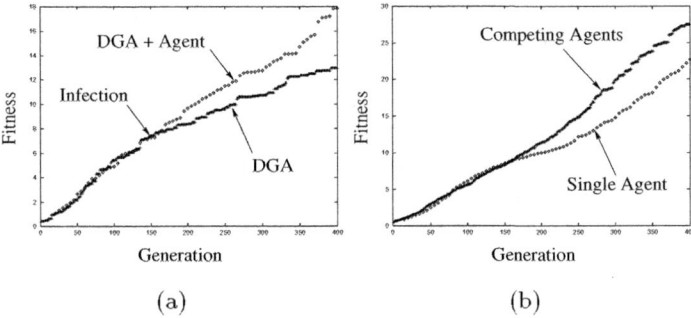

Fig. 8. Agent supervision: (a) Comparison of DGA and simple agent supervised DGA;
(b)Comparison of simple agent and competing agent supervised DGA. (Note difference
in fitness scales; results in (b) are also averaged over more trials)

6   Conclusion

The distributed genetic algorithm (DGA) has been applied to a number of aeronautical design optimisation problems. Earlier results indicated that the method
has better convergence behaviour than the single population GA; the present
work outlines agent supervision strategies to improve the DGA further. A learning classifier scheme applied to the single population GA showed some improvement, but appears very costly at present. In contrast, the use of competing
agents which evolve appears quite promising for further investigation. Applica-

tions of the DGA to low speed airfoil optimisation, demonstrate that the method is straightforward to implement, and can be easily applied to different problems.

References

1. Goldberg D E, Genetic Algorithms in Search, Optimisation and Machine Learning, Addison-Wesley, 1988.
2. Bäck T, Evolutionary Algorithms in Theory and practice, Oxford, 1996.
3. Schwefel H P, Evolution and Optimum Seeking, Wilrey New York, 1995.
4. Davis L, Handbook of Genetic Algorithms, Van Nostrand Reinhold, New York 1991.
5. Obayashi, S., Yamaguchi Y., and Nakamura, T. Multiobjective genetic algorithm for multidisciplinary design of transonic wing planfoem, J. Aircraft 34, 5, pp 690–693, 1997
6. Doorly D J, *Ch. 13* of Genetic Algorithms in Engineering and Computer Science, ed. G. Winter et al., Wiley, 1995.
7. Quagliarella D and DellaCioppa A, Genetic Algorithms Applied to the Aerodynamic Design of Transonic Airfoils, J. Aircraft **32**, 889–891, 1995.
8. Poloni C, *Ch. 20* of Genetic Algorithms in Eng. and Comp. Sci., ed. G. Winter et al., Wiley, 1995.
9. Yamamoto K, and Inoue O, Applications of Genetic Algorithms to Aerodynamic Shape Optimisation, AIAA-95-1650-CP, 1995
10. Tanese R, Distributed Genetic Algorithms, PhD thesis, U. Michigan, 1989.
11. Doorly D J, Peiró J, Kuan T, and Oesterle J-P, Optimisation of Airfoils Using Parallel Genetic Algorithms, in Proc. 15th Int. Conf. Num. Meth. Fluid Dyn., Monterey, 1996.
12. Nang J and Matsuo K, A Survey of Parallel Genetic Algorithms, J. SICE **33**, 6, 500–509, 1994.
13. Doorly D J, Peiró J, and Oesterle J-P, Optimisation of Aerodynamic and Coupled Aerodynamic-Structural Design using Parallel Genetic Algorithms, in Proc. Sixth AIAA/NASA/ISSMO Symposium on Multidisciplinary Analysis and Optimization, 401–409, 1996.
14. Holland J H, Adaptation in Natural and Artificial Systems, MIT Press, 1992.
15. Doorly D J and Peiró , Supervised parallel genetic algorithms in Aerodynamic Optimisation, AIAA paper 97-1852, 1997.
16. Oesterle J-P, Aeronautical optimisation using parallel genetic algorithms, MSc thesis, Aeronautics Dept.,Imperial College London, 1996.

An Evolutionary Algorithm for Large Scale Set Covering Problems with Application to Airline Crew Scheduling

Elena Marchiori[1] and Adri Steenbeek[2]

[1] Free University
Faculty of Sciences, Department of Mathematics and Computer Science
De Boelelaan 1081a, 1081 HV Amsterdam, The Netherlands
elena@cs.vu.nl
[2] CWI,
P.O. Box 94079, 1090 GB Amsterdam, The Netherlands
adri@cwi.nl

Abstract. The set covering problem is a paradigmatic NP-hard combinatorial optimization problem which is used as model in relevant applications, in particular crew scheduling in airline and mass-transit companies. This paper is concerned with the approximated solution of large scale set covering problems arising from crew scheduling in airline companies. We propose an adaptive heuristic-based evolutionary algorithm whose main ingredient is a mechanism for selecting a small core subproblem which is dynamically updated during the execution. This mechanism allows the algorithm to find covers of good quality in rather short time. Experiments conducted on real-world benchmark instances from crew scheduling in airline companies yield results which are competitive with those obtained by other commercial/academic systems, indicating the effectiveness of our approach for dealing with large scale set covering problems.

1 Introduction

The set covering problem (SCP) is one of the oldest and most studied NP-hard problems (cf. [14]).

Given a m-row, n-column, zero-one matrix (a_{ij}), and an n-dimensional integer vector (w_j), the problem consists of finding a subset of columns covering all the rows and having minimum total weight. A row i is covered by a column j if the entry a_{ij} is equal to 1. This problem can be formulated as a constrained optimization problem as follows:

$$\text{minimize} \quad \sum_{j=1}^{n} w_j x_j$$

$$\text{subject to the constraints} \quad \begin{cases} x_j \in \{0,1\} & j = 1, \ldots, n, \\ \\ \sum_{j=1}^{n} a_{ij} x_j \geq 1 & i = 1, \ldots, m. \end{cases}$$

S. Cagnoni et al. (Eds.): EvoWorkshops 2000, LNCS 1803, pp. 367–381, 2000.
© Springer-Verlag Berlin Heidelberg 2000

The variable x_j indicates whether column j belongs to the solution ($x_j = 1$) or not ($x_j = 0$). The m constraint inequalities are used to express the requirement that each row be covered by at least one column. The weight w_j is a positive integer that specifies the cost of column j. When all w_j's are equal to 1, then the SCP is called unicost SCP.

Relevant practical applications of the SCP include crew scheduling [1, 2, 12, 15]: find a set of pairings having minimum-cost which covers a given set of trips, where a pairing is a sequence of trips that can be performed by a single crew. A widely used approach to crew scheduling works as follows. First, a very large number of pairings is generated. Next, a SCP is solved, having as rows the trips to be covered, and as columns the pairings generated. When this approach is used in mass-transit applications, very large scale SCP instances may arise, involving thousands of rows and millions of columns.

The most successful heuristic algorithms for large scale SCP's are based on Lagrangian relaxation [13]. Lagrangian relaxation is used to compute the score of a column according to its likelihood to be selected in an optimal solution. These scores are employed in simple greedy heuristics for computing a solution. A very effective heuristic algorithm for large scale SCPs based on this approach is [7]. We refer the reader to [8] for a recent survey on exact and heuristic algorithms for SCP. All effective heuristics for large scale SCP's act on a subset of the columns, called core, which is selected before the execution of the algorithm. In the static approach the core remains the same during the execution (cf. [6, 11]), while in the dynamic approach it is updated using an adaptive mechanism (e.g. [7, 10, 9]).

In this paper we propose a novel heuristic algorithm for large scale SCPs arising from crew sheduling problems in airline companies. At each iteration a near optimal cover is constructed using the information provided by the previous iterations to guide the search. The final solution is the best cover found in all the iterations.

Given a problem instance, the algorithm extracts an initial core from the set of columns given in the input. Then the algorithm consits of the iterated application of the following three steps: 1) First, an approximated solution to the actual SCP core is constructed by means of a novel greedy heuristic. 2) Next, a local search optimization algorithm is applied to the resulting solution. 3) Finally, some columns that occur in the best solution found in all iterations up to now are selected for forming the initial partial solution for the next iteration.

The size of the core is determined by an adaptive size parameter, while the selection of a column is specified by a suitable merit criterion. During the execution, the score of the columns is modified as well as the size parameter, and the core is dynamically updated.

This algorithm can be viewed as a hybrid $(1 + 1)$ steady-state evolutionary algorithm, where at each iteration a child is generated from the parent using the above described heuristic, and the best between the parent and the child survives.

In order to assess the performance of the algorithm, we conduct extensive experiments on real-world problem instances arising from crew scheduling in airlines, as well as on other benchmark instances from the literature. The results of the experiments are rather satisfactory: our algorithm is able to find covers of very good quality in a short amount of time, yielding results which are competitive with those reported by the best industrial as well as academic methods for solving large set covering problems.

The rest of the paper is organized as follows. In the next subsections we briefly discuss some related work, and set up the notation and terminology used throughout the paper. In Section 2 we introduce the overall method and present in detail the four main modules of the algorithm. In Section 3 we report the results of extensive computational experiments. We conclude with some final remarks on the present investigation and on future work.

1.1   Related Work

An experimental comparison of the most effective exact and heuristic algorithms for the weighted SCP is given in a recent paper by Caprara et al [8].

A rather effective heuristic algorithm based on Lagrangian relaxation is the CFT algorithm [7] by Caprara et al. This algorithm has been tested also on large scale problem instances arising from crew scheduling in railway, yielding rather satisfactory results. In [15] a approximation algorithm for solving large 0-1 integer programming problems is proposed. This algorithm is used in the CARMEN system for airline crew scheduling, a industrial system used by several major airlines.

Research based on evolutionary computation includes the following two papers.

Beasley and Chu in [6] introduce a genetic algorithm for the SCP. The authors employ a representation where a chromosome is a bit string of lenght equal to the number of columns, one bit for each column, representing the set of columns whose bit in the string are equal to 1. The algorithm employs a heuristic repair mechanism for transforming infeasible chromosomes into solutions. Moreover, a core is used for constructing the chromosomes of the initial population.

A genetic algorithm based on a non-binary representation has been proposed by Eremeev in [11]. Here a chromosome is a string of lenght equal to the number of rows, where the i-th entry contains (the index of) a column covering the i-th row. As a consequence, all chromosomes are feasible solutions, thus they do not need to be repaired as in [6]. Moreover, heuristics are used for eliminating redundant columns as well as for defining the crossover operator.

In Section 3 we will compare experimentally the above mentioned algorithms with the algorithm introduced in this paper.

1.2   Notation and Terminology

In order to describe our method, we use the following terminology and notation.

In the sequel, the indexes i, j denote a generic row and column, respectively. A column will also be denoted by c, and a row by r, possibly subscripted. Moreover, S denotes a set of columns.

Let $cov(S)$ be the set of rows that are covered by the columns in S:

$$cov(S) = \{i \mid a_{ij} = 1 \text{ for some } j \in S \}.$$

For simplicity, we write $cov(j)$ instead of $cov(\{j\})$. We say that a column j is *redundant* with respect to S if $cov(S \setminus \{j\}) = cov(S)$.

A *partial cover* (also called partial solution) is a set of columns containing no redundant column.

Let $cov(j, S)$ be the set of rows which are covered by column j, but are not covered by any column in $S \setminus \{j\}$:

$$cov(j, S) = \{i \mid a_{ij} = 1 \text{ and } a_{ij'} = 0 \text{ for all } j' \in S \setminus \{j\} \}.$$

Moreover, let $min\_weight(i)$ be the minimum weight of the columns that cover i:

$$min\_weight(i) = minimum \ \{w_j \mid i \in cov(j) \}.$$

We can now define the function cov_val, called *cover value*, which is used to evaluate a column j with respect to a partial cover S in order to select a column to be added (resp. removed) to (resp. from) S:

$$cov\_val(j, S) = \sum_{i \in cov(j,S)} min\_weight(i).$$

A convenient property of cov_val is that $cov\_val(j, S) = cov\_val(j, S \setminus \{j\})$. This allows one to compute the cover value of a column without taking into account whether it belongs to the partial solution S or not. Moreover, we can characterize the redundancy of a column by means of the condition $cov\_val(j, S) = 0$.

The cover value is used to define the *selection value* $sel\_val(j, S)$ of a column j with respect to the partial cover S:

$$sel\_val(j, S) = \begin{cases} Lim & \text{if } j \text{ redundant wrt } S, \\ w_j / cov\_val(j, S) & \text{otherwise.} \end{cases}$$

The selection value of redundant columns is set to a very big constant Lim. In this way, redundant columns do not have any chance of being selected.

2   The Overall Method

The algorithm we propose consists of an iterated procedure, where each iteration generates an approximated solution using only columns from the actual core.

Roughly, at each iteration a greedy heuristic is used to construct incrementally a cover starting from a partial cover: in the first iteration the partial cover is empty, while in the following iterations the partial cover is a proper subset of the best cover found in all iterations up to now. The cover found after the application of the greedy heuristic is given as input to an optimization procedure which tries to improve the partial solution. The core is updated from time to time during the execution. The final result is the best cover found in all iterations.

The corresponding algorithm WSCP (Weighted Set Covering Problem) is illustrated below in pseudo-code, where Sbest represents the best cover found so far, S denotes the actual partial cover, and value(S) is the sum of the weights of the columns of S, that is, $\sum_{j \in S} w_j$. Therefore the optimal cover is the cover S having minimum value(S).

```
FUNCTION WSCP()
BEGIN
    RECOMPUTE_CORE();
    Sbest <- { 1..ncol };
    S <- { };
    FOR 1 .. param.number_of_iterations DO
        IF ( core_selection() ) RECOMPUTE_CORE(); ENDIF;
        S <- GREEDY(S);
        S <- OPTIMIZE(S);
        IF ( value(S) <= value(Sbest) ) THEN Sbest <- S; ENDIF;
        S <- SELECT_PARTIAL_COVER(Sbest);
    ENDFOR
    RETURN Sbest
END
```

2.1   Greedy Heuristic

Our greedy heuristic GREEDY is described in pseudo-code below. Lines starting with "//" are comments. The algorithm constructs a solution (a cover), starting from a (possibly empty) partial cover S. Columns are added (resp. removed) to (resp. from) S until S covers all the rows.

```
// extend S until it is a  cover:
FUNCTION GREEDY( var S )
BEGIN
    WHILE ( S is not a cover ) DO
        // select and add one column to S
        S <- S + select_add();
        // remove 0 or more columns from S
        WHILE ( remove_is_okay() ) DO
            S <- S - select_rmv();
        ENDWHILE
    ENDWHILE
    // S is a cover, without redundant columns
    return S;
END
```

The function *select_add* selects a column j not in S having minimum selection value *sel_val* (j, S).

The test *remove_is_okay* determines whether columns should be removed from S. If S is empty it returns *false*; if S contains at least one redundant column then it returns *true*; otherwise, with probability *param.p_rmv* (typical value 0.3) it returns *true*, otherwise *false*.

Finally, the function *select_rmv* selects a column in S having maximum selection value.

2.2 Local Optimization

The local optimization procedure OPTIMIZE is based on the following idea. Given a cover S, suppose there is a column $j \notin S$ such that $S \cup \{j\}$ contains at least two columns other than j, say $j_1, \ldots, j_l$, with $l \geq 2$ that are redundant, and such that the sum of their weights is greater than the weight of j, that is, $\sum_{k=1,\ldots,l} w_{j_k} > w_j$. Then $(S \setminus \{j_1, \ldots, j_l\}) \cup \{j\}$ is a better cover than S. In this case we call j a *superior column*. The gain of j is defined by

$$gain(j) = \sum_{k=1,\ldots,l} w_{j_k} - w_j.$$

So a best superior column is the one having highest gain. Note that the optimization procedure operates on a cover containing no redundant columns. The optimization algorithm OPTIMIZE in pseudo-code is given below.

```
// S is a cover, without redundant columns
FUNCTION OPTIMIZE( var S )
BEGIN
    Sup <- select_superior();
    WHILE ( Sup not empty) DO
        // select best column from Sup
        best <- select_best();
        Sup <- Sup - best;
        // add superior and remove redundant columns from S
        IF ( best superior )
            S <- S + best;
            S <- S - select_redundant();
        ENDIF
    ENDWHILE
    // S is a cover, without redundant columns
    return S;
END
```

First, the function *select_superior*() is used, which generates the list *Sup* consisting of all the superior columns ordered in decreasing order according to their gain. Next, the list *Sup* is scanned in the WHILE loop. At each iteration, the head of *Sup* is removed and memorized in the variable *best* using the function *select_best*(). If the selected column is still superior (that is if the test `best superior` is satisfied) then it is added to S, and the set of redundant columns are removed from the resulting partial cover S using the function *select_redundant*().

2.3 Restoring Part of the Actual Best Solution

In the first iteration of WSCP the heuristic GREEDY constructs a cover starting from the empty set; in the following iterations, GREEDY builds a cover starting from a subset of the best cover found so far. For a column j, we keep track of the number $chosen(j)$ of times that j has been part of a best solution. The function SELECT_PARTIAL_COVER considers the set E of so-called *elite columns*, consisting of those columns j of the best solution Sbest such that $cov\_val(j, Sbest) > w_j$. Then SELECT_PARTIAL_COVER selects from E the set of columns having low $chosen(j)$ (in our implementation $chosen(j)$ has to be smaller than $\sum_j chosen(j)/(neli * 10)$, where $neli$ is the number of elements of E) while the remaining columns of E are selected with a probability that is set to a random value between 0.1 and 0.9.

2.4 Selecting the SCP Core

This is a fundamental step in the design of an algorithm for dealing with large SCP instances. We introduce the following method for constructing an SCP core, which has been implemented in the function RECOMPUTE_CORE.

The SCP cover is constructed from the empty set by incrementally adding columns according to the following criterion. Columns are selected in increasing order according to their selection value. Suppose column j has been selected:

1. if j is an elite column then with probability close to 1 it is added to the actual SCP core;
2. otherwise, j is added if there exists a row i such that j covers i and $w_j < min\_weight(i) \cdot K_0$, with K_0 a given constant real value greater or equal than 1;
3. otherwise, j is added if there exists a row i such that j covers i and i is covered by less than K_1 columns of the actual SCP core, with K_1 a given constant integer value greater or equal than 1.

Note that K_0, K_1 are parameters which are chosen depending on the class of problems one considers. Condition 3 implies that the SCP core contains for each row, at least the first K_1 best columns (according to the ordering induced by the selection value function) that cover that row.

The function *core_selection*() determines when the actual SCP core has to be recomputed. In our implementation, we recompute the SCP core every 100 iterations of WSCP.

During the execution of GREEDY, when ninety per cent of a cover has been constructed, the *min_weight* of those rows that are not yet covered is increased by a small quantity (in our implementation $min\_weight(i)$ is multiplied by 1.1). This affects the selection value of the columns, hence their order of selection in the construction of the SCP core changes during the execution of the overall algorithm WSCP.

3   Experimental Evaluation

The algorithm WSCP has been tested on large set covering problems arising from crew scheduling applications in various airline companies. Moreover, we have considered the weighted SCP instances from the OR library maintained by J.E. Beasley [1]. These instances are considered standard benchmarks for testing the effectiveness of exact and heuristic algorithms for the SCP. In particular, they have been used in [8] for comparing experimentally various exact and heuristic algorithms for SCP.

WSCP has been implemented in C++. The algorithm was run on a Sun Ultra 10 (UltraSPARC-IIi 300MHz).

The results of the experiments are based on 10 runs on each problem instance of the OR Library, and on 5 runs on the other instances. In each table, the entry labeled **Id** contains the name of the problem instance. The label **BK** denotes the best known solution for that instance; **Bst** denotes the best result found by the algorithm; **Fbst** indicates the frequency of obtaining the best solution in the performed runs; **Apd** denotes the average percentage deviation $\sum_{k=1}^{10}(z_k - z^*)/(10 \cdot z^*) \cdot 100\%$, where z_k is the solution found in the k-th run, and z^* is the optimal or best known solution. **Tbst** denotes the average cpu time for obtaining the best solution Bst, while **Tsol** denotes the average cpu time for finding a solution. Finally, **Ibst** and **Isol** denote the average number of iterations of obtaining the best solution Bst, and a solution, respectively.

3.1   Experiments on Airline Crew Scheduling Problem Instances

We consider three sets of benchmark instances from real-world airline crew scheduling problems. A set of instances from a major airline company, here called AIR instances, the airline scheduling instances from Wedelin [15], and the instances from Balas and Carrera [3]. The characteristics of these problems are reported in Tables 2, 3, and 7, respectively. Observe that in many instances, like, e.g., the Wedelin instances, the weights of the columns are very large numbers, because the weight represents the cost of a pairing and takes into account several factors.

We compare experimentally WSCP with the industrial system used by an airline company on the AIR instances, with the CFT algorithm by Caprara et al [7], and with the Wedelin algorithm [15]. The results of the experiments are given in Tables 2, 4, 5, and 7. Note that the results for the Wedelin and CFT algorithms are taken from the paper [7], where the cpu time is estimated in DECstation

---

[1] see http://mscmga.ms.ic.ac.uk/jeb/orlib/scpinfo.html

5000/240 CPU seconds. Only the value of the best solution is reported. For the CFT algorithm, the time for finding the best solution is given, while for the Wedelin algorithm, only the overall execution time **Texe** of the algorithm is reported. The authors do not specify the setting of the various parameters in their algorithms, and the total number of trials performed.

Id	Rows	Columns	Density (%)	Weight Range
A01	5265	258303	0.167	1319-35302
A02	3878	19441	0.135	1437-37206
A03	4965	40580	0.092	1337-37148
A04	4916	79481	0.123	1460-37142
A05	4656	72377	0.126	1411-37251
A06	1971	23741	0.135	1437-37037
A07	4203	32363	0.15	1319-36370
A08	4320	45286	0.18	1345-36370
A09	4287	50047	0.19	1361-36370
A10	4369	49525	0.18	1344-36370
A11	150	389388	5.55	1800-18768
A12	682	642613	1.45	1630-19000

Table 1. Characteristics of AIR instances

Id	Industry	WSCP						
	Bst	Bst	Fbst	Apd	Tbst	Tsol	Ibst	Isol
A01	16351667	16351667	1.0	0.0	550.9	550.9	919.8	919.8
A02	12879297	12879297	1.0	0.0	131.0	131.0	822.4	822.4
A03	15663720	**15663688**	1.0	0.0	254.2	254.2	1004.4	1004.4
A04	16110608	16110608	1.0	0.0	363.1	363.1	1085.0	1085.0
A05	16315241	**16315070**	0.3	0.0001	923.5	848.1	3501.6	3203.2
A06	13162511	13162511	1.0	0.0	156.9	156.9	907.6	907.6
A07	13301520	13301520	1.0	0.0	200.9	200.9	945.6	945.6
A08	13510606	**13510584**	1.0	0.0	254.4	254.4	946.2	946.2
A09	13489489	13489489	1.0	0.0	235.4	235.4	944.2	944.2
A10	13571530	13571530	1.0	0.0	237.3	237.3	933.8	933.8
A11	247775	247775	1.0	0.0	224.2	224.2	1087.8	1087.8
A12	732587	732587	0.3	0.11	1064.9	896.8	1460.6	1167.4

Table 2. Results for AIR instances

On three AIR instances WSCP found a solution which is better than the best solution found by the industrial system, while on the other instances WSCP found solutions of equal value as those found by the industrial system.

Id	Rows	Columns	Density (%)	Weight Range
B727scratch	29	157	8.2	1600-11850
ALITALIA	118	1165	3.1	2200-2110900
A320	199	6931	2.3	1600-2111450
A320coc	235	18753	1.9	1900-1812000
SASjump	742	10.370	0.6	4720-55849
SASD9imp2	1366	25032	0.3	3860-35200

Table 3. Characteristics of Wedelin instances

Id	BK	CFT		Wedelin	
		Bst	Tbst	Bst	Texe
B727scratch	94400	94.400	0.3	94400	4.7
ALITALIA	27258300	27258300	6.2	27258300	37.2
A320	1262100	1262100	79.5	1262100	216.9
A320coc	14495500	14495600	577.8	14495500	1023.7
SASjump	7338844	7339537	396.3	7340777	806.8
SASD9imp2	5262190	5263640	2082.1	5262190	1579.7

Table 4. Results of CFT and Wedelin on Wedelin instances

Id	BK	WSCP						
		Bst	Fbst	Apd	Tbst	Tsol	Ibst	Isol
B727scratch	94400	94400	1.0	0.0	0.018	0.018	38.4	38.4
ALITALIA	27258300	27258300	1.0	0.0	0.63	0.63	106.8	106.8
A320	1262100	1262100	1.0	0.0	17.34	17.34	326.2	326.2
A320coc	14495500	14495500	0.2	0.0006	651.08	446.20	3494.5	2402.0
SASjump	7338844	7339541	0.1	0.02	269.3	200.98	4635.0	3454.6
SASD9imp2	5262190	5263590	0.1	0.04	741.9	608.452	4603.0	3671.4

Table 5. Results of WSCP on Wedelin instances

Id	Rows	Columns	Density (%)	Weight Range
AA03	106	8661	4.05	91-3619
AA04	106	8002	4.05	91-3619
AA05	105	7435	4.05	91-3619
AA06	105	6951	4.11	91-3619
AA11	271	4413	2.53	35-2966
AA12	272	4208	2.52	35-2966
AA13	265	4025	2.60	35-2966
AA14	266	3868	2.50	35-2966
AA15	267	3701	2.58	35-2966
AA16	265	3558	2.63	35-2966
AA17	264	3425	2.61	35-2966
AA18	271	3314	2.55	35-2966
AA19	263	3202	2.63	35-2966
AA20	269	3095	2.58	35-2966
BUS1	454	2241	1.89	120-877
BUS2	681	9524	0.51	120-576

Table 6. Characteristics of Balas and Carrera instances

	CFT		WSCP						
Id	Bst	Tbst	Bst	Fbst	Apd	Tbst	Tsol	Ibst	Isol
AA03	33155	61.0	33155	1.0	0.0	1.26	1.266	40.0	40.0
AA04	34573	3.6	34573	1.0	0.0	1.73	1.73	74.6	74.6
AA05	31623	3.1	31623	1.0	0.0	0.48	0.48	9.6	9.6
AA06	37464	5.2	37464	1.0	0.0	2.67	2.67	128.2	128.2
AA11	35384	193.7	35384	1.0	0.0	19.11	19.11	755.4	755.4
AA12	30809	53.8	30809	1.0	0.0	7.88	7.88	350.8	350.8
AA13	33211	8.3	33211	1.0	0.0	2.32	2.32	103.8	103.8
AA14	33219	30.3	33219	1.0	0.0	11.74	11.74	557.8	557.8
AA15	34409	18.8	34409	1.0	0.0	8.92	8.92	485.6	485.6
AA16	32752	33.6	32752	1.0	0.0	4.63	4.63	257.4	257.4
AA17	31612	10.9	31612	1.0	0.0	4.69	4.69	262.2	262.2
AA18	36782	13.5	36782	0.1	0.01	17.1	6.94	1108.0	433.0
AA19	32317	5.9	32317	1.0	0.0	2.73	2.73	175.4	175.4
AA20	34912	13.6	34912	1.0	0.0	4.76	4.76	318.4	318.4
BUS1	27947	5.0	27947	1.0	0.0	8.19	8.19	382.6	382.6
BUS2	67760	19.2	67760	1.0	0.0	37.24	37.24	616.2	616.2

Table 7. Results of CFT and WSCP on Balas and Carrera instances

On the instances from Wedelin the performance of WSCP is comparable to the one of the CFT and Wedelin algorithms.

Finally, on the instances from Balas and Carrera, both WSCP and CFT are always able to find the optimal solution. In the AA instances WSCP is faster that CFT, while in the BUS instances CFT finds the optimum in a shorter time.

The results of the experiments indicate that WSCP is a rather powerful tool for solving large real-life airline crew scheduling problems.

3.2   Experiments on the OR Library SCP Instances

We consider the families A-D from [4], and the NRE-NRH from [5], consisting of randomly generated SCP instances. Each class contains 5 instances. The values of the characteristic parameters of these problem classes, like number of rows and columns, are given in Table 8.

We compare experimentally WSCP with the genetic algorithms by Beasley and Chu [6], and by Eremeev [11], and with the CFT algorithm by Caprara et al [7]. The results of the experiments are summarized in Tables 9, 10, and 11.

The results for the CFT, Beasley Chu, and Eremeev algorithms are from [11]. In particular, the cpu time is estimated in 100MHz Pentium CPU seconds.

All the algorithms are able to solve the instances of the classes A-D. On these instances, WSCP seems to have a more robust behaviour that the two genetic algorithms, finding the optimum in each of the 10 trials. The performance of WSCP on the other problem instances of classes E-H is rather satisfactory, both in terms of quality of the solutions as well as running time. On each instance, WSCP is able to find the optimum or best known solution, while the two genetic algorithms BC and Er do not find the optimum value on instances H1 and H2. Moreover, WSCP finds the solutions for instances in the harder classes G and H in a much shorter time than all the other algorithms.

Id	A	B	C	D	E	F	G	H
Rows	300	300	400	400	500	500	1000	1000
Columns	3000	3000	4000	4000	5000	5000	10000	10000
Density (%)	2	5	2	5	10	20	2	5
Weight Range	1-100	1-100	1-100	1-100	1-100	1-100	1-100	1-100

Table 8. Characteristics of Classes A, B, C, D

4   Conclusion

In this paper we have introduced a novel heuristic method for solving large weighted set covering problems. The results of the experiments indicate that WSCP is able to find covers of satisfactory quality in short running time.

Id	CFT Tbst	Beasley Chu			Eremeev				WSCP					
		Fbst	Apd	Tsol	Fbst	Apd	Tbst	Tsol	Fbst	Apd	Tbst	Tsol	Ibst	Isol
A	47.15	0.86	0.20	65.98	0.44	0.35	82.00	71.8	1.00	0.00	0.98	0.98	108.0	108.0
B	3.34	1.00	0.00	68.63	1.00	0.00	20.80	20.80	1.00	0.00	0.30	0.30	7.8	7.8
C	29.23	0.68	0.41	87.93	0.74	0.26	53.50	52.40	1.00	0.00	0.76	0.76	72.6	72.6
D	7.64	0.96	0.06	101.70	0.94	0.08	26.62	23.33	1.00	0.00	0.40	0.40	26.0	26.0

Table 9. Results for Classes A, B, C, D

Id	BK	CFT		Beasley Chu				Eremeev				
		Bst	Tbst	Bst	Fbst	Apd	Tsol	Bst	Fbst	Apd	Tbst	Tsol
E1	29	29	11.5	29	1.0	0.0	16.9	29	1.0	0.0	1.0	1.0
E2	30	30	180.5	30	0.4	2.0	266.9	30	1.0	0.0	94.8	94.8
E3	27	27	41.7	27	0.3	2.6	85.1	27	1.0	0.0	23.1	23.1
E4	28	28	11.6	28	1.0	0.0	238.5	28	1.0	0.0	11.0	11.0
E5	28	28	16.2	28	1.0	0.0	15.5	28	1.0	0.0	2.1	2.1
F1	14	14	14.7	14	1.0	0.0	33.8	14	1.0	0.0	7.9	7.9
F2	15	15	13.8	15	1.0	0.0	34.5	15	1.0	0.0	1.3	1.3
F3	14	14	110.0	14	1.0	0.0	117.9	14	1.0	0.0	55.4	55.4
F4	14	14	13.7	14	1.0	0.0	92.6	14	1.0	0.0	20.4	20.4
F5	13	13	89.0	13	0.3	5.4	67.1	13	0.3	5.4	497.4	151.2
G1	176	176	65.0	176	0.2	1.0	451.3	176	0.7	0.3	115.0	96.0
G2	154	154	346.6	155	0.5	1.5	159.3	154	0.5	0.65	318.3	226.6
G3	166	166	432.7	166	0.1	1.1	312.1	166	0.1	0.8	627.6	319.1
G4	168	168	105.0	168	0.2	1.4	665.4	168	0.4	0.7	160.0	172.5
G5	168	168	105.0	168	0.2	0.8	242.6	168	0.7	0.05	161.2	170.4
H1	63	63	642.1	64	1.0	1.6	743.0	64	1.0	1.6	90.5	90.5
H2	63	63	392.5	64	1.0	1.6	234.3	64	1.0	1.6	34.7	34.7
H3	59	59	690.4	59	0.9	0.2	796.6	59	1.0	0.0	493.2	493.2
H4	58	58	105.1	58	0.4	91.6	62.9	58	1.0	0.0	218.2	218.2
H5	55	55	68.8	55	0.9	0.2	198.6	55	1.0	0.0	25.2	25.2

Table 10. Results of CFT, Beasley and Chu, and Eremeev on Classes E, F, G, H

Id	BK	WSCP						
		Bst	Fbst	Apd	Tbst	Tsol	Ibst	Isol
E1	29	29	1.0	0.0	1.8	1.8	2	2
E2	30	30	1.0	0.0	2.7	2.7	62.9	62.9
E3	27	27	1.0	0.0	2.3	2.3	48.3	48.3
E4	28	28	1.0	0.0	2.1	2.1	31.1	31.1
E5	28	28	1.0	0.0	1.8	1.8	5.0	5.0
F1	14	14	1.0	0.0	3.6	3.6	19.6	19.6
F2	15	15	1.0	0.0	3.6	3.6	9.0	9.0
F3	14	14	1.0	0.0	6.2	6.2	153.2	153.2
F4	14	14	1.0	0.0	3.6	3.6	13.1	13.1
F5	13	13	1.0	0.0	34.1	34.1	2061.1	2061.5
G1	176	176	1.0	0.0	2.2	2.2	29.7	29.7
G2	154	154	0.5	1.1	8.9	4.0	315	107.9
G3	166	166	0.2	0.6	30.2	14.9	1433.5	640.3
G4	168	168	0.4	0.8	18.4	25.0	812.3	1114.7
G5	168	168	0.9	0.6	5.9	5.7	207.7	197.4
H1	63	63	0.2	1.1	9.5	11.7	161.5	269.2
H2	63	63	1.0	0.0	50.4	50.4	1872	1872
H3	59	59	0.3	1.1	25.2	21.6	778.3	678.9
H4	58	58	0.8	0.3	28.1	23.9	1016.3	834.1
H5	55	55	1.0	0.0	5.5	5.5	64.1	64.1

Table 11. Results of WSCP on Classes E, F, G, H

In all the experiments we have worked with a core which is a proper subset of the set of all columns. The size of the core depends on the problem instance. However, in general a small fraction (which varies from 10 per cent to 50 per cent) of the set of columns is used as core. Using small covers helps the efficiency of the algorithm. Moreover, extensive experiments with different core sizes have revealed a somehow counter intuitive phenomenon: in many instances, the quality of the results become worse by using a larger core, even if the same number of iterations is used. This seems to indicate that the merit criterion used in WSCP is not the best possible, because it can make the wrong decision when all the columns are present in the core. We are actually investigating the use of alternative merit criteria and their relationship with the selection of the core.

Future work concerns the investigation of how to tune automatically the parameters K_0, K_1 for determining the core problem, and how the value of these can be adaptively change during the execution.

Acknowledgements

We would like to thank Thomas Baeck and Martin Schuetz for interesting discussions on the subject of this paper.

References

1. E. Andersson, E. Housos, Kohl, and D. Wedelin. Crew pairing optimization. In *Operation Research in the Airline Industry*. Kluwer Scientific Publishers, 1997.
2. J.P. Arabeyre, J. Fearnley, F.C. Steiger, and W. Teather. The airline crew scheduling problem: A survey. *Transportation Science*, (3):140–163, 1969.
3. E. Balas and M.C. Carrera. A dynamic subgradient-based branch-and-bound procedure for set covering problem. *Operations Research*, 44:875–890, 1996.
4. J.E. Beasley. An algorithm for set covering problem. *European Journal of Operational Research*, 31:85–93, 1987.
5. J.E. Beasley. A lagrangian heuristic for set covering problems. *Naval Research Logistics*, 37:151–164, 1990.
6. J.E. Beasley and P.C. Chu. A genetic algorithm for the set covering problem. *European Journal of Operational Research*, 94:392–404, 1996.
7. A. Caprara, M. Fischetti, and P. Toth. A heuristic method for the set covering problem. In W.H. Cunningham, T.S. McCormick, and M. Queyranne, editors, *Proc. of the Fifth IPCO Integer Programming and Combinatorial Optimization Conference*. Springer-Verlag, 1996.
8. A. Caprara, M. Fischetti, and P. Toth. Algorithms for the set covering problem. Technical report, DEIS Operation Research Technical Report, Italy, 03 1998.
9. S. Ceria, P. Nobili, and A. Sassano. A Lagrangian-based heuristic for large-scale set covering problems. *Mathematical Programming*, 1995. to appear.
10. H.D. Chu, E. Gelman, and E.L. Johson. Solving large scale crew scheduling problems. *European Journal of Operational Research*, 97:260–268, 1997.
11. A.V. Eremeev. A genetic algorithm with a non-binary represenation for the set covering problem. In *Proc. of OR'98*, pages 175–181. Springer-Verlag, 1998.
12. M.M. Etschmaier and D.F. Mathaisel. Airline scheduling: An overview. *Transportation Science*, (19):127–138, 1985.
13. M.L. Fisher. An application oriented guide to Lagrangian relaxation. *Interfaces*, 15(2):10–21, 1985.
14. M.R. Garey and D.S. Johnson. *Computers and Intractability: A Guide to the Theory of NP-completeness*. Freeman, San Francisco, 1979.
15. D. Wedelin. An algorithm for large scale 0-1 integer programming with application to airline crew scheduling. *Annals of Operational Research*, 57:283–301, 1995.

Design, Implementation, and Application of a Tool for Optimal Aircraft Positioning

J. Pfalzgraf [1], K. Frank [1], J. Weichenberger [1], S. Stolzenberg [2]

[1]Department of Computer Science,University of Salzburg (jpfalz@cosy.sbg.ac.at),
[2]Deutsche Lufthansa AG, Frankfurt/Main (Siegfried.Stolzenberg@dlh.de)

Abstract. Optimal positioning of aircraft at a specific airport is a very difficult problem involving the modeling of many constraints. Lufthansa AG formulated this problem field for the airport Frankfurt/Main. In this contribution we describe the development of a tool for finding solutions to positioning problems automatically. Our approach consists of two parts. A generic airport model is developed where the notion of logical fiberings plays a basic role. The optimization task is treated by application of modified and extended genetic algorithms. A system has been implemented which is capable of computing concrete positioning plans that can be used by a human operator for further processing. This leads to a considerable speed up in the generation of positioning plans for aircraft in comparison with the former method. The application of the aircraft positioning tool to a real world scenario (airport Frankfurt) is briefly presented.

Keywords: optimal aircraft positioning, logical fiberings, genetic algorithms, hybrid problem solving

1 Introduction

This contribution deals with the general problem of optimal positioning of aircraft at an airport. The problem formulation has been provided by Lufthansa AG for the concrete case of airport Frankfurt/Main. The main task in the field of aircraft positioning is to find an optimal schedule for all incoming and outgoing aircraft with respect to their position at corresponding gates of an airport. One has to take into account many constraints and requests, such as neighborhood relationships, runway crossings, critical passenger connections, aircraft types, airline requests, special gates, security constraints, and others, depending on particular situations. The basic task is to fulfill all these constraints and requests in an optimal way so that the yield converges to a predefined maximum. Our work consists of two main parts. The development of a "generic airport model" not depending on the choice of a specific airport – and thus reusable – and the application of extended and modified genetic algorithms to work on the optimization problem. A rather complicated fitness function (cost function)

S. Cagnoni et al. (Eds.): EvoWorkshops 2000, LNCS 1803, pp. 382–393, 2000.

has been developed in cooperation with Lufthansa. In establishing the airport model, the concept of logical fiberings plays a basic role. This is a logical modeling approach which allows to describe what we call the logical state space of a virtual airport (distribution of logics). Thus, for example, a gate and a position has its own "local" logic, respectively, and they can communicate via non-classical logical operations called transjunctions. In principle, it would be possible to apply methods from multi-agent systems (MAS). In our case a gate or an airplane position could be interpreted as an agent, respectively. Application of MAS techniques is intended future work. In terms of logical fiberings an "agent" corresponds to a "local fiber". For the treatment of the optimization problem we develop and apply modified and extended genetic algorithms. A suitable fitness function has been devised which encodes the basic information and constraints underlying the positioning task. A prototype of an aircraft positioning tool has been implemented. A detailed description of the generic airport model, the evolutionary optimization approach, the system design and real world applications can be found in [FW99].

As far as we know no such approach to treat the problem field as described above has been tried before. This has been confirmed by somebody from the EvoNet project. And no such implemented system existed before, as Lufthansa states. There is a considerable increase of performance by our system in comparison with the procedure in use so far. Our PC-based system needs about one hour to calculate a positioning plan which can then be used by a human operator to process it further. In contrast to that, with the former method, the human planner needed about one week to establish a similar plan "by hand" supported by a graphical computing system, interactively.

2   On logical fiberings

The purpose of this section is twofold: to give a very brief introduction to the elementary notions of logical fiberings for later use and to bring them to the attention of the reader. The concept of logical fiberings originates in an industrial project – a case study on so-called polycontextural logics and their possible applications to complex communication and information systems. Subsequently we present only the elementary notion of a logical fibering and point to practical applications in the area of cooperating agents (in particular cooperating robots) scenarios. A detailed introduction to logical fiberings with background information and motivating comments can be found in [Pfa91].

The notion of a logical fibering is inspired by the mathematical modeling language of fiber bundles. This very expressive and powerful notion integrates different structures, namely geometric, topological, and algebraic structures. A fiber bundle consists of a whole bunch of fibers which form a so-called total space ditributed over a base manifold. Each fiber is mapped via a projection map onto its base point in the base space. Thus, for example, in a vector bundle the typical fiber is a vector space of a given dimension. So we can say that in a fiber bundle vertically one does algebra and horizontally geometry, topology.

Concerning a logical fibering, now the idea is to use a fiber bundle and take logics as fibers, thus, vertically one does logic. In [Pfa91] we take as typical fiber a classical two valued logic. Thus a logical fibering is an abstract fiber space (or bundle) with typical fiber $F = L$, a classical first order logical space or a (sub-)space generated by a set of formulas. The base space B of the fibering will often be denoted by $B = I$, the indexing set, the total space is denoted by E, and $\pi : E \longrightarrow I$ is the corresponding projection map. For $i \in I$ we have the fiber $\pi^{-1}(i)$ over i, namely $\pi^{-1}(i) = \{x \in E | \pi(x) = i\}$. It has the structure of a logic as mentioned above. We note that E is decomposed into the fibers $\pi^{-1}(i)$ for all $i \in I$.

The simplest form of a fibering is the "trivial fibering" having total space $E = I \times L$, base space I and $\pi(i,l) = i$. Therefore the fiber over $i \in I$ is $L_i := \pi^{-1}(i) = \{i\} \times L$. Such a trivial fibering is a "parallel system" of logics L_i over the index set I as base space. We can think of reasoning processes running in parallel within each fiber $L_i = \pi^{-1}(i)$. A fiber L_i is interpreted as a local logical system (a subsystem of the whole fibering). Transition (communication) betweeen fibers is described with the help of suitable maps (cf. [Pfa91]). Such a trivial fibering as previously considered will also be called "free parallel system" with total space the disjoint union of the fibers L_i. Each subsystem L_i has local classical truth values $\Omega_i = \{T_i, F_i\}$. The global set of truth values is denoted by Ω^I. In a free parallel system Ω^I is just the disjoint union of the Ω_i for $i \in I$.

Logical connectives can be introduced by taking "fiberwise" logical operations. For example, one can form logical expressions like the following for a system with three fibers (using vector notation): $(x_1 \wedge y_1, x_2 \Rightarrow y_2, x_3 \vee (\neg y_3))$, etc.. For more details we refer to [Pfa91].

A special nonclassical bivariate operation arises naturally: a local pair (x_i, y_i) in $L_i \times L_i$, $i \in I$, can be mapped into different subsystems $L_\alpha, L_\beta, ...$ Taking truth values as input we can observe that for the four possible input pairs in $\Omega_i \times \Omega_i$ for a locally defined bivariate operation there can be maximally four different subsystems where that function can be evaluated. We can say that the values will be distributed over the subsystems. Such an operation is called transjunction. Below we give an example of a "conjunctional" transjunction – we just display the truth table

$$
\begin{array}{c||c|c}
 & T_0 & F_0 \\
\hline\hline
T_0 & T_\alpha & F_\beta \\
F_0 & F_\gamma & F_\delta
\end{array}
$$

Every transjunction can be described by such a table or "T-F-pattern" together with the indices $\{\alpha, \beta, \gamma, \delta\}$. If $\{\alpha, \beta, \gamma, \delta\} \neq \{i\}$ then we obtain a transjunction – in the previous example we have the type of a conjunction as can be seen by omitting the indices. If $\{\alpha, \beta, \gamma, \delta\} = \{i\}$ then we have a classical conjunction remaining in subsystem L_i. In work cited below, we introduced generalized transjunctions having more than 4 input pairs. The example which we present in section 3 deals with such a generalized trunsjunction.

We applied transjunctions for the logical control of cooperating robots scenarios. Cf.[Pfa97] for a brief discussion of such an application. For further information on the subject we refer to [PSS96a], [PSS96b]. In the framework of this contribution here we apply methods from logical fiberings to support logically the modeling of the aircraft positioning system as described subseqently.

Generally spoken, we consider the concept of logical fiberings as a natural logical modeling approach for multi-agent systems (MAS). This has been discussed in [Mei99]. Future work is planned, especially with respect to an extended generic airport model using MAS techniques.

3   Development of an aircraft positioning system

3.1   Design concept of the positioning tool

First of all, we want to give a short problem description and explain what we mean by aircraft positioning. The main task of aircraft positioning is to find schedules for all incoming and outgoing aircraft at an airport. Difficult constraints have to be considered, like runway crossings, (optimal) passenger connections, aircraft types, special flights, particular gates, airline requests, security problems, and others. The main problem in the field of aircraft positioning is to find an optimal schedule for all incoming and outgoing aircraft with respect to their positions and corresponding gates. Three subtasks have to be distinguished: long-time, short-time and the actual day scheduling. To each subtask corresponds an individual knowledge about the aircraft which influences the positioning. The knowledge changes rapidly during scheduling. The external state of the airport can change rapidly too, caused by construction work, for example. Thus an important design objective is to build a tool which is able to react to rapidly changing situations. As previously mentioned, we decided to choose logical fiberings as a logical modeling approach. This decision was naturally motivated by our problem analysis. Figure 1 shows the global system design.

Our system includes various agencies and data areas, namely the kernel agency, the airport agency and the external data areas, like airport database, temporary flightplan. Another part of our system is the output unit and the output communication unit, which will be used to visualize an airport utilization.

The communication between the various parts of our model will be handled by a negotiation protocol. This allows us to handle the basic communication in our system for the short-time planning task. An augmented final version of the tool will be able to treat the two remaining cases too, namely long time planning and the actual day.

3.2   External data areas

The external data areas are specialized data storages for, e.g. airport description, flight characteristics, airport characteristics , airline characteristcs. We use an airport database, which includes basic airport information like the number of

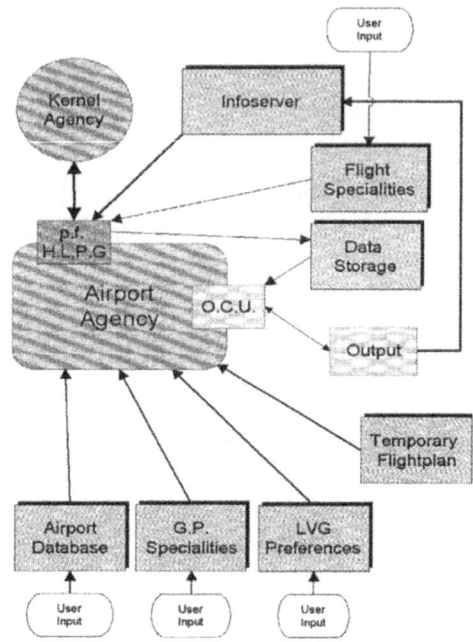

Fig. 1. Global system design

positions and gates. Besides that there are the infoserver, which is our main data source for flight information, the airline preferences, the flight specialities and the gate and position specialities. All these areas contain positioning relevant data which are necessary to calculate an optimal solution for the positioning problem.

3.3   Airport Agency

The airport agency is used to model a virtual airport. The virtual airport model can be applied to an existing airport, in our case airport Frankfurt/Main. The design of the virtual airport is based on the concept of logical fiberings. The airport agency, respectively the virtual airport, consists of serveral units. Two units are interfaces for communication with the enviroment and all others are internal units. First, there is the precalculating filtering hierarchical list of positions and gates (p.f.H.L.P.G) and second the output communication unit (O.C.U) which will not be discussed in this contribution. These two units manage the data transfer with the enviroment and prepare input data for the internal use in the airport agency and the kernel agency which will be decribed later in this contribution. Figure 2 gives an overview of the design of the airport agency with their units and the internal communication paths.

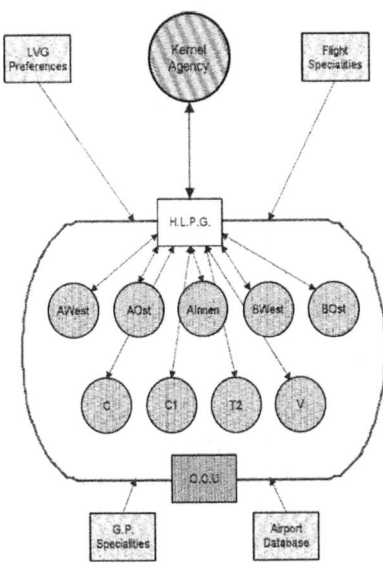

Fig. 2. Design of the Airport Agency

The airport agency, shown in Figure 2, includes a number of virtual clusters, e.g. "AWEST", "AOST" or "BWEST". These clusters include the fibers of our airport. We use two different types of fibers one for positions and the other one for gates because they have a lot of different characteristics. Some typical characteristics of a position fiber are the maximum valid wing code or the buffertime, on the other side some typical characteristics of a gate fiber are the maximum number of allowed passengers in the gate area or the time to bring the passengers to the aircraft.

In terms of logical fiberings to each "agent" corresponds a "local" fiber. These fibers are connected via communication paths with the p.f.H.L.P.G. and they can have further connections with their neighbors. The connections (communication) with neighbors are modeled by transjunctions. A transjunction can be used to control the state spaces of neighbored (connected) fibers. A special effect in an application of a transjunction can be described as follows: if a local fiber A corresponds to, for example, an aircraft or group of passengers (GoPax), and there is a transjunction from A to a fiber B, then the state space of fiber B will be downgraded by the transjunction. This downgrade of the state space of fiber B will be cancelled again as soon as fiber A is no longer used by ("attached to") an aircraft or GoPax.

Example: *Transjunction rule 'A10' to 'A12'*
if 11 then 5
The effect of this transjunction is, that the state space of 'A12' will be downgraded to SWC 5, if the current SWC of 'A10' is 11.

Here SWC is the short notation for "wing code". One has to take into account that different wing codes have to be distinguished – there exist priorities which must be taken into consideration.

The complete transjunction corresponding to this example has 3 truth values in the local fiber A and 12 values in local fiber B and therefore represents a generalized transjunction in the sense of section 2. The complete truth table of the transjunction is displayed in [FW99].

Figure 3 shows a special cluster with its communication paths and neighborhood relationships. Moreover, the complexity of the state space of an agent depends on the number of allowed aircraft types for that agent.

3.4   Kernel Agency

The kernel agency includes currently 4 different algorithms, the adaptive long-time scheduling algorithm, the genetic short-time scheduling algorithm, the random scheduling algorithm, the conflict solving algorithm. Another part of the kernel is the kernel communication unit. In this contribution we discuss the genetic short-time scheduling algorithm. Treatment of the other scheduling algorithms is planned as future work.

Genetic short-time scheduling algorithm This genetic algorithm [Mic96], [Hof96] is designed to solve the problems occurring in short time scheduling. It uses an already pre-optimized season plan (generated by a human operator or by

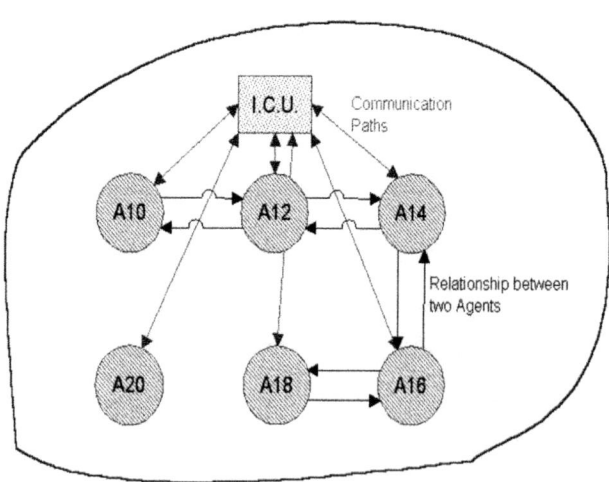

Fig. 3. Internal structure of cluster 'AWest'

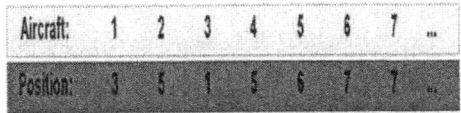

Fig. 4. First attempt

a long time scheduler) together with actual changes and additional information (including, among others, actually flying passengers, planes available, passengers that need to get a connection). We decided to use evolutionary computing to solve this problem.

The original idea was to treat the positioning problem with a usual genetic algorithm, encoding the positions of a given aircraft in a standard way (figure 4).

Although this approach works it causes a crucial problem: almost no valid (valid for a final solution) individuals are created this way since most solutions will position one of the planes either on an impossible position (plane is too large, for example) or will put planes at the same time on the same position. Of course, such solutions would get a very low fitness, so they are not likely to produce offspring. Still we get the problem that the resulting algorithm spends most of the time searching for a solution where the planes actually fit and not for an optimal positioning of the planes – thus wasting valuable time.

Position:	1	2	3	4	5	6	7	...
Aircraft:	3		1		2	5	6	...
					4		7	

Fig. 5. Second attempt

Therefore, another approach was chosen. First we changed the coding such that the aircraft are mapped onto positions (figure 5).

This enables us to use the complete rule set which is provided by the airport agency. So we are now able to allow only such genetic operators which produce a valid individuum. For example, if the aircraft number one would be a B747 and the only positions which were allowed for such a plane would be position one, two, and three, then a mutation on plane number one could only mutate it towards position one, two, or three.

This second model works very well with the mutation operator, but new problems arise when using a crossover operator. The loss of data or the duplication of data can happen. These problems are well known since they also occur when trying to find solutions to a traveling salesman problem with the help of GAs.

When doing a crossover not only the parts that are actually selected are exchanged, but also other parts which are necessary to maintain consistency. To this end we have to check for each exchanged plane whether its "counterpart" is also moved. If not, we have to set this aircraft also on the exchange list. On first view this seems to be a very useful trick to solve the problem, but it also produces a problematic side effect. Every time we wish to do a crossover at a certain point we also exchange planes on quite randomly chosen other positions. This can destroy building blocks, especially in a problem as big as the positioning task. For this reason we decided not to implement a crossover operator in the classical way.

The Condense Operator (Mutation 1) This is the simplest mutation operator which processes all planes that are not yet at valid positions. It aims at positioning such planes correctly.

The Replace Operator (Mutation 2) This second mutation operator is almost as simple as the first one, but works in the opposite way. It is mainly used to maintain the diversity in a population. To do this it moves a plane from one position to another one. Technically this is also done in three steps. First one aircraft is randomly selected. This aircraft is moved to the Temp place. Temp denotes the set of planes not yet positioned. Then the algorithm selects randomly a position where the plane fits. Then the selected plane is force-positioned on this selected position. It will be positioned there in any case removing disturbing planes if necessary. Later the removed planes will be reordered using the random scheduler.

Chromosome Repositioning (Mutation 3) This is the most 'advanced' mutation operator that will be used in our implementation of the GA. It does not operate on single planes like the previous ones, but on complete positions.

Guided Replace (Crossover 1) As already mentioned, many difficulties can arise when using crossover operators in a given problem. Therefore we decided not to use them in a classical way, but in form of a new class of genetic operators. This class works like a mixture of crossover and mutation. It behaves like a normal crossover functionally, but it has a probability of occurrence like a mutation. So this kind of crossover operators can be considered as guided mutations. The Guided Replace operator is the simplest case of an operator of this type.

Guided Chromosome Repositioning (Crossover 2) This crossover-like operator has similarity to the operator "Mutation 3". The idea is to make a transfer of a perfect distribution of planes corresponding to a certain position in an individual to another individual. The exact algorithm becomes a rather complicated ruleset. A detailed description can be found in [FW99].

Cluster Crossover (Crossover 3) This is the most advanced crossover operator. Since the airport contains groups (clusters) of positions it is plausible to use these in the optimization process. The internal algorithm of the operator has close links with the second crossover operator. In fact, we can use this operator for the cluster crossover if we select more than one position (all positions in the given cluster).

Fitness Function To find an appropriate fitness function was one of the main problems of the complete optimization process. We designed the function in close cooperation with the experts from Lufthansa. The optimization process is influenced by many factors which have to be taken into account. Furthermore, the approach should be flexible and it should be possible to cope with an optimization task which depends on selected factors only. The complete fitness function F consists of several constituents. $F(i)$ is based on four different main optimization criteria with respect to an individual i:

$$F(i) = P(i) + C(i) + S(i) + Q(i).$$

$P(i)$ denotes the part that tries to optimize the number of passengers. $C(i)$ is responsible for reducing the connecting time. $S(i)$ describes the service aspects and $Q(i)$ the quality of the solution.

The Greek letters in the subsequent formulas denote parameters which can be tuned problem dependent.

$$P(i) = \alpha * IP + \beta * OP + \psi * IB + \delta * OB + \varepsilon * IY + \phi * OY$$

The passenger part takes care that a maximum amount of passengers can leave or enter a plane directly from the gate (IP and OP). Additionally, also the yield of passengers can be taken into account (IY and OY). $P(i)$ models the request that a maximum amount of aircraft are well positioned right in front of the building (IB and OB).

$$C(i) = \gamma * CT + \eta * GD.$$

The connexe (connecting passengers) part optimizes the time needed to get from one plane to the other. To this end two options are possible. Either the buffer time (maximal possible time to reach the plane - walking time) is maximized (CT) or special gates are used for planes with many connecting passengers (GD).

$$S(i) = \iota * BN + \kappa * BY + \lambda * PD.$$

$S(i)$ represents the service part of the fitness function. Three factors are summarized here. BN and BY reduce the amount of bus transfers or the amount of bus transfers for valuable passengers, respectively.

The third factor, PD, makes the time between boarding and take off as small as possible.

$$Q(i) = \mu * PC + \nu * GC + \pi * CC.$$

Finally, $Q(i)$ is responsible for the fact that the solution is admissible at all. PC and GC model the constraints that all positions and gates are valid. CC is responsible for the request that all connecting passengers are able to catch their planes.

4   Test of the tool in a real world scenario

In [FW99] a real world example is presented which corresponds to a typical scenario to be handled every day by Lufthansa and FAG (the Frankfurt airport operating center). The initial input to our system is a concrete schedule elaborated by the long-term scheduling team of FAG. The syntax of the input data is of the following type (example) – we display three lines only.

inbound LH 00201 sta 07:45 in_type I in_flight_type P inpaxcode 3
outbound LH 03720 std 08:40 out_type S outpaxcode 3 out_flight_type P
dest VIE air_type A321 swc 11 pos V123 out_gate B13

The complete input comprises about 1200 such 3-line units. Evaluating the performance of the tool in its application to a real problem situation it turned out that the system found better solutions than the human expert. In order to produce an optimized solution to the position scheduling task the human operator needs about one week, whereas, working on the same task, our system needs about one hour (on a standard modern PC) and it produces even better positioning plans. More details and an example of an airport resource utilization plan produced by the tool can be found in [FW99].

5   Conclusions

In the previous sections we presented work on the hard problem of optimal positioning of aircraft at an airport. Many real world constraints have to be considered. The original problem has been described by Lufthansa AG focusing on airport Frankfurt/Main. We developed a general (generic) airport model using, among others, the concept of logical fiberings. The optimization problem was treated by modified and extended genetic algortihms. On the basis of these approaches an aircraft positioning tool was developed and implemented, especially tailored for computing positioning configurations of the airport Frankfurt/Main. A prototypic first version of the system is currently being tested with Lufthansa at Frankfurt airport. Our new system achieves much better performance than the methods applied before.

In addition to the previously described methods, in future work we intend to use also methods from artificial neural networks (for modeling position constraints and optimization), make systematic applications of multi-agent systems techniques and rule based systems.

References

[FW99]   K. Frank and J. Weichenberger. Design and implementation of an aircraft positioning tool using hybrid problem-solving methods. Master's thesis, Institut für Computerwissenschaften, Universität Salzburg, Austria, 1999.

[Hof96]   Frank Hoffmann. *Automatischer Entwurf von Fuzzy-Reglern mit genetischen Algorithmen*. PhD thesis, Mathematisch-Naturwissenschaftliche Fakultät, Christian-Albrechts Universität zu Kiel, Germany, 1996.

[Mei99]   W. Meixl. Logical fiberings. a general decomposition method for many-valued logics and a modeling approach for multi-agent systems. Master's thesis, Institut für Computerwissenschaften, Universität Salzburg, Austria, 1999.

[Mic96]   Zbigniew Michalewicz. *Genetic Algorithms + Data Structures = Evolution Programs*. Springer–Verlag Berlin Heidelberg, 1996.

[Pfa91]   J. Pfalzgraf. Logical fiberings and polycontextural systems. In *Fundamentals of Artificial Intelligence Research, Ph.Jorrand, J.Kelemen (eds.). Lecture Notes in Computer Science 535, Subseries in AI, Springer Verlag*, 1991.

[Pfa97]   J. Pfalzgraf. On geometric and topological reasoning in robotics. *Annals of Mathematics and Artificial Intelligence*, 19:279–318, 1997.

[PSS96a] J. Pfalzgraf, U. Sigmund, and K. Stokkermans. Towards a general approach for modeling actions and change in cooperating agents scenarios. special issue of *IGPL (Journal of the Interest Group in Pure and Applied Logics)*, IGPL **4 (3)** 445-472, 1996.

[PSS96b] J. Pfalzgraf, V. Sofronie, and K. Stokkermans. On a semantics for cooperative agents scenarios. In *Proceedings 13th European Meeting on Cybernetics and Systems Research (EMCSR'96), Vienna, April 9-12, 1996*, 1996.

Author Index

Lecture Notes in Computer Science

For information about Vols. 1–1713
please contact your bookseller or Springer-Verlag

Vol. 1745: P. Banerjee, V.K. Prasanna, B.P. Sinha (Eds.), High Performance Computing – HiPC'99. Proceedings, 1999. XXII, 412 pages. 1999.

Vol. 1746: M. Walker (Ed.), Cryptography and Coding. Proceedings, 1999. IX, 313 pages. 1999.

Vol. 1747: N. Foo (Ed.), Adavanced Topics in Artificial Intelligence. Proceedings, 1999. XV, 500 pages. 1999. (Subseries LNAI).

Vol. 1748: H.V. Leong, W.-C. Lee, B. Li, L. Yin (Eds.), Mobile Data Access. Proceedings, 1999. X, 245 pages. 1999.

Vol. 1749: L. C.-K. Hui, D.L. Lee (Eds.), Internet Applications. Proceedings, 1999. XX, 518 pages. 1999.

Vol. 1750: D.E. Knuth, MMIXware. VIII, 550 pages. 1999.

Vol. 1751: H. Imai, Y. Zheng (Eds.), Public Key Cryptography. Proceedings, 2000. XI, 485 pages. 2000.

Vol. 1752: S. Krakowiak, S. Shrivastava (Eds.), Advances in Distributed Systems. VIII, 509 pages. 2000.

Vol. 1753: E. Pontelli, V. Santos Costa (Eds.), Practical Aspects of Declarative Languages. Proceedings, 2000. X, 327 pages. 2000.

Vol. 1754: J. Väänänen (Ed.), Generalized Quantifiers and Computation. Proceedings, 1997. VII, 139 pages. 1999.

Vol. 1755: D. Bjørner, M. Broy, A.V. Zamulin (Eds.), Perspectives of System Informatics. Proceedings, 1999. XII, 540 pages. 2000.

Vol. 1757: N.R. Jennings, Y. Lespérance (Eds.), Intelligent Agents VI. Proceedings, 1999. XII, 380 pages. 2000. (Subseries LNAI).

Vol. 1758: H. Heys, C. Adams (Eds.), Selected Areas in Cryptography. Proceedings, 1999. VIII, 243 pages. 2000.

Vol. 1759: M.J. Zaki, C.-T. Ho (Eds.), Large-Scale Parallel Data Mining. VIII, 261 pages. 2000. (Subseries LNAI).

Vol. 1760: J.-J. Ch. Meyer, P.-Y. Schobbens (Eds.), Formal Models of Agents. Poceedings. VIII, 253 pages. 1999. (Subseries LNAI).

Vol. 1761: R. Caferra, G. Salzer (Eds.), Automated Deduction in Classical and Non-Classical Logics. Proceedings. VIII, 299 pages. 2000. (Subseries LNAI).

Vol. 1762: K.-D. Schewe, B. Thalheim (Eds.), Foundations of Information and Knowledge Systems. Proceedings, 2000. X, 305 pages. 2000.

Vol. 1763: J. Akiyama, M. Kano, M. Urabe (Eds.), Discrete and Computational Geometry. Proceedings, 1998. VIII, 333 pages. 2000.

Vol. 1764: H. Ehrig, G. Engels, H.-J. Kreowski, G. Rozenberg (Eds.), Theory and Application of Graph Transformations. Proceedings, 1998. IX, 490 pages. 2000.

Vol. 1765: T. Ishida, K. Isbister (Eds.), Digital Cities. IX, 444 pages. 2000.

Vol. 1767: G. Bongiovanni, G. Gambosi, R. Petreschi (Eds.), Algorithms and Complexity. Proceedings, 2000. VIII, 317 pages. 2000.

Vol. 1768: A. Pfitzmann (Ed.), Information Hiding. Proceedings, 1999. IX, 492 pages. 2000.

Vol. 1769: G. Haring, C. Lindemann, M. Reiser (Eds.), Performance Evaluation: Origins and Directions. X, 529 pages. 2000.

Vol. 1770: H. Reichel, S. Tison (Eds.), STACS 2000. Proceedings, 2000. XIV, 662 pages. 2000.

Vol. 1771: P. Lambrix, Part-Whole Reasoning in an Object-Centered Framework. XII, 195 pages. 2000. (Subseries LNAI).

Vol. 1772: M. Beetz, Concurrent Ractive Plans. XVI, 213 pages. 2000. (Subseries LNAI).

Vol. 1773: G. Saake, K. Schwarz, C. Türker (Eds.), Transactions and Database Dynamics. Proceedings, 1999. VIII, 247 pages. 2000.

Vol. 1774: J. Delgado, G.D. Stamoulis, A. Mullery, D. Prevedourou, K. Start (Eds.), Telecommunications and IT Convergence Towards Service E-volution. Proceedings, 2000. XIII, 350 pages. 2000.

Vol. 1776: G.H. Gonnet, D. Panario, A. Viola (Eds.), LATIN 2000: Theoretical Informatics. Proceedings, 2000. XIV, 484 pages. 2000.

Vol. 1777: C. Zaniolo, P.C. Lockemann, M.H. Scholl, T. Grust (Eds.), Advances in Database Technology – EDBT 2000. Proceedings, 2000. XII, 540 pages. 2000.

Vol. 1778: S. Wermter, R. Sun (Eds.), Hybrid Neural Systems. IX, 403 pages. 2000. (Subseries LNAI).

Vol. 1780: R. Conradi (Ed.), Software Process Technology. Proceedings, 2000. IX, 249 pages. 2000.

Vol. 1781: D.A. Watt (Ed.), Compiler Construction. Proceedings, 2000. X, 295 pages. 2000.

Vol. 1782: G. Smolka (Ed.), Programming Languages and Systems. Proceedings, 2000. XIII, 429 pages. 2000.

Vol. 1783: T. Maibaum (Ed.), Fundamental Approaches to Software Engineering. Proceedings, 2000. XIII, 375 pages. 2000.

Vol. 1784: J. Tiuryn (Ed.), Foundations of Software Science and Computation Structures. Proceedings, 2000. X, 391 pages. 2000.

Vol. 1785: S. Graf, M. Schwartzbach (Eds.), Tools and Algorithms for the Construction and Analysis of Systems. Proceedings, 2000. XIV, 552 pages. 2000.

Vol. 1786: B.H. Haverkort, H.C. Bohnenkamp, C.U. Smith (Eds.), Computer Performance Evaluation. Proceedings, 2000. XIV, 383 pages. 2000.

Vol. 1790: N. Lynch, B.H. Krogh (Eds.), Hybrid Systems: Computation and Control. Proceedings, 2000. XII, 465 pages. 2000.

Vol. 1793: O. Cairo, L.E. Sucar, F.J. Cantu (Eds.), MICAI 2000: Advances in Artificial Intelligence. Proceedings, 2000. XIV, 750 pages. 2000. (Subseries LNAI).

Vol. 1795: J. Sventek, G. Coulson (Eds.), Middleware 2000. Proceedings, 2000. XI, 436 pages. 2000.

Vol. 1794: H. Kirchner, C. Ringeissen (Eds.), Frontiers of Combining Systems. Proceedings, 2000. X, 291 pages. 2000. (Subseries LNAI).

Vol. 1801: J. Miller, A. Thompson, P. Thomson, T.C. Fogarty (Eds.), Evolvable Systems: From Biology to Hardware. Proceedings, 2000. X, 286 pages. 2000.

Vol. 1802: R. Poli, W. Banzhaf, W.B. Langdon, J. Miller, P. Nordin, T.C. Fogarty (Eds.), Genetic Programming. Proceedings, 2000. X, 361 pages. 2000.

Vol. 1803: S. Cagnoni et al. (Eds.), Real-World Applications and Evolutionary Computing. Proceedings, 2000. XII, 396 pages. 2000.

GPSR Compliance

The European Union's (EU) General Product Safety Regulation (GPSR) is a set of rules that requires consumer products to be safe and our obligations to ensure this.

If you have any concerns about our products, you can contact us on ProductSafety@springernature.com

In case Publisher is established outside the EU, the EU authorized representative is:

Springer Nature Customer Service Center GmbH
Europaplatz 3
69115 Heidelberg, Germany

Batch number: 09624486

Printed by Printforce, the Netherlands